Springer Series on

Atoms+Plasmas

24

Editor: G. Ecker

Springer-Verlag Berlin Heidelberg GmbH

Physics and Astronomy ONLINE LIBRARY

http://www.springer.de/phys/

Yu. M. Aliev   H. Schlüter   A. Shivarova

Guided-Wave-Produced Plasmas

With 126 Figures

 Springer

Professor Dr. Yuri M. Aliev

Russian Academy of Sciences, Lebedev Physical Institute
53, Leninskii Prospect, 117924 Moscow, Russia

Professor Dr. Hans Schlüter

Ruhr-Universität Bochum, Institut für Experimentalphysik II
NB 5/67, 44780 Bochum, Germany

Professor Dr. Antonia Shivarova

Sofia University, Faculty of Physics
5, James Bourchier Blvd., 1164 Sofia, Bulgaria

Series Editors:

Professor Dr. Günter Ecker

Ruhr-Universität Bochum, Fakultät für Physik und Astronomie,
Lehrstuhl Theoretische Physik I, Universitätsstrasse 150,
44801 Bochum, Germany

Professor Peter Lambropoulos, Ph. D.

Max-Planck-Institut für Quantenoptik,
85748 Garching, Germany, and
Foundation for Research and Technology – Hellas (FO.R.T.H.),
Institute of Electronic Structure & Laser (IESL),
University of Crete, PO Box 1527, Heraklion, Crete 71110, Greece

Professor Jürgen Mlynek

Universität Konstanz, Universitätsstrasse 10
78434 Konstanz, Germany

Professor Dr. Herbert Walther

Sektion Physik der Universität München, Am Coulombwall 1,
85748 Garching/München, Germany

Library of Congress Cataloging-in-Publication Data

ISSN 0177-6495
ISBN 978-3-642-62982-2   ISBN 978-3-642-57060-5 (eBook)
DOI 10.1007/978-3-642-57060-5

Cover concept: eStudio Calamar Steinen
Cover production: *design & production* GmbH, Heidelberg
Typesetting: Camera ready copy from the authors
Printed on acid-free paper   SPIN 10098916   57/3144/tr - 5 4 3 2 1 0

Preface

The topic of this book is a class of gas discharges which has become important over the last twenty-five years and has attracted much interest and stimulated increased research activities and applicational work. The fields that maintain this class of discharges are those of travelling electromagnetic waves, in the radio-frequency and microwave ranges. Since the surface waves which maintain these discharges are waveguided modes of gas-discharge plasmas, these discharges could be also called "waveguided discharges" and the created plasmas are, thus, "guided-wave-produced plasmas".

Discharges generated by travelling electromagnetic surface waves are currently being investigated very actively in view of their advantages as electrodeless discharges with a high efficiency in the transfer of the wave power to plasma production. This is the reason why they are considered promising for various applications such as plasma technology (surface treatment, microelectronics, plasma chemistry), lighting (as light sources for spectroscopy and elemental analysis), laser physics, gas detoxification and the medical industry.

Regardless of the wide spectrum of applications in which surface-wave-produced plasmas can be involved, it is often stated that the theory of high-frequency discharges which has been developed over the years of research on surface-wave-sustained discharges presents the foremost "application" of these discharges. This theory is the main subject of the book.

Surface-wave-sustained discharges are structures which unify wave-fields and gas-discharge plasmas. The creation of these discharges is a clear example of a self-consistent nonlinear problem: the wave produces the plasma and the plasma produced maintains the wave propagation. Wave characteristics and plasma properties are self-consistently related to each other. The discharge behaviour is simultaneously governed by electrodynamics and gas-discharge physics. This is stressed in the book.

Various complicated theoretical and experimental problems are met in studies on this type of plasma. The self-consistent models of gas discharges maintained by guided electromagnetic waves discussed in this volume are completely based on nonlinear plasma theory. The ionization nonlinearity, with local and nonlocal Joule heating of the electrons in the wave field as well as with noncollisional heating, results in a description of the self-consistent structure of the discharge. Different nonlinear mechanisms are considered, depending on the gas-discharge conditions, as sources of the interrelation between the plasma and wave properties which establish the surface-wave-sustained discharge as a united structure of wave fields and plasmas.

Besides stressing the gas-discharge physics and, in particular, the ionization nonlinearity, the book gives a broad but detailed discussion of quite a number of problems in the physics of waves in plasmas. Results on wave propagation properties in inhomogeneous collisional plasmas, although investigated with a view to their appearance in the mechanisms of the maintenance of the discharge, are presented in their own right and are not limited to the aspects needed to solve the particular problem of the discharge under consideration. Methods of treatment of the wave behaviour in plasmas with plasma-density inhomogeneities in the longitudinal and transverse directions with respect to the wave propagation are discussed. These treatments concern wave propagation in weakly nonuniform media as well as wave behaviour in resonance regions of strongly inhomogeneous plasmas. Although the main content of the book is in the field of gas-discharge physics, those chapters where wave propagation behaviour is presented extend the scope of the book to more general aspects of waves in plasmas, encompassing neighbouring disciplines such as hot-plasma research and ionospheric physics.

The experimental and theoretical work in the field by many researchers in various countries has achieved a good basic understanding of the phenomena encountered. The authors have worked for a long time on a wide range of plasma physics problems and they have used their experience in attempting to achieve the desired uniform presentation. No restriction of the generality of the subject results from employing surface waves as a guided mode maintaining the discharge. The essential features treated are characteristic of all types of wave-produced plasmas.

The book is addressed to physicists and engineers active in research on gas discharges and their industrial applications and is intended to provide material suited also for graduate courses. It could also be of value to researchers working in the fields of plasma physics, plasma chemistry and technology, radio communications and space physics.

The authors are greatly indebted to colleagues in the field for stimulating and generous discussions and are very grateful to all members of their research groups in Bochum, Moscow and Sofia for cooperation and contributions. To Mr. B. Lehnhoff they express sincere thanks for electronical assistance and the realization of the manuscript. Much of the work contributing to the book would not have been possible without the support of the Deutsche Forschungsgemeinschaft. Two of the authors (A.Sh., H.Sch.) are grateful for valuable support from the Volkswagen Foundation and the NATO Office of Scientific Affairs. Additionally, Yu.M.A. acknowledges support by INTAS-96-0235 and by the Russian Foundation of Basic Research (project no.: 98-02-16435), A.Sh. is deeply obliged to the Alexander von Humboldt Foundation for a most helpful reinvitation.

The authors wish to express their appreciation to Prof. Dr. G. Ecker, Dr. H. Lotsch and Dr. C. Ascheron for their encouragement of and editorial assistance for this book. The editorial contribution of Mrs. P. Treiber is also highly acknowledged.

December 1999

Yu.M. Aliev
H. Schlüter
A. Shivarova

Contents

List of Symbols and Abbreviations

A — amplitude of the E_z SW field component at the plasma–dielectric interface

A_{mk} — radiative transition probability

$\tilde{A}$ — defined by (3.67)

$\boldsymbol{A}$ — defined by (5.119d)

$\mathcal{A}$ — defined by (3.71a)

$a(\xi)$ — introduced in (5.45a)

a_n^{p} — coefficients in solutions (4.169)

a_n^{v} — coefficients in solutions (4.169)

B — $\equiv B_y(x)$, magnetic-field component of SW in semi-infinite plasma according to (4.1)

$\tilde{B}$ — defined by (3.68)

$\boldsymbol{B}$ — magnetic induction

$\mathcal{B}$ — defined in (4.2)

$\mathcal{B}$ — defined by (3.71b)

B_0 — magnetic-field intensity at the interface in a planar waveguide (single interface)

$B^{(1,2)}$ — $\equiv B_y^{(1,2)}$, magnetic-field components of SW in media denoted by "1" and "2"

b_{e} — coefficient of electron mobility

C — constant introduced in (5.37b)

$\tilde{C}$ — defined by (3.68)

$\mathcal{C}$ — defined by (3.72c)

C_{a} — characteristics of associative ionization

C_k^m — rate coefficient for collisional excitation from initial state k to final state m

c — speed of light in vacuum

$\boldsymbol{D}$ — electrical displacement

$\mathcal{D}$ — defined by (3.72d)

$D(v,x)$ — diffusion coefficient in velocity space

D_{A} — ambipolar diffusion coefficient

D_{e} — electron diffusion coefficient

D_m — diffusion coefficient of atoms in excited state m

D_{ql} — diffusion coefficient in energy space for noncollisional (quasi-linear) energy transfer

D_r — diffusion coefficient in energy space for Joule heating through the E_r field component

$D_{r_{\text{total}}}$ — diffusion coefficient containing both collisional and noncollisional effects, defined by (6.18b)

D_z — diffusion coefficient in energy space for Joule heating through the E_z field component

D_ϵ — total diffusion coefficient in energy space according to (6.20)

D_{e}^{T} — thermal diffusion coefficient

$\mathcal{D}(\omega, \boldsymbol{k})$ — the function involved in the dispersion relation

$\bar{D}_W$ — defined by (6.5)

DC — direct current

d — semi-width of a plasma slab and thickness of a plasma/dielectric layer in planar and cylindrical waveguides

d — defined by (6.13)

E_{A} — ambipolar electric field

E_{eff} — effective field intensity defined by (2.41b)

E_{i} — normalizing field determining the efficiency of the thermal nonlinearity according to (5.7)

$E_{\text{i}_{(1,2)}}$ — normalizing fields determining the efficiency of the thermal nonlinearity according to (5.119c)

$E_{n(r)}$ — normalizing field amplitude defined by (5.110b)

E_{Th} — threshold-field intensity for SW-produced plasma

E_{th} — threshold-field intensity according to Schottky theory

$E_\parallel$ — maintenance field in the positive column of a DC discharge

E_0 — $(= |E_z(r = R, z = 0)|)$ axial field component at the onset of the discharge

$E_{(\text{r})}^2$ — maintenance-field intensity for electron heating in regions of resonance absorption of SWs

$\langle E^2 \rangle$ — maintenance-field intensity in nonlocal regime of electron heating according to (5.136)

$\boldsymbol{E}(\boldsymbol{r}, z)$ — electric-field strength

$\boldsymbol{E}_{\text{HF}}$ — high-frequency electric field

EDF — electron distribution function

EEDF — electron energy distribution function

EM — electromagnetic

e — electron charge (absolute value)

$F(\boldsymbol{r}, \boldsymbol{v}, t)$ — electron distribution function in phase space

$F_0(\boldsymbol{r}, \boldsymbol{v}, t)$ — isotropic part of the distribution function

$F_0(\boldsymbol{r}, u)$	electron energy distribution function
$F_0^{(0)}(\mathcal{W})$	introduced by (6.2)
$F_0^{(1)}(\mathcal{W}, r)$	introduced by (6.2)
$\bar{F}$	defined by (3.69)
$\bar{F}_1$	defined by (3.67)
$\bar{F}_2$	defined by (3.68)
$\bar{F}_3$	introduced by (5.117a)
$\bar{F}_3'$	introduced by (5.117b)
$\boldsymbol{F}_1(\boldsymbol{r}, v, t)$	anisotropic part of the distribution function
FRD	field radial decay
f	$\equiv \omega/2\pi$
$f(\boldsymbol{r}, v, t)$	electron velocity distribution function
f_0	isotropic part of $f(\boldsymbol{r}, v, t)$
$\bar{f}$	defined by (3.73)
$\langle \bar{f} \rangle$	mean value of $\bar{f}$

$G(z)$	introduced by (4.138)
$G_c(z)$	introduced by (4.155a)
$G_{\{{}^s_a\}}(z)$	introduced by (4.144)
$\tilde{G}$	introduced in (3.69b)
g	defined by (5.48c)
$g(\bar{n})$	introduced by (6.41)
$g(r)$	introduced in (5.80)
$g(u)$	defined by (2.46)
g_{mk}	escape factor
g_0	statistical weight of the ground state
$\bar{g}(\mathcal{W})$	introduced in (6.7)

$H(|\varepsilon|)$	defined by (5.84b)
$\mathrm{H}_1^{(1,2)}(x)$	Hankel functions of the first order, of the first and second kind

H	magnetic-field strength
HF	high frequency
$h(u)$	defined by (2.46)
$\bar{h}(\mathcal{W})$	introduced by (6.8)

$I(n_{1_0})$	defined by (5.22b)
$I(n_{2_0})$	defined by (5.31c)
$I(\gamma_{\mathrm{NL}})$	defined by (5.37e)
I_L	spectral line intensity
$I_{\mathrm{p, v}}$	notations introduced in (6.40)
$\mathrm{I}_{0,1}(x)$	modified Bessel functions of the first kind
$I^*(\gamma_{\mathrm{NL}})$	defined by (5.38b)
$\tilde{I}(n_{1_0})$	defined by (5.23c)
$\tilde{I}(n_{2_0})$	defined by (5.32a)

$\mathrm{J}_{0,1}(x)$	Bessel functions of the first kind
$\boldsymbol{j}$	current density

K_{c}	introduced by (4.155b)
$\mathrm{K}_{1,2}(x)$	modified Bessel functions of the second kind
$K_{\{{}^s_a\}}$	abbreviation introduced by (4.144)
$\mathcal{K}$	coefficient of temperature conductivity
k	wave number
k_{p}	rate coefficient for ionization by collisions between two excited atoms
k_{v}	vacuum wave number

k_r — Fourier harmonics in the series expansion of the transverse distribution of the wave field in cylindrical coordinates

k_x — Fourier harmonics in the series expansion of the transverse distribution of the wave field in rectangular coordinates

k_0 — zero-order approximation to k (in terms of the geometrical-optics approximation), i.e. the local wave number

$k_{1,2}$ — wave numbers in different space regions (denoted by "1" and "2")

$\bar{k}$ — $\equiv i\,k$

$\boldsymbol{k}$ — wave vector

L_D — diffusion length

$L_{D(m)}$ — diffusion length of metastable atoms (in excited state m)

L_E — scale length of the heating field

L_N — characteristic length of plasma density inhomogeneity in the longitudinal direction in the entire system

L_T — scale length of temperature variation

L_n — characteristic length of plasma density inhomogeneity across the discharge cross-section

L_ε — electron energy relaxation length (see also L_χ)

L_χ — scale length of thermal conduction

L_{eff} — dimensionless effective length of the discharge

$L_N^{(1,2)}$ — characteristic length of plasma density inhomogeneity in the longitudinal direction in regions "1" and "2"

$L_n^{(\mathrm{r})}$ — characteristic length of plasma-density inhomogeneity in the transverse direction around the resonant point defined in terms of variation of the plasma permittivity

$L_{n(\mathrm{r})}$ — characteristic length of plasma density inhomogeneity in the transverse direction around the resonance absorption point defined in terms of variation of the plasma density

$\tilde{L}$ — defined by (3.66)

l — introduced in (5.47)

l_n — eigenvalues of the solutions of (4.166)

$\ln(\Lambda)$ — Coulomb logarithm

M — atom mass and ion mass

$M_{\mathrm{res}}(r)$ — step function

$\tilde{M}$ — defined by (3.66)

m — electron mass

N — ($\equiv n(z)/n_c$) normalized plasma density

N_k — population density of excited atoms (in state k)

N_0	neutral-gas density
N_{1_0}	defined by (5.26d)
N_{2_0}	defined by (5.34b)
N_{3_0}	defined by (5.41a)
$\mathrm{N}_{0,1}(x)$	Bessel functions of the second kind
N'_{2_0}	defined by (5.35b)
$\bar{N}$	$\equiv \bar{n}/n_\mathrm{c}$
n	plasma density, electron density
$n(r)$	plasma density, electron density
n_c	critical density
n_e	electron density
n_w	plasma density at the walls
n_0	plasma density at the discharge axis
n_1	normalized plasma density as introduced in (5.20a)
n_2	normalized plasma density as introduced in (5.30b)
n_3	normalized plasma density as introduced in (5.37a)
n_c^*	surface wave cut-off frequency
$\bar{n}$	plasma density averaged over the slab width or the column cross-section
$\bar{n}_\mathrm{r}$	plasma density (averaged over the discharge cross-section) at the position along the discharge where the resonance absorption starts
$\bar{n}_0$	plasma density (averaged over the discharge cross-section) at a given z_0-position
$P(r)$	introduced in (4.57)
P_r	introduced in (4.73c)
P_0	externally applied power
P	axial component of the SW flux after integration over the discharge cross-section
P_i	Poynting vector component over an unit length integrated over the waveguide cross-section
$\boldsymbol{P}$	wave-energy flux (Poynting vector)
$\boldsymbol{P}^\mathrm{v}$	that part of the wave-energy flux which is in the vacuum
Pr	principal value of an integral
p	gas pressure
Q	Joule losses of wave power; Joule heating of electrons
Q^h	total energy absorbed in the region of resonant absorption
Q	Q defined over a unit length of the discharge after integration over the waveguide cross-section
$\overline{Q}$	Q divided by the cross-sectional area
Ω	defined by (4.111)
$\boldsymbol{q}_\mathrm{c}$	heat-flux vector

R	radius of the plasma column; inner radius of the gas discharge tube
R_m	radius of the shielding metal cylinder
$\bar{R}$	$= \varkappa_d R$ (see (3.76c))
$\overline{R+d}$	$= \varkappa_d(R+d)$ (see (3.76c))
Ry	Rydberg constant
r	radial coordinate
r_D	Debye length of the electrons
r_r	radial coordinate of the point of resonance absorption of SWs
r_0	arbitrary point in a region of strong inhomogeneity (inhomogeneity in transverse direction, cylindrical geometry)
$r^k(\mathcal{W})$	introduced in (6.9)
$r^*(\mathcal{W})$	introduced in (6.5)
$\boldsymbol{r}$	radius vector
$\boldsymbol{r}_\mathrm{r}$	radius vector of the point of resonance absorption of SWs
$S(F)$	collisional integral
S_m	rate coefficient for collisional ionization from the mth state
S_0	velocity-angle-averaged collisional integral
$S_\perp$	cross-sectional area of the discharge
SW	surface wave
s	$= U_\mathrm{i}/U_*$
s_{0j}	cross-section (integral) for excitation
$s^{\mathrm{i}(j)}$	cross-section (integral) for ionization from a metastable state
$\bar{s}$	defined by (6.14)
T	period of the HF field
T_e	electron temperature (in energy units)
T_g	gas temperature
$\tilde{T}$	defined by (3.66)
$\boldsymbol{T}$	flux of coherent-motion energy of charged carriers
TM	transverse magnetic
t	time
U_i	ionization energy
U_j	excitation energy for atomic level j
U_*	excitation energy of the first atomic excited level
U	introduced in (5.37c)
$\mathcal{U}(r)$	introduced by (4.53)
u	$(= mv^2/2e)$ kinetic energy of the electrons (scaled in volts)
$\bar{u}$	mean energy
u	$(\equiv |E_z^2|/E_\mathrm{th}^2)$ normalized intensity of the maintenance field for Joule heating in the plasma volume
$\bar{V}_\mathcal{W}$	introduced by (6.6)
$\mathcal{V}(z)$	introduced by (4.89)
v	electron velocity
v_E	electron velocity in the electric field
v_gr	wave group velocity

v_r	radial component of v
$v_{\rm ph}$	wave phase velocity
$v_{\rm th}$	thermal velocity of electrons
$\boldsymbol{v}$	velocity
$\boldsymbol{v}_{\rm A}$	ambipolar velocity
$\boldsymbol{v}_\alpha$	velocity of particles of type α ($\alpha = $ e, i)
W	wave energy
W	wave energy per unit length integrated over the waveguide cross-section
$\mathcal{W}$	total energy of electrons
WKB	approximation: Wentzel, Kramers, Brillouin (geometrical-optics) approximation
w	heat flux
w	($= E_{(r)}^2/E_{\rm th}^2$) normalized intensity of the maintenance field for Joule heating in regions of resonance absorption
$\mathcal{X}(x)$	introduced in (4.164)
$\mathcal{X}(r)$	introduced in (4.170)
x	transverse coordinate in planar waveguide
$x_{\rm r}$	coordinate of the point of resonance absorption in planar waveguide
x_0	arbitrary point in a region of strong inhomogeneity (inhomogeneity in transverse direction in planar waveguide)
$\tilde{x}$	$= \varkappa_{\rm p} R$

$\mathcal{Y}(z)$	introduced in (4.164) and (4.170)
y	coordinate in rectangular coordinate frame
$\tilde{y}$	$= \varkappa_{\rm v} R$
Z	$= -1$ or ≥ 1, number of charges per particle
$\mathcal{Z}$	impedance
$\mathcal{Z}_{\rm d}$	impedance of dielectric region
$\mathcal{Z}_{\rm p}$	impedance of plasma region
$\mathcal{Z}_{\rm inh}^{\rm p}$	impedance of region of inhomogeneous plasma
z	axial coordinate (direction of wave propagation)
$\tilde{z}$	$= \beta R$
α	(space) damping rate
α	($=$ e, i, n) index indicating electrons, ions and atoms
$\alpha_{\rm c}$	defined by (2.52d)
α_L	damping rate at the end of the plasma column
$\alpha_{\rm c(exc)}$	defined by (2.74a)
β	real part of the wave number
$\beta_{\rm i}$	coefficient in the cross-section for direct ionization
$\beta_{\rm r}$	defined by (5.109b)
$\beta_{\rm s}$	defined in (5.42b)

β — real part of the wave vector

Γ — $= 0.577\ldots$

$\Gamma(a, x)$ — incomplete gamma function

Γ_1 — defined in (2.59b)

Γ' — $= 1.781$

γ — time damping rate

$\gamma(a, x)$ — incomplete gamma function

γ_{NL} — introduced by (5.37b)

γ_r — introduced by (5.43)

γ_{si} — introduced by (5.43)

Δ — space interval covering resonance regions

Δ_n — deviation of the maintenance-field intensity from the value corresponding to the Schottky condition

Δ_{n_1} — introduced in (5.54a)

Δ_r — introduced in (5.65b)

Δ — defined by (3.70c)

δ — proportion of energy transferred by an elastic collision

$\delta(x)$ — delta function

δ_{en}^{h} — proportion of energy lost by an electron in inelastic collisions with large energy loss

δ_{en}^{l} — proportion of energy lost by an electron in inelastic collisions with low energy loss

δ_{en}^{s} — proportion of energy loss on average in a collision

δk — first-order correction to the wave number

ε — plasma permittivity

$\varepsilon(\omega)$ — plasma permittivity

$\varepsilon(\omega, \mathbf{k})$ — relative permittivity of a medium

$\varepsilon(\omega, \mathbf{r})$ — plasma permittivity

ε_d — relative permittivity of a medium

ε_{eff} — effective plasma permittivity introduced in (6.35)

ε_i — imaginary part of the plasma permittivity

ε_r — real part of the plasma permittivity

ε_s — permittivity of a dielectric/plasma layer

ε_v — $(= 1)$ relative permittivity of vacuum

ε_0 — vacuum permittivity

$\varepsilon_{1,2}$ — plasma permittivities in different space regions denoted by "1" and "2"

$\varepsilon_i(\omega, \mathbf{k})$ — imaginary part of the plasma permittivity

$\varepsilon_r(\omega, \mathbf{k})$ — real part of the plasma permittivity

ε^{l} — longitudinal part of the plasma permittivity

ε^{tr} — transverse part of the plasma permittivity

$\bar{\varepsilon}$ — plasma permittivity averaged over the discharge cross-section

$\tilde{\varepsilon}$ — as introduced by (3.10c), with the meaning of an "effective" permittivity of a single-interface waveguide

$\tilde{\varepsilon}_i$ — imaginary part of $\tilde{\varepsilon}$

$\tilde{\varepsilon}_r$ — real part of $\tilde{\varepsilon}$

ζ — normalized longitudinal coordinate according to the definition given (Chap. 5)

η' — small quantity

Θ — power absorbed on average by one electron

Θ_B — Brewster angle

$\Theta_{\text{coll-less}}$ — (Θ_{cl}) the part of Θ related to collisionless heating

Θ_r — the part of Θ related to Joule heating through the E_r-field component

Θ_{total} — (total) power absorbed on average by one electron

Θ_z — the part of Θ related to Joule heating through the E_z-field component

κ — coefficient specifying the discharge regime in the results for the axial density profile

$\varkappa$ — defined by (3.43)

$\varkappa_d$ — introduced by (3.9b) for describing the SW-field distribution in the transverse direction in the dielectric region

$\varkappa_p$ — introduced by (3.9a) for describing the SW-field distribution in the transverse direction in the plasma region

$\varkappa_s$ — introduced by (3.39) for describing the SW-field distribution in the transverse direction in the layer region

$\varkappa_v$ — as $\varkappa_s$ with $\varepsilon_d = \varepsilon_v = 1$

$\varkappa_x$ — Fourier harmonics in the series expansion of the transverse distribution of the SW field

$\varkappa_{1,2}$ — quantities describing the field distribution in the transverse direction of two plasma media denoted by "1" and "2"

$\varkappa_d^{(i)}$ — imaginary part of $\varkappa_d$

$\varkappa_d^{(r)}$ — real part of $\varkappa_d$

$\varkappa_p^{(i)}$ — imaginary part of $\varkappa_p$

$\varkappa_p^{(r)}$ — real part of $\varkappa_p$

$\bar{\varkappa}_p$ — introduced in (6.33) for describing the SW-field distribution in the transverse direction in the plasma region

Λ — argument of the Coulomb logarithm $(\ln \Lambda)$

Λ_1 — defined in (2.59b)

$\bar{\Lambda}$ — defined in (2.60b)

λ — SW wavelength

$\lambda_{\text{f.p.}}$ — electron mean free path

$\lambda_{\text{f.p. (i)}}$ — ion mean free path

λ_{sk}	skin depth for weak collisions	ν_{m}	electron–neutral (atom) elastic collision frequency
λ_{v}	vacuum wavelength	ν_{ε}	energy relaxation frequency
λ'_{sk}	skin depth for strong collisions	ν_{jk}	frequency of electron–neutral (atom) collisions for stepwise excitation of the state j to the state k
μ	parameter which characcterizes the transverse density profile ($0 < \mu < \pi/2$ for a plasma slab and $0 < \mu < 2.405$ for a plasma column)	$\nu_{0\mathrm{i}}$	ionization frequency
		ν_{0k}	collision frequency for inelastic processes (excitation) starting from the ground state to state k
μ_{n}	defined by (4.109)		
$\mu_{\alpha\mathrm{n}}$	reduced mass, introduced in (2.51b)	ν_{*}	frequency of excitation from the ground state to the first excited state
μ_0	vacuum susceptibility		
μ_{*}	defined by (5.87b)	ν^{h}	inelastic electron–neutral (atom) collision frequency with large energy losses
μ'	relative magnetic susceptibility		
$\mu'(\omega, \boldsymbol{k})$	relative magnetic susceptibility	ν^{l}	inelastic electron–neutral (atom) collision frequency with small energy losses
$\tilde{\mu}$	parameter characterizing saturation in step ionization, introduced in (2.63)	$\bar{\nu}_{0_k}(\mathcal{W})$	introduced by (6.9)
		$\mathring{\nu}_{\mathrm{i}}$	slowly varying function of T_{e} in ν_{i}
$\tilde{\mu}_0$	normalized $\tilde{\mu}$, introduced in (5.46)	$\mathring{\nu}_{*}$	slowly varying function of T_{e} in ν_{*}
		ξ	normalized transverse coordinate according to the definition given (Chap. 5)
ν	electron–neutral (atom) elastic collision frequency		
ν_{eff}	effective collision frequency	ξ_{w}	value of ξ at the wall
ν_{en}	electron–neutral (atom) elastic collision frequency	$\tilde{\xi}$	proportion of molecular ions
ν_{i}	ionization frequency		
ν_{in}	ion–neutral collision frequency	Π	$(= \varrho_{\mathrm{r}}\beta_{\mathrm{s}}n_{\mathrm{c}}/\mathring{\nu}_{\mathrm{i}})$ parameter characterizing effects

	related to recombination losses
π	$= 3.14159\ldots$
ϱ	charge density
ϱ_{NL}	defined by (5.36b)
ϱ_{r}	coefficient of recombination
ϱ_{si}	rate coefficients of step ionization
ϱ'_{si}	rate coefficients of step ionization
$\mathring{\varrho}_{\mathrm{si}}$	slowly varying function of T_{e} in the rate coefficients ϱ_{si} of step ionization
$\mathring{\varrho}'_{\mathrm{si}}$	slowly varying function of T_{e} in the rate coefficients ϱ'_{si} of step ionization
Σ	$(= \mathring{\varrho}_{\mathrm{si}} n_{\mathrm{c}}/\mathring{\nu}_{\mathrm{i}})$ parameter characterizing effects related to step ionization
Σ_{i}	partition function of the ions
σ	parameter of phase diagram of SWs in planar ($\sigma \equiv \omega d/c$) and cylindrical ($\sigma \equiv \omega R/c$) waveguides
σ_{c}	cross-section for elastic collisions
σ_{i}	cross-section for collisions with ionization
σ_{p}	plasma conductivity
$\sigma_{\mathrm{p}(r)}$	real part of the plasma conductivity
$\sigma_{\mathrm{p_{eff}}}$	real part of the plasma conductivity defined using ν_{eff} according to (4.25)

σ_*	total cross-section for excitation
σ_k^j	cross-section for inelastic collisions with stepwise excitation of the jth atomic state from the kth state
$\sigma_0^{(\mathrm{i})}$	cross-section for collisions with ionization
$\sigma_0^j(u)$	cross-section for electron collisions with excitation to atomic state j
τ	transit time for a single pass of an electron through the region of resonance absorption of SWs
τ_{c}	time interval between two elastic electron–neutral (atom) collisions
τ_j	lifetime of the metastable state j
Υ	defined by (5.119b)
Φ	potential of the ambipolar field
Φ_{sh}	potential at the plasma–sheath boundary
Φ_{wall}	potential at the discharge wall
φ	azimuthal coordinate in cylindrical geometry
χ	coefficient of thermal conductivity

$\bar{\chi}$	$= 2.405\ldots$	ω_{r}	real part of the frequency
		$\omega_{\mathrm{p}}(r)$	plasma frequency
Ψ_1	defined by (4.155c)	$\overline{\omega_{\mathrm{p}}}$	plasma frequency presented in terms of the averaged density
Ψ_2	defined by (4.155d)		
Ω	angle in velocity space	$\overline{\omega_{\mathrm{p}}^2}$	squared plasma frequency presented in terms of the averaged, over the discharge cross-section, density
ω	frequency		
ω_{p}	plasma frequency		
$\omega_{\mathrm{p_w}}$	plasma frequency with electron density at the wall		
ω_{p_0}	plasma frequency with electron density at the discharge axis	$\wp$	introduced in (5.67a)

1. Introduction

The wide spectrum of possible investigations which gas-discharge physics offers has invigorated interest in the field, resulting in a long history of research. Different stages mark this activity: the simple accumulation of experimental results and the derivation of empirical laws, followed by a period of understanding the creation and maintenance of the discharge as a process in the self-organization of dynamic nonlinear systems, and thereafter proceeding to a kinetic modelling of discharges. The latter trend – with the rising capability of modern numerical facilities – has led again to a new accumulation of results, obtained by accounting for the specific details of the atomic and molecular structure of different gases. Though seemingly eclectic to some degree, these activities are motivated to a large extent by applicational aspects. Indeed, the realization and rising potential of technological applications have strengthened and continue to stimulate growing interest in research on gas-discharge phenomena. However, whichever objectives motivate the investigations – modelling, diagnostics, applications – it is always necessary to be aware that gas discharges are self-organized systems governed by a strong ionization nonlinearity.

The research during the last twenty-five years on the so-called "surface-wave-sustained discharges" established a new branch in gas-discharge physics and stimulated further development in the field. On the one hand, since these discharges are electrodeless and exclude processes and phenomena associated with pre-electrode regions, the area of research is narrowed down within the field of plasma physics. On the other hand, the nature of this type of discharge – plasmas created by propagating waves – requires a combined understanding of concepts in the physics of gas-discharge plasmas and of plasma electrodynamics and the physics of waves in plasmas. Work in this area provides opportunities for new contributions and injects fresh trends into the development of gas-discharge physics.

Discharges sustained by surface waves (SWs) are prime members of a group of discharges which are maintained by the fields of travelling waves. The present status of research justifies regarding them as a new type of gas discharge within the more general class of high-frequency (HF) discharges, which are here meant to encompass both radio-frequency and microwave plasmas. In general, in HF discharges the plasma is produced by the energy which electrons, oscillating in the field, gain owing to phase decorrelations by collisions and equivalent effects. However, whereas in other types of HF discharges the plasmas are created and maintained essentially by electric fields not exhibiting a wave nature, the maintenance of SW-produced discharges is in the field of travelling waves. This leads to some characteristic and interesting differences. The plasma extends far away from the zone of the wave launcher. Though this launcher is essential for the efficiency of power transfer from the source to the discharge system, the self-consistent field–plasma behaviour is virtually independent of the wave coupler. This self-consistent behaviour acts over the complete length of the discharge and manifests itself in the resulting longitudinal structure. The self-consistent variation of wave and plasma characteristics along the discharge length is in the very nature of these discharges. In SW-sustained discharges the electrodynamic boundary conditions are clearly and simply defined by the electrodynamic continuity conditions at the interfaces between the plasma and the outside.

In the following the travelling-wave-produced plasma is represented by discharges maintained by the propagation of SWs. Since the surface mode is a proper mode of a waveguide system, these discharges can also be called *waveguided* discharges. In these discharges it is the wave which creates the discharge. On the other hand, the plasma which is produced by the wave energy is a part of the waveguide structure which maintains the wave propagation. The wave and the plasma behave in a self-consistent way in all details: their properties are mutually determined. The characteristics of the wave (only its frequency ω being given externally) – its wave number k, damping rate α and electric-field intensity $|E(r, z)|^2$ – and the plasma parameters – electron temperature T_e and density n or the electron energy distribution function in general, and all characteristics of the discharges determined by those quantities – are related to each other in a self-consistent way. This self-consistent behaviour of the wave and plasma occurs on a nonlinear level. This might be regarded as obvious since the creation of any discharge is a nonlinear phenomenon. However, it is appropriate to stress the nonlinearity, because for many years of research on SW-produced plasmas it has been argued that the discharge maintenance – at least that in a diffusion-controlled regime – operates under linear conditions, thus constituting a linear problem. It has been claimed that, firstly, there is no relation between the plasma density n and the wave field intensity $|E|^2$ and, secondly, the linear properties of the wave determine the maintenance of the discharge. However, subsequent attention to the nonlinear aspects has invalidated these claims.

The physical picture of the creation of an SW-sustained discharge can be described schematically by Fig. 1.1. Arrows symbolize the transfer of energy to the Poynting flux in the vacuum outside, which is radially coupled step by step to the plasma inside. A high-amplitude signal of frequency ω is applied to launch the wave. To ensure an efficient energy transfer of the applied power into energy of an SW the design of the launcher should create a con-

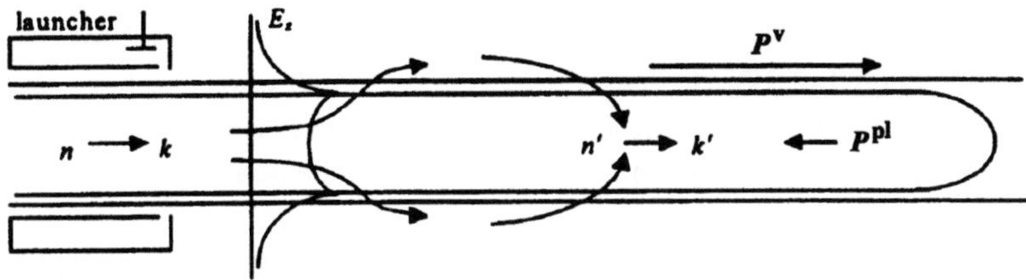

Figure 1.1. Schematic picture visualizing the maintenance of SW-sustained discharges

figuration of the field pattern in the launcher which can be transferred into that of an SW (e.g. the longitudinal field component E_z of the SWs, the field in the direction z of wave propagation is given schematically in Fig. 1.1). The breakdown of the gas inside the launcher determines the plasma density n (and the electric-field intensity) and the value of the wave number k with which the wave starts its propagation. Usually the waveguide structure consists of the plasma column created by the wave itself, a dielectric (the glass tube containing the gas discharge) and a vacuum (the surrounding free space). The wave energy flux P moves along the waveguide with its main part positioned in free space (P^{v}). Part of the wave energy is lost in creating the plasma. Because of this energy loss, the density n' of the plasma at some distance away from the launcher along the plasma column is lower. The new value n' of the plasma density determines a new value k' of the wave number, with which the wave continues its propagation. Therefore, variation of the plasma characteristics – electron temperature $T_{\mathrm{e}}(z)$, density $n(z)$, etc. – and of the wave characteristics – wave number $k(z)$, energy flux $P(z)$ and field intensity $|E(z)|^2$ – along the discharge length z is a fundamental physical feature of the discharge maintenance. The z-variations of all these quantities are self-consistently coupled, and the ionization nonlinearity which couples the plasma characteristics $T_{\mathrm{e}}(|E|^2)$ and $n(|E|^2)$ with the variation of the wave field intensity $|E|^2$ is the basis of the axial structure of the discharge.

A self-consistent interdependence of the field and plasma properties, of course, exists in all HF discharges and indeed in all discharges in general. However, in the case of discharges sustained by travelling SWs, the situation is more clearly and uniquely defined. This makes SW-sustained discharges a kind of prototype of HF discharges that is very suitable for studying the basis of self-consistency and nonlinearity in discharge formation.

1.1 The Structure of the Book

In Chap. 2 some basic relations frequently used in the other chapters are described. Chapters 3 and 4, in which the properties of the propagation of

SWs in homogeneous and inhomogeneous plasmas are summarized, contain the basic information necessary for describing the "electrodynamic part" of a discharge sustained by SW propagation. Chapters 5, 6 and 7, where the present status of the theory of these discharges and experiments on their diagnostics, respectively, are summarized, constitute the main part of this book.

Chapter 2 outlines the basic procedure of a kinetic approach to discharges, in particular the treatment of Boltzmann's equation for the electron (energy) distribution function. The "noncollisional" (stochastic) heating effects which are gaining importance in modern studies of HF discharges are encompassed. These effects are based on the Landau-type damping; in the case of SWs, however, they largely stem from components of large wave numbers present in the pronounced radial structures of nonuniform electric-field amplitudes. Such spatial amplitude structures of the (radial) field component are dominant when a strong radial density inhomogeneity is encountered. The procedure used is that of a quasi-linear treatment known from previous studies of plasma turbulence. A brief introduction is included to the different scenarios, the nonlocal and the local ones, involved in the formation of the distribution function, a subject of much interest in recent years and still of interest in current work. Chapter 2 also introduces a fluid treatment, a much used and useful approximation to gain insight into dependences and trends, where details of the kinetic effects are not specifically addressed. Basic electrodynamic laws are listed, in addition.

The theory of SW propagation in guiding structures with different geometries which contain a homogeneous plasma is outlined in Chap. 3. The SW propagates along the waveguide structure and its amplitude decays in the transverse direction away from the interface to which the field is attached. Since the SW field variation is in both the longitudinal and the transverse directions, the description of its propagation is a two–dimensional wave problem. A classification of the SW branches which exist in different regions of variation of the plasma permittivity ε is discussed. The characteristics (wave number, damping rate and field distribution in the transverse direction) of the wave behaviour for weak ($\nu < \omega$) and strong ($\nu > \omega$) collisions (where ν is the electron–atom collision frequency) are summarized. Effects associated with dielectric and plasma layers between the plasma and vacuum, as well as thermal effects, are also mentioned.

However, gas-discharge plasmas are almost always inhomogeneous. The type of transverse density profile is connected to the gas-discharge regime, which is usually defined by the type of charged-particle losses: the free fall of charged particles to the discharge walls at low pressures, losses through ambipolar diffusion in the range of medium pressures and recombination losses at higher pressure. In addition to this, the plasmas in SW-produced discharges are also inhomogeneous in the longitudinal direction (the direction along the wave propagation). The main properties of SW behaviour in inhomogeneous plasmas are included in Chap. 4. The propagation of SWs in SW-sustained plasmas represents a problem of propagation of waves with a two-dimensional field variation in a medium with two-dimensional inhomogeneity. The cases of both weak and strong inhomogeneities are considered analytically and numerically. The behaviour of the SWs in the regions of plasma resonance absorption, which is due to the inhomogeneity of the plasma density in the transverse direction, and in the region of the quasi-static resonance of the waves associated with the inhomogeneity in the longitudinal direction is analysed. The geometrical-optics (WKB) approximation, applied to analysing effects of weak inhomogeneity, is described, including the extension of its application to propagation of waves with a two-dimensional field variation. Numerical solutions of the electrodynamic field equations in the presence of two-dimensional density nonuniformity are also discussed.

Chapter 5 deals with the fluid plasma model of discharges sustained by SW propagation. The treatment is completely analytical. Since the nature of the SW-sustained discharges combines plasma and wave behaviour in a self-consistent way, basic relations of gas-discharge physics, which describe the electron heating and charged-particle production, and basic electrodynamic laws, which describe the wave propagation, are both involved in the model. The maintenance of stationary discharges without an external magnetic field is described. Local and nonlocal regimes of electron heating in the SW field are considered. The determination of the electron temperature $T_\mathrm{e}(|\boldsymbol{E}|)$ in terms of the field intensity, giving the thermal nonlinearity, is considered as a stage in the development of the ionization nonlinearity. The charged-particle production is considered as due to direct and step ionization, which compensates the losses of charged particles due to diffusion to the tube walls and recombination. It is shown that the mechanisms of nonlinearity – step ionization and/or recombination – provide possibilities to couple the plasma density to the field intensity and to obtain the relation for $n(|\boldsymbol{E}|^2)$, which is the final result of the ionization nonlinearity. The permittivity of the plasma sustained by the field is nonlinear. It enters the electrodynamic laws by which the SW propagation is described. In such a scheme the description of the problem of sustaining discharges by SW propagation involves a closed set of nonlinear equations which gives, at its first stage, results for the effect of the nonlinearity – $T_\mathrm{e}(|\boldsymbol{E}|^2)$, $n(|\boldsymbol{E}|^2)$ – which are valid for discharge maintenance in an inhomogeneous field in principle, and, at its second stage, gives results for a self-consistent axial variation of all the characteristics of the wave and the plasma, which is a specific feature of SW-sustained discharges. The self-consistency can also be related to the fact that the Joule losses of wave power, which enter the electrodynamic part of the problem, are the same quantity as that which enters the gas-discharge part as a Joule heating term. Results for the axial structure of discharges maintained by Joule heating in the plasma volume for both weak ($\nu < \omega$) and strong ($\nu > \omega$) collisions are obtained. The contribution of the resonance absorption of SWs to the electron heating is analysed. In all these cases both a diffusion-controlled regime and a recombination-controlled regime are treated.

The results from numerical kinetic modelling of the discharges are presented in Chap. 6. Both nonlocal and local regimes for the determination of electron-energy-distribution functions (EEDFs) in SW-sustained discharges are discussed, and the nonlocal regime is applied extensively in numerical treatments. On the basis of solutions for the spatially dependent EEDF, the conditions for the applicability of these two approaches are discussed. The Joule heating in the plasma volume and heating by the enhanced field in the regions of resonance absorption of the SW field are included. The collisionless heating, as an effect of the particle–wave kinetics, is introduced as a third mechanism of electron heating. It is interpreted as a kind of "transit time" heating in the presence of sharp field inhomogeneities. The contributions of the three channels of energy transfer to the formation of the discharge structure are compared. The role of two-dimensional density (and field) nonuniformity is considered in detail, and further influences on the interdependence of electron density and field intensity and their interrelation are discussed.

In Chap. 7 connection is made between the theory and modelling treated in the previous chapters to the experimental background. Actually, there is a tremendous body of observations and experimental findings, on which the interest in waveguided discharges is based and which stimulated the analysis of these discharges. However, in the context of this book Chap. 7 is a survey. The main results included are experimental results which reflect features typical of waveguided discharges. Some more recent observations which have a particular bearing on the theoretical aspects stressed in the book are included and commented on. The chapter starts, though, with a brief description of the diagnostic methods used, with some stress on methods typically used for SW discharges. At the end applicational aspects are listed; these are expected to be of growing importance in the future expansion of the work in this field.

All the results discussed concern the problem of discharge maintenance in the absence of static magnetic fields, under stationary conditions and at comparatively low pressure. The presentation of theoretical and experimental results is in clear imbalance. This is done on purpose, since many experimental results have been collected and discussed in excellent review papers, while the theoretical results are scattered in separate papers. In this monograph, the aim has been to try to convincingly show that SW-produced plasmas or, in general, waveguided discharges represent an interesting subject for fundamental research in the field of gas-discharge physics.

2. Basic Relations

Some relations used frequently and widely in the further presentation are briefly commented on in this chapter. As has been mentioned before (Chap. 1), the nature of waveguided discharges combines wave properties and plasma characteristics. As a consequence of this, their description is a problem which unifies concepts from the fields of electrodynamics and gas-discharge physics. Therefore, appropriately here at the very beginning, a summary is given of the electrodynamic laws, equations modelling the plasma behaviour and relations used for describing the gas-discharge regimes.

2.1 Basic Laws of Plasma Electrodynamics

In the description of electromagnetic (EM) fields in plasmas [2.1–3], Maxwell's equations are the starting point and are employed in their ordinary form:

$$\operatorname{curl} \boldsymbol{E} = -\frac{\partial \boldsymbol{B}}{\partial t} \tag{2.1}$$

$$\operatorname{curl} \boldsymbol{H} = \frac{\partial \boldsymbol{D}}{\partial t} \quad \text{or} \quad \operatorname{curl} \boldsymbol{H} = \varepsilon_0 \frac{\partial \boldsymbol{E}}{\partial t} + \boldsymbol{j}\,, \tag{2.2}$$

$$\operatorname{div} \boldsymbol{B} = 0\,, \tag{2.3}$$

$$\operatorname{div} \boldsymbol{D} = 0 \quad \text{or} \quad \operatorname{div} \boldsymbol{E} = \frac{\varrho}{\varepsilon_0}\,. \tag{2.4}$$

Here $\boldsymbol{E}$ and $\boldsymbol{H}$ are the electric- and magnetic-field strengths, respectively, $\boldsymbol{D}$ and $\boldsymbol{B}$ are the electric displacement and magnetic induction, $\boldsymbol{j}$ and ϱ are the current and charge densities, and ε_0 is the vacuum permittivity. The time (t) and space $(\boldsymbol{r})$ variations of the quantities describing the field are taken in the subsequent chapters in the form

$$\boldsymbol{E}(t,\boldsymbol{r}) = \frac{1}{2}\left[\boldsymbol{E}(\boldsymbol{r},\omega,\boldsymbol{k})\exp(-\mathrm{i}\omega t + \mathrm{i}\,\boldsymbol{k}\cdot\boldsymbol{r}) + \text{c.c.}\right] \tag{2.5}$$

for the electric field, and analogously for the other quantities $\boldsymbol{H}$, $\boldsymbol{B}$, $\boldsymbol{D}$; ω and $\boldsymbol{k}$ are the frequency and the wave vector, and $\boldsymbol{E}(\boldsymbol{r},\omega,\boldsymbol{k})$ is the wave amplitude taken for stationary cases.

The electric displacement and magnetic induction

$$\boldsymbol{D}(\omega,\boldsymbol{k}) = \varepsilon_0\varepsilon(\omega,\boldsymbol{k})\boldsymbol{E}(\omega,\boldsymbol{k})\,, \tag{2.6}$$

$$\boldsymbol{B}(\omega,\boldsymbol{k}) = \mu_0\mu'(\omega,\boldsymbol{k})\boldsymbol{H}(\omega,\boldsymbol{k}) \tag{2.7}$$

involve the relative permittivity $\varepsilon(\omega,\boldsymbol{k})$ of the medium and its relative magnetic susceptibility $\mu'(\omega,\boldsymbol{k})$. A value of $\mu' = 1$ is taken for the media of the waveguided systems under consideration; μ_0 is the vacuum susceptibility.

The expression used for the plasma permittivity is that for cold inhomogeneous, weakly ionized collisional plasmas:

$$\varepsilon(\omega,\boldsymbol{r}) = 1 - \frac{\omega_{\mathrm{p}}^2(\boldsymbol{r})}{\omega(\omega + \mathrm{i}\nu)}\,, \tag{2.8a}$$

where

$$\omega_{\mathrm{p}}(\boldsymbol{r}) = \left(\frac{e^2 n(\boldsymbol{r})}{\varepsilon_0 m}\right)^{1/2} \tag{2.8b}$$

is the plasma frequency, and $\nu \equiv \nu_{\mathrm{en}} \equiv \nu_{\mathrm{m}}$ is the electron–neutral (atom) collision frequency for momentum transfer; e and m are the electron charge (absolute value) and mass, respectively, and $n(\boldsymbol{r})$ is the electron density. The real and imaginary parts of the permittivity are

$$\varepsilon_{\mathrm{r}}(\omega,\boldsymbol{r}) = 1 - \frac{\omega_{\mathrm{p}}^2(\boldsymbol{r})}{\omega^2 + \nu^2}\,, \tag{2.9a}$$

$$\varepsilon_{\mathrm{i}}(\omega,\boldsymbol{r}) = \frac{\nu}{\omega}\,\frac{\omega_{\mathrm{p}}^2(\boldsymbol{r})}{\omega^2 + \nu^2}\,. \tag{2.9b}$$

The scale of the space dispersion effects, when they are incorporated in the plasma permittivity $\varepsilon(\omega,\boldsymbol{k})$, is determined with respect to the Debye length r_{D} of the electrons, i.e.

$$r_{\mathrm{D}} = \left(\frac{\varepsilon_0 T_{\mathrm{e}}}{e^2 n}\right)^{1/2}\,, \tag{2.10}$$

where T_{e} is the electron temperature in energy units.

With varying wave frequency ω or electron density n (which is equivalent to varying the plasma frequency ω_{p}), the plasma permittivity $\varepsilon(\omega)$ could cover regions of positive and negative values. When the plasma frequency $\omega_{\mathrm{p}}(r)$ is equal to the wave frequency, the plasma density takes its critical value

$$n_{\mathrm{c}} = \frac{\varepsilon_0 m \omega^2}{e^2}\,. \tag{2.11}$$

The plasma conductivity σ_{p} is defined by

$$\boldsymbol{j} = \sigma_{\mathrm{p}}\boldsymbol{E}\,, \tag{2.12a}$$

where

$$\boldsymbol{j} = \sum_\alpha e_\alpha n_\alpha \boldsymbol{v}_\alpha \tag{2.12b}$$

is the current density (with $\alpha = $ e, i as an index denoting the electron and ion plasma components, and $\boldsymbol{v}_\alpha$ and $e_\alpha = Ze$ being the mean particle velocity and the particle charge, respectively; $Z = -1$ for electrons and $Z \geq 1$ for ions). In the case of cold isotropic plasmas its real part is determined by

$$\sigma_{\mathrm{p(r)}} = \frac{\varepsilon_0 \nu \omega_{\mathrm{p}}^2}{\omega^2 + \nu^2} \,. \tag{2.13}$$

In general, the relation between the plasma conductivity and the permittivity is

$$\varepsilon(\omega, \boldsymbol{k}) = 1 + \frac{\mathrm{i}}{\varepsilon_0 \omega} \sigma_{\mathrm{p}}(\omega, \boldsymbol{k}) \,. \tag{2.14}$$

The wave field equations

$$\Delta \boldsymbol{E} - \mathrm{grad}\ \mathrm{div}\ \boldsymbol{E} + \frac{\omega^2 \varepsilon(\omega, \boldsymbol{r})}{c^2} \boldsymbol{E} = 0 \,, \tag{2.15}$$

$$\Delta \boldsymbol{H} + \frac{1}{\varepsilon(\omega, \boldsymbol{r})} \mathrm{grad}\ \varepsilon(\omega, \boldsymbol{r}) \times \mathrm{curl}\ \boldsymbol{H} + \frac{\omega^2 \varepsilon(\omega, \boldsymbol{r})}{c^2} \boldsymbol{H} = 0 \tag{2.16}$$

are used in the form commonly employed for arbitrary inhomogeneous dielectric media; here $c = (\varepsilon_0 \mu_0)^{-1/2}$ is the speed of light in vacuum.

The solution of problems of EM wave propagation in bounded media – the cases of waveguided systems considered in this book – involves application of the boundary conditions for the fields. Conditions for the continuity of the tangential components (with respect to the interface between the two media, denoted by the indices 1 and 2) of $\boldsymbol{E}$ and $\boldsymbol{B}$,

$$\{E_{\mathrm{t}}\}|_{\text{interface}} = 0 \,, \tag{2.17a}$$

$$\{B_{\mathrm{t}}\}|_{\text{interface}} = 0 \,, \tag{2.17b}$$

are applied. Here $\{f\}|_{\text{interface}}$ is a shorthand for the difference $f_2 - f_1$, and surface currents and energy absorption at the interface are assumed to be absent. Continuity of the impedances

$$\mathcal{Z} = \frac{E_{\mathrm{t}}}{H_{\mathrm{t}}} \tag{2.18a}$$

of the two media at the interface

$$\{\mathcal{Z}\}|_{\text{interface}} = 0 \tag{2.18b}$$

is also used as a boundary condition.

The dispersion relation

$$\mathcal{D}(\omega, \boldsymbol{k}) = 0 \tag{2.19}$$

appears as a condition for the existence of a solution of a set of homogeneous equations. It is obtained when the solutions of Maxwell's equations for the different media of the system are inserted into the boundary conditions. The function $\mathcal{D}(\omega, \boldsymbol{k})$ is complex, in general.

In initial-value problems, the solution of the dispersion relation gives the dependence of the frequency $\omega \to \omega_r + i\gamma$ on the wave vector k, i.e. $\omega(k)$; the slow time evolution of the field is characterized by the time damping rate γ. When the problem is considered as a boundary value problem, the solution gives the dependence of the wave number $k \to \beta + i\alpha$ on the frequency, i.e. $k(\omega)$, and the space damping rate α describes the slow space evolution of the wave.

In the case of weakly damped guided modes k is along the waveguide axis (the z direction from here on), i.e. $k(0, 0, k)$. The real part of (2.19),

$$\mathrm{Re}\left\{\mathcal{D}[\omega(k), k]\right\} = 0 \qquad \text{or} \qquad \mathrm{Re}\left\{\mathcal{D}[\omega, k(\omega)]\right\} = 0 \,, \tag{2.20}$$

giving the wave dispersion behaviour, determines both the phase velocity

$$v_{\mathrm{ph}} = \frac{\omega}{\beta} \tag{2.21a}$$

and the group velocity

$$v_{\mathrm{gr}} = \frac{\partial \omega_r}{\partial \beta} = -\frac{\partial \mathrm{Re}\{\mathcal{D}\}/\partial k|_{k=\beta}}{\partial \mathrm{Re}\{\mathcal{D}\}/\partial \omega|_{\omega=\omega_r}} \tag{2.21b}$$

of the waves. When v_{ph} and v_{gr} point in the same direction, the waves are propagating forward; the case of oppositely directed v_{ph} and v_{gr} corresponds to backward waves.

To obtain the damping rates of waves with arbitrary damping, including cases of anomalous dispersion [2.4], the complex equation (2.19) should be solved by separating its real and imaginary parts. For strongly damped waves, this is the only way to obtain a solution. In the case of weakly damped waves ($\gamma \ll \omega_r$, $\alpha \ll \beta$) a Taylor expansion can be applied to $\mathrm{Re}\{\mathcal{D}(\omega, k)\}$ and, with the imaginary part of (2.19) separated, expressions for the time and space damping rates can be obtained as follows:

$$\gamma = -\frac{\mathrm{Im}\{\mathcal{D}[\omega_r(k), k]\}}{\partial \mathrm{Re}\{\mathcal{D}[\omega(k), k]\}/\partial \omega|_{\omega_r}} \,, \tag{2.22a}$$

$$\alpha = -\frac{\mathrm{Im}\{\mathcal{D}[\omega, \beta(\omega)]\}}{\partial \mathrm{Re}\{\mathcal{D}[\omega, k(\omega)]\}/\partial k|_{k=\beta}} \,, \tag{2.22b}$$

where ω_r and β are the real parts of the frequency and the wave number, respectively.

With the field variation taken in the form (2.5), $\gamma < 0$ and $\alpha > 0$ yield damped waves considered in terms of initial- and boundary-value problems, respectively, whereas $\gamma > 0$ and $\alpha < 0$ correspond to instabilities in the evolution of the wave field. The time and space damping rates are related to each other through the group velocity, according to

$$\alpha = -\frac{\gamma}{v_{\mathrm{gr}}} \,. \tag{2.23}$$

The bulk waves in isotropic infinite, spatially uniform plasmas are either purely longitudinal ($k \parallel E$) or purely transverse ($k \perp E$), and they propagate for $\varepsilon \geq 0$, i.e. for $\omega/\omega_{\mathrm{p}} \geq 1$.

In cold plasmas the longitudinal waves degenerate into plasma oscillations (bulk plasmons) at the frequency

$$\omega = \omega_{\mathrm{p}} \tag{2.24a}$$

and the dispersion law of the EM bulk waves is

$$k = \frac{\omega}{c}\sqrt{\varepsilon} \,. \tag{2.24b}$$

The spatial dispersion introduced by thermal motion transforms the plasma oscillations (2.24a) into waves, the collisional time damping rate of which,

$$\gamma = -\frac{\nu}{2}, \tag{2.25a}$$

is almost constant because of the weak plasma dispersion. The variation of their space damping rate is due to the variation of the group velocity along the dispersion curve. The collisionless Landau damping of the longitudinal waves is exponentially small. The space damping rate (2.23) of the transverse waves in transparent plasmas ($\omega \gg \omega_{\mathrm{p}}$, where the group velocity is almost constant and equal to the vacuum light speed c) changes because of variation of their time damping rate

$$\gamma = -\frac{\nu}{2}\frac{\omega_{\mathrm{p}}^2}{\beta^2 c^2} \,. \tag{2.25b}$$

At frequencies ω below the plasma frequency (i.e. for $\varepsilon < 0$), the penetration depth of EM waves in cold plasmas over the skin depth λ_{sk}, which in the cases of weak ($\nu < \omega$) and strong ($\nu > \omega$) collisions is given by

$$\lambda_{\mathrm{sk}} = \frac{c}{\omega_{\mathrm{p}}} \tag{2.26a}$$

and

$$\lambda'_{\mathrm{sk}} = \frac{c}{\omega_{\mathrm{p}}}\left(\frac{2\nu}{\omega}\right)^{1/2}, \tag{2.26b}$$

respectively.

The wave energy conservation law is

$$\frac{\partial W}{\partial t} + \operatorname{div} P = -Q \tag{2.27}$$

and the quantities involved in it are as follows:

$$W = \frac{1}{4}\left(\mu_0 |H|^2 + \varepsilon_0 \left.\frac{\partial}{\partial \omega}[\omega \varepsilon_{\mathrm{r}}(\omega, k)]\right|_{\omega_{\mathrm{r}}} |E|^2\right) \tag{2.28a}$$

is the wave energy, consisting of the contributions from the magnetic- and electric-field energies and the average kinetic energy of the particle oscillations,

$$P = \frac{1}{4}\left(E \times H^* + E^* \times H\right) \tag{2.28b}$$

is the Poynting vector (or the wave energy flux through a unit surface), and

$$Q = \frac{1}{2}\varepsilon_0\omega_r\varepsilon_i(\omega, k)|E|^2 \tag{2.28c}$$

describes the wave losses in the medium. In the case of a space-dispersive medium the flux

$$T = -\frac{\omega}{4}\frac{\partial\varepsilon_r(\omega, k)}{\partial k}|E|^2 \tag{2.28d}$$

of the coherent-motion energy of charged carriers should be added to the wave energy flux P. By using (2.14), the wave power losses (2.28c) can be represented in the form

$$Q = \frac{1}{2}\sigma_{p(r)}|E|^2\,, \tag{2.29}$$

where $\sigma_{p(r)}$ is the real part of the plasma conductivity (in the case of cold plasmas, see (2.13)).

The group velocity (2.21b) can be expressed in a more general form by the ratio of the total energy flux to the total energy density:

$$v_{gr} = \frac{P + T}{W}. \tag{2.30}$$

The form in which the Poynting theorem is usually applied to guided-wave propagation comes out after integration of (2.27) over a volume enclosed between two planes $S_\perp$ parallel to the waveguide cross-section extended to infinity:

$$\int\limits_{(V)}\frac{\partial W}{\partial t}\,d\boldsymbol{r} + \oint\limits_{S}\boldsymbol{P}\cdot d\boldsymbol{S} = -\int\limits_{(V)}Q\,d\boldsymbol{r} \tag{2.31a}$$

or, equivalently,

$$\frac{\partial}{\partial t}\int\limits_{S_\perp}W\,dS_\perp + \frac{\partial}{\partial r_i}\int\limits_{S_\perp}P_i\,dS_\perp = -\int\limits_{S_\perp}Q\,dS_\perp\,, \tag{2.31b}$$

where P_i is the Poynting vector component parallel to the direction of propagation r_i. Therefore, the Poynting theorem finally takes the form

$$\frac{\partial W}{\partial t} + \frac{\partial P_i}{\partial r_i} = -Q, \tag{2.32a}$$

where

$$W = \int\limits_{S_\perp} W \, dS_\perp, \quad P_i = \int\limits_{S_\perp} P_i \, dS_\perp \text{ and } Q = \int\limits_{S_\perp} Q \, dS_\perp. \tag{2.32b}$$

It can be shown that W is given by (2.28a), in which the volume plasma permittivity ε_r is replaced by the function $\mathrm{Re}\{\mathcal{D}(\omega, \beta)\}$ involved in the dispersion relation of the guided waves.

2.2 Basic Equations for Modelling Gas Discharges

2.2.1 Kinetic-Model Equations

Boltzmann's Equation. The electron distribution function (EDF) $F(r, v, t)$ obeys the kinetic equation

$$\frac{\partial F}{\partial t} + v \cdot \frac{\partial F}{\partial r} - \frac{e}{m}(E + v \times B) \cdot \frac{\partial F}{\partial v} = S(F), \tag{2.33}$$

where the evolution of the EDF connected with particle collisions is described by the collisional integral $S(F)$, and $E(r, t)$ and $B(r, t)$ are the electric and magnetic fields, respectively (self-consistent with the plasma).

The solution of (2.33) gives the EDF $F(r, v, t)$ in phase space or the electron velocity distribution function $f(r, v, t)$ defined by $F(r, v, t) = n(r, t)f(r, v, t)$, where $n(r, t) = \int_{(v)} F(r, v, t) \, d^3r$ is the plasma density. (The normalizing condition for $f(r, v, t)$ is $\int_{(v)} f(r, v, t) \, d^3v = 1$.)

The procedure commonly used [2.5–7] for solving the kinetic equation (2.33) in gas-discharge physics is the so-called Lorentz approximation, i.e. a series expansion of the EDF in Legendre polynomials, resulting in

$$F(r, v, t) = F_0(r, v, t) + \frac{v}{v} \cdot F_1(r, v, t). \tag{2.34a}$$

Here F_0 is the isotropic component of the EDF, and F_1 is its anisotropic part. They both depend on the absolute value of the velocity vector $v = |v|$. Since the anisotropy of the EDF is small, even in strong electric fields, the series expansion (2.34a) in a parameter characterizing the anisotropy can be restricted to the first-order term. For example, with elastic collisions as the main mechanism for losses of electron energy, the ratio of the directed velocity v_E of the electrons to their thermal velocity v_{th}

$$\frac{v_E}{v_{\mathrm{th}}} = \sqrt{\delta} \tag{2.34b}$$

is the small parameter in (2.34a); $\delta = 2m/M$ is the proportion of energy transferred by an elastic collision (i.e. the so-called energy transfer coefficient for elastic collisions), and M is the atom mass.

In quasi-stationary situations, for most of the cases considered, the EDF F_0 is almost independent of time, when the frequency ω of the applied high-frequency electric field

$$E_{\mathrm{HF}} = \frac{1}{2}\left[E\exp(-\mathrm{i}\,\omega t) + \mathrm{c.c.}\right]$$

is larger than the energy relaxation frequency ν_ε, i.e.

$$\omega > \nu_\varepsilon\,, \tag{2.35a}$$

where

$$\nu_\varepsilon = \delta\nu + \sum \nu_{0k}\,. \tag{2.35b}$$

By ν_{0k} the collision frequencies for the inelastic processes starting from the ground state are denoted.

When effects of the ambipolar field are not taken into account, the equation for the isotropic part of the distribution function F_0 in the one-dimensional case is

$$\frac{\partial}{\partial x}\left(\frac{v^2}{3\nu}\frac{\partial F_0}{\partial x}\right) + \frac{1}{v^2}\frac{\partial}{\partial v}\left[v^2 D(v,x)\frac{\partial F_0}{\partial v}\right] = -S_0\,, \tag{2.36}$$

where $D(v,x)$ is the diffusion coefficient in the velocity space, and S_0 is the velocity-angle-averaged collisional integral. Usually the assumption is made that $\nabla_r\nu = 0$, simplifying (2.36) somewhat.

The case of nonlocal heating is realized when the scale length L_E of the heating field is much less than the electron-energy relaxation length L_ε. Then, the diffusion coefficient $D(v,x)$ can be represented as

$$D(v,x) = \frac{1}{v^2}\frac{\pi e^2}{m^2}\,\delta(x - x_0)\int\frac{\mathrm{d}\Omega}{4\pi}\int \mathrm{d}k_x\,\left|vE(k_x)\right|^2\frac{\nu}{(\omega - k_x v_x)^2 + \nu^2}\,, \tag{2.37a}$$

where an integration over the angle Ω is performed in order to cover all the directions of the velocity vector and

$$E(k_x) = \int_{-\infty}^{\infty} E(x)\exp(-\mathrm{i}\,k_x x)\,\frac{\mathrm{d}x}{2\pi} \tag{2.37b}$$

is a Fourier component of the heating field $E(x)$. The appearance of the delta function $\delta(x - x_0)$ in (2.37a) is related to a space localization of the heating field in the region $x = x_0$ on scale lengths smaller than L_ε. Since the magnetic field does not perform work on the electrons and does not heat them, the magnetic-field component of the wave field does not enter into the angle-averaged coefficient of the electron diffusion in energy space [2.8]. In the long-wavelength range of the Fourier spectrum ($k_x < \nu/v_x$), the electron heating occurs through the ordinary mechanism of Joule heating via collisions and, correspondingly, the diffusion coefficient (2.37a) reduces to

$$D(v,x) = \frac{1}{6}\frac{e^2}{m^2}\,\delta(x - x_0)\,\frac{\nu}{\omega^2 + \nu^2}\int_{-\infty}^{\infty}\left|E(x)\right|^2\,\mathrm{d}x\,. \tag{2.38a}$$

In the opposite limiting case, i.e. the short-wavelength range $k_x \gg \nu/v_x$, the collisions do not play a role, the heating becomes collisionless and, correspondingly, the diffusion coefficient (2.37a) becomes

$$D(v, x) = \frac{1}{v^2} \frac{\pi^2 e^2}{m^2} \delta(x - x_0) \int \frac{\left| v\boldsymbol{E}\left(k_x = \omega/v_x\right)\right|^2}{|v_x|} \frac{\mathrm{d}\Omega}{4\pi} . \tag{2.38b}$$

In obtaining (2.38b) from (2.37a) the relation

$$\lim_{\nu \to +0} \frac{\nu}{(\omega - k_x v_x)^2 + \nu^2} = \pi\delta(\omega - k_x v_x) \tag{2.38c}$$

has been employed. For more details on collisionless heating in different types of discharges, see [2.8].

In the opposite case $L_E \gg L_\varepsilon$, the heating regime is local, and the usual collisional Joule's heating is the only heating mechanism which can arise. Then, the diffusion coefficient is

$$D(v, x) = \frac{1}{6} \frac{e^2}{m^2} \frac{\nu}{\omega^2 + \nu^2} \left|\boldsymbol{E}(x)\right|^2 . \tag{2.39}$$

When the heating field has different scale lengths for its components, e.g. its longitudinal and transverse components, as may happen in the case of SWs, the total diffusion coefficient is a sum of the partial diffusion coefficients defined with respect to each field component, taking into account the predominant mechanism of heating associated with it (Chap. 6).

However, even in the case of the local regime of plasma heating $L_E \gg L_\varepsilon$, another reason for a nonlocality in the kinetic description can appear. This is connected to the plasma inhomogeneity and arises when the characteristic length L_n of the density inhomogeneity (and, correspondingly, of the ambipolar electric field $E_A = -\partial\Phi/\partial x$) becomes less than L_ε. In this case the EDF cannot be represented (in the general case) in the form $F_0(x, u) = n(x)f(u)$ and is nonlocal. However, over the scale length L_n, the electrons do not lose energy through collisions and, in a first approximation, the distribution function $F_0(u)$ – in the energy scale $u = (mv^2)/(2e)$ – depends only on the total energy $\mathcal{W}$ [2.9], in which the potential energy in the ambipolar field is included, i.e.

$$F_0(x, u) = F_0^{(0)} \left[\mathcal{W} \equiv u - \Phi(x)\right] . \tag{2.40}$$

This aspect of nonlocality in the determination of the EDF is taken up in more detail in Chap. 6, where a nonlocal scenario is used, followed by a discussion of an opposite local scenario disregarding inhomogeneity and space-charge effects.

In the space of the energy u, the equation equivalent to (2.36), but including the space-charge effects of the ambipolar field $\boldsymbol{E}_A$ to be used later, is

$$\frac{2}{3} \frac{e}{m} u^{3/2} \left\{ \boldsymbol{\nabla} \cdot \left[\frac{1}{\nu} \boldsymbol{\nabla} F_0(u, \boldsymbol{r})\right] - \boldsymbol{\nabla} \cdot \left[\boldsymbol{E}_A \frac{\partial F_0(u, \boldsymbol{r})}{\partial u}\right]\right\}$$

$$+ \frac{2}{3} \frac{e}{m} \frac{\partial}{\partial u} \left\{\frac{u^{3/2}}{\nu} \left[(-\boldsymbol{E}_A \cdot \boldsymbol{\nabla}) F_0(u, \boldsymbol{r}) + (E_A^2 + E_{\text{eff}}^2) \frac{\partial F_0(u, \boldsymbol{r})}{\partial u}\right]\right\}$$

$$= S_0 , \tag{2.41a}$$

where

$$E^2_{\text{eff}} = \frac{1}{2}\,\frac{\nu^2}{\omega^2 + \nu^2}\,|\boldsymbol{E}|^2 \tag{2.41b}$$

is the effective field intensity, and the collisional term S_0 is redefined and specified below. It includes [2.5–7] the following.

(1) Electron–atom elastic collisions with

$$S_0|_{\text{elastic}} = -\frac{\partial}{\partial u}\left(u^{3/2}\delta\nu F_0\right)\,, \tag{2.42}$$

taken in a form in which the term containing the gas temperature T_{g} is neglected.

(2) Electron–atom inelastic collisions with direct excitation from the ground state:

$$S_0|^{(1)}_{\text{inelastic}} = \sum_j \nu_{0j}\sqrt{u}F_0(u) \equiv \left(\frac{2e}{m}\right)^{1/2} N_0\left[\sum_j \sigma^j_0(u)\right] uF_0(u)\,, \tag{2.43a}$$

where $\nu_{0j} = (2e/m)^{1/2}N_0\sigma^j_0(u)u^{1/2}$ is the electron collision frequency for excitation of the jth atomic level, $\sigma^j_0(u)$ is the corresponding cross-section and N_0 is the neutral-gas density. The term which enters into S_0 together with (2.43a) to describe the transfer of electrons from the high-energy range into the low-energy one after this type of inelastic collision is

$$S_{0r}|^{(1)}_{\text{inelastic}} = -\sum_j \sqrt{u + U_j}\,\nu_{0j}(u + U_j)\,F_0(u + U_j)\,, \tag{2.43b}$$

where U_j is the excitation energy for the jth atomic level.

(3) Electron–atom inelastic collisions with direct ionization from the ground state:

$$S_0|^{(2)}_{\text{inelastic}} = \nu_{0i}(u)\sqrt{u}F_0(u) \equiv \left(\frac{2e}{m}\right)^{1/2} N_0\sigma^{(i)}_0(u)uF_0(u)\,, \tag{2.44a}$$

where $\nu_{0i} = (2e/m)^{1/2}N_0\sigma^{(i)}_0(u)u^{1/2}$ is the ionization frequency and $\sigma^{(i)}_0$ is the corresponding cross-section. The arrival of electrons in the low-energy range after collisions causing ionization is described by

$$S_{0r}|^{(2)}_{\text{inelastic}} = -4\sqrt{2u + U_i}\,\nu_{0i}(2u + U_i)\,F_0(2u + U_i)\,, \tag{2.44b}$$

with U_i being the ionization energy. See later on step ionization.

(4) Electron–atom inelastic collisions with stepwise excitation and de-excitation (i.e. superelastic collisions) to the jth atomic state from the kth state :

$$S_0\big|^{(3)}_{\text{inelastic}} = \sum_j \sum_k \nu_{jk}\sqrt{u}\,F_0(u) \equiv \sqrt{\frac{2e}{m}}\left(\sum_k N_k \sigma^j_k\right) u F_0(u)\,,$$

$$(2.45\text{a})$$

where N_k is the density of excited atoms in the kth level. The corresponding terms for the arrival of electrons in other energy ranges are

$$S_{0\text{r}}\big|^{(3)}_{\text{inelastic excitation}}$$
$$= -\sum_j \sum_{k<j} \sqrt{u + U_k - U_j}\,\nu_{jk}(u + U_k - U_j)\,F_0(u + U_k - U_j)\,,$$

$$(2.45\text{b})$$

after inelastic collisions with excitation which transfer comparatively high-energy electrons into the lower-energy range, and

$$S_{0\text{r}}\big|^{(3)}_{\text{superelastic}}$$
$$= -\sum_j \sum_{k>j} \sqrt{u + U_j - U_k}\,\nu_{jk}(u + U_j - U_k)\,F_0(u + U_j - U_k)\,,$$

$$(2.45\text{c})$$

after superelastic collisions which transfer comparatively low-energy electrons into a higher-energy range.

(5) Electron–electron collisions:

$$S_{0\text{e–e}} = -\frac{e^2\sqrt{e}}{4\pi\sqrt{2}\varepsilon_0^2\sqrt{m}}\ln(\Lambda)\,\frac{\mathrm{d}}{\mathrm{d}u}\left[\frac{2}{3}g(u)\frac{\partial F_0}{\partial u} + h(u)F_0(u)\right]\,, \quad (2.46)$$

where

$$g(u) = \int_0^u F_0(u')u'^{\,3/2}\,\mathrm{d}u' + u^{3/2}\int_u^\infty F_0(u')\,\mathrm{d}u'\,,$$
$$h(u) = \int_0^u F_0(u')u'^{\,1/2}\,\mathrm{d}u'$$

and

$$\ln(\Lambda) = \ln\left[12\pi\frac{(^2\!/_3\,\bar{u}\varepsilon_0 e)^{3/2}}{e^3 n^{1/2}}\right]$$

is the Coulomb logarithm, with $\bar{u}$ being the mean value of u. Λ is given by the ratio of the Debye length to the minimum binary-collision length.

The influence of the different types of collisions on the EDF is as follows:

- The electron–atom elastic collisions are the main contributor to the formation of the body of the distribution function and lead, with $\nu = \text{const.}$, to its Maxwellization. Their frequency is

$$\nu = N_0\langle\sigma_c v\rangle = N_0\sqrt{\frac{2e}{m}}\int_0^\infty \sigma(u)\sqrt{u}\,F_0(u)\,\mathrm{d}u\,, \qquad (2.47\text{a})$$

where $\sigma_{\rm c}$ is the cross section for elastic collisions, and $f_0(u)$ is involved through $F_0(u) = n f_0(u)$.

- The inelastic collisions with direct excitation and ionization lead to the depopulation of the high-energy part of the distribution function and play an important role in forming the tail of the distribution. Their frequencies are obtained by integration over the tail of the distribution:

$$\nu_{0j,0i} = N_0 \langle \sigma_{0j,i} v \rangle = N_0 \sqrt{\frac{2e}{m}} \int\limits_{U_j, U_{\rm i}}^{\infty} \sqrt{u}\, \sigma_{0j,i}(u) \sqrt{u} F_0(u)\, {\rm d}u \ . \qquad (2.47\mathrm{b})$$

- The inelastic collisions with excitation to a higher level from an excited state could be important for determining the EDF in discharges in gases with metastable states.
- The electron–electron elastic collisions, which are important in gas discharges with relatively high electron densities, thermalize the electron plasma component and cause Maxwellization of the EDF. In a sense, since they lead to an increase of the population in the tail of the distribution function, they can compensate, to a certain extent, the effect of the inelastic collisions. In fact, the competition between the inelastic process above and electron–electron collisions essentially determines whether or not the tail of the distribution function is lowered or whether the Maxwellian distribution extends into the tail. Since the rate of inelastic processes from the ground state is proportional to nN_0, whereas the rate of electron–electron collisions is proportional to n^2, the degree of ionization n/N_0 is decisive in the formation of the distribution.

In weakly ionized plasmas the elastic electron–atom collision frequency has the largest value. However, because of the small value of the energy transfer in elastic collisions (of the order of $\delta \equiv 2m/M$), the inelastic collision frequency for excitation is usually the largest among the frequencies for energy transfer.

The anisotropic part $F_1(r, v, t)$ of the EDF (see (2.34a)) determines the coefficients of the transport processes: mobility, diffusion, thermal diffusion and thermal conductivity. These coefficients are as follows:

- electron mobility coefficient

$$b_{\rm e} = -\frac{4\pi}{3} \frac{e}{m} \int_0^{\infty} \frac{v^3}{\nu_{\rm en}^s} \frac{{\rm d}f_0}{{\rm d}v}\, {\rm d}v \qquad (2.48\mathrm{a})$$

- electron diffusion coefficient

$$D_{\rm e} = \frac{4\pi}{3} \int_0^{\infty} \frac{v^4}{\nu_{\rm en}^s} f_0\, {\rm d}v \qquad (2.48\mathrm{b})$$

- thermal diffusion coefficient

$$D_{\rm e}^{\rm T} = T_{\rm e} \frac{\partial D_{\rm e}}{\partial T_{\rm e}}, \qquad (2.48\mathrm{c})$$

where $\nu_{\mathrm{en}}^{\mathrm{s}} = \nu_{\mathrm{en}} + \nu^{\mathrm{l}} + \nu^{\mathrm{h}}$ is a "summary frequency" of the electron–atom collisions which includes the frequencies of elastic collisions (ν_{en}) and of inelastic collisions with small (ν^{l}) and large (ν^{h}) energy losses.

Collisional–Radiative Model. The atomic-physics database required for solving Boltzmann's equation (2.36) essentially accounts for the collision term S_0, the details of which have just been commented on. If collisional and radiative processes involving excited states are taken into account, additional rate equations for the population densities N_k (which are involved in the corresponding excitation frequencies ν_{jk}) are needed to close the system of equations to be solved simultaneously. This requires, in general, knowledge of a lot of atomic (and possibly molecular) data, the quality of which can have a significant influence on the accuracy of the results finally obtained. Judgement is needed as to how many excited levels may be important and required. (For highly-excited levels the Saha–Boltzmann population densities can sometimes be assumed.)

The collisional–radiative model for argon, given for example in [2.10], can be used for orientation. Though in molecular situations extensions are, of course, necessary, the most important processes commonly considered in the case of inert gases are commented on. These are inelastic and superelastic collisions, ionizing collisions, radiative transitions, diffusion to the wall and ionization by collisions between excited atoms. This results in the following rate equation system, if four 4s states and a block of 4p states are considered:

$$\sum_{k=0,\,k\neq m}^{5} n N_k C_k^m + \sum_{k=m+1}^{5} N_k A_{mk} g_{mk}$$

$$= \sum_{k=0,\,k\neq m}^{5} n N_m C_m^k + \sum_{k=0}^{m-1} N_m A_{km} g_{km} + n N_m S_m + N_m \sum_{k=1}^{4} N_k \overline{k_{\mathrm{p}}}$$

$$+ N_m D_m / L_{\mathrm{D}(m)}^2 \text{ (for the metastable states } m = 1,\,3)\,.$$

$$(2.49)$$

On the left-hand side the productions term are specified, while on the right-hand side the destruction terms are gathered. The rate coefficients for collisional excitation and for superelastic processes caused by electrons, from the initial state k to the final state m, are denoted by C_k^m. The radiative processes from state k to state m are expressed by the product of the radiative transition probability A_{mk} and the escape factor g_{mk}. The escape factors account for possible trapping of radiation and usually differ noticeably from unity for two resonant transitions. Collisional ionization by electrons is described by the rate coefficient S_m and the probability of ionization by collision between two excited atoms is represented by the rate coefficient $\overline{k_{\mathrm{p}}}$. The diffusion loss of metastables is written in terms of the diffusion coefficient D_m with the diffusion length $L_{\mathrm{D}(m)} = R/2.4$, where R is the radius of the gas-discharge tube. The rate coefficients for excitation, superelastic processes and ionization by electron collisions are calculated with the EDF obtained from Boltzmann's equation, solved simultaneously.

2.2.2 Fluid-Model Equations

The fluid equations [2.2,6] used later in presenting the fluid-model theory of discharges are those for the first three moments of the EDF. The first two equations are

$$\frac{\partial n_\alpha}{\partial t} + \text{div}\,(n_\alpha v_\alpha) = \frac{\delta n_\alpha}{\delta t} \tag{2.50}$$

for the plasma density and

$$m_\alpha n_\alpha \left[\frac{\partial v_\alpha}{\partial t} + (v_\alpha \cdot \nabla) v_\alpha \right]$$

$$= Z e n_\alpha E + Z e n_\alpha (v_\alpha \times B) - \nabla p_\alpha + n_\alpha m \frac{\delta v_\alpha}{\delta t} \tag{2.51a}$$

for the charged-particle velocity.

The term on the right-hand side of (2.50) and the last term in (2.51a),

$$m_\alpha n_\alpha \frac{\delta v_\alpha}{\delta t} \simeq -\mu_{\alpha n} \nu_{\alpha n} v_\alpha \,, \tag{2.51b}$$

come from the collisional integral in Boltzmann's equation. Here $\mu_{\alpha n} = m_\alpha m_n/(m_\alpha + m_n)$ are the reduced masses ($\mu_{en} \approx m$ for electrons and $\mu_{in} \approx M/2$ for ions) and the index "n" denotes the neutrals. The elastic-collision frequency $\nu_{\alpha n}$ is estimated for discharges in argon gas [2.11], for example, as follows:

$$\nu_{en} = 2 N_0 \alpha_c \left(\frac{2U_*}{\pi m} \right)^{1/2} \left[\left(\frac{T_e}{U_*} \right) \gamma \left(3, \frac{U_*}{T_e} \right) + \Gamma \left(\frac{3}{2}, \frac{U_*}{T_e} \right) \right] , \tag{2.52a}$$

which, for $T_e \ll U_*$, gives

$$\nu_{en}$$
$$= 1.84 \times 10^{-8} T_e^{3/2} N_0 \left[1 + \frac{1}{2} \left(\frac{U_*}{T_e} \right)^2 \left(1 + \frac{T_e}{2U_*} \right) \exp \left(-\frac{U_*}{T_e} \right) \right] \tag{2.52b}$$

with ν_{en} in s^{-1}, T_e in eV and N_0 in cm^{-3}.

Here the assumption of a Maxwellian electron-energy distribution

$$f(u) = \left(\frac{m}{2\pi T_e} \right)^{3/2} \exp \left(-\frac{eu}{T_e} \right) \tag{2.52c}$$

and the following approximations for the cross-section for elastic collisions [2.12]

$$\sigma_c = \begin{cases} \alpha_c (u/U_*) & \text{for } u \leq U_* \\ \alpha_c (u/U_*)^{-1/2} & \text{for } u \geq U_* \end{cases} \tag{2.52d}$$

with $\alpha_c = 1.59 \times 10^{-15}\,\mathrm{cm}^2$ have been made. U_* is the excitation energy of the first atomic excited state and $\gamma(a,x) = \int_0^x t^{a-1}\exp(-t)\mathrm{d}t$ and $\Gamma(a,x) = \int_x^\infty t^{a-1}\exp(-t)\mathrm{d}t$ are the incomplete gamma functions. For example, for gas pressures $p = 10$–40 Pa, the collision frequency is of the order of $\nu_{\mathrm{en}} = (2$–$3) \times 10^8\,\mathrm{s}^{-1}$. In discharges in helium gas, the elastic collision frequency is simply given by [2.13]

$$\nu_{\mathrm{en}} = 1.8p \times 10^7 \tag{2.52e}$$

with ν_{en} in s^{-1} and p in Pa, and for $p \approx 40$ Pa, for example, $\nu_{\mathrm{en}} \approx 7 \times 10^8\,\mathrm{s}^{-1}$.

The stationary form of the linearized equation of motion (2.51),

$$\boldsymbol{v}_\alpha = -b_\alpha \boldsymbol{E} - D_\alpha \frac{\boldsymbol{\nabla} n}{n} - D_\alpha^T \frac{\boldsymbol{\nabla} T_\alpha}{T_\alpha}\,, \tag{2.53a}$$

valid for

$$\nu_{\alpha n}\tau \gg 1\,, \tag{2.53b}$$

i.e. when the characteristic time τ of the variation of the plasma parameters exceeds the time $1/\nu_{\alpha n}$ between collisions, leads to the definition of the following transport coefficients

- mobility coefficient

$$b_\alpha = \frac{e}{\mu_{\alpha n}\nu_{\alpha n}} \tag{2.54a}$$

- diffusion coefficient

$$D_\alpha = \frac{T_\alpha}{\mu_{\alpha n}\nu_{\alpha n}} \tag{2.54b}$$

- thermal diffusion coefficient

$$D_\alpha^T \equiv D_\alpha = \frac{T_\alpha}{\mu_{\alpha n}\nu_{\alpha n}}\,. \tag{2.54c}$$

The quantities involved in the description of the ambipolar-controlled regime of the discharges are the

- ambipolar diffusion coefficient

$$D_{\mathrm{A}} = \frac{b_e D_i + b_i D_e}{b_e + b_i} \approx \frac{2T_e}{M\nu_{\mathrm{in}}} \tag{2.55a}$$

- ambipolar velocity

$$\boldsymbol{v}_{\mathrm{A}} = -\frac{1}{n}\mathrm{grad}(D_{\mathrm{A}} n) \tag{2.55b}$$

- ambipolar field

$$\boldsymbol{E}_{\mathrm{A}} = -\frac{1}{en}\mathrm{grad}(nT_e) \approx -\frac{T_e}{e}\frac{\boldsymbol{\nabla} n}{n}\,. \tag{2.55c}$$

The particle balance equation obtained from (2.50) and (2.51a) is

$$\frac{\partial n}{\partial t} - \Delta(D_A n) = \nu_i n + \varrho_{si} n^2 - \varrho_r n^2 \,.$$
(2.56)

Here, ν_i is the frequency of direct ionization, ϱ_{si} is the rate of step ionization and ϱ_r is the coefficient of recombination. On the basis of [2.14], the frequency of direct ionization can be expressed in the form

$$\nu_i = \mathring{\nu}_i \exp(-U_i/T_e)$$
(2.57)

where $\mathring{\nu}_i$ is a slowly varying function of T_e, which might be considered as a constant, and U_i is the ionization energy. For instance, with the assumption of a Maxwellian distribution and with the approximation $\sigma_i = \beta_i(U - U_i)$ for the ionization cross-section [2.15,16], the frequency of direct ionization is [2.11,15]

$$\nu_i = 2N_0 \left(\frac{2}{\pi m}\right)^{1/2} \frac{\beta_i U_i^3}{T_e^{3/2}} \left[\left(\frac{T_e}{U_i}\right)^3 \Gamma\left(3, \frac{U_i}{T_e}\right) - \left(\frac{T_e}{U_i}\right)^2 \Gamma\left(2, \frac{U_i}{T_e}\right)\right] \,.$$
(2.58a)

Therefore, the slowly varying function $\mathring{\nu}_i$ has the form

$$\mathring{\nu}_i = 6.69 \times 10^7 N_0 \beta_i U_i T_e^{1/2}$$
(2.58b)

with $\mathring{\nu}_i$ in s^{-1}, N_0 in cm^{-3}, β_i in cm^2/eV and U_i and T_e in eV. With $\beta_i = 2 \times 10^{-17}\, cm^2/eV$ [2.15] for discharges in argon gas, $\mathring{\nu}_i$ is estimated as $\mathring{\nu}_i \approx 10^8\, s^{-1}$ at $p \approx 30$ Pa and $T_e \approx 2.5$ eV, which gives a value of the order of $\nu_i \approx 10^5\, s^{-1}$ for the frequency of direct ionization.

When attention need not be paid to the nonlinearity in the term for the step ionization, for instance, in situations of saturation [2.17] in the step ionization, this term can be represented by a term which has the same form $(\varrho_{si} n^2 \to \nu_{si} n)$ as that for the direct ionization. If direct and step ionization can be considered in a unified way, then the formulae derived in [2.14] are convenient for analytical treatments. These formulae are as follows.

- At comparatively high temperatures

$$T_e > \frac{2}{3}(U_i - U_*)\,,$$
(2.59a)

the ionization frequency is

$$\nu_i^{(1)} = \mathring{\nu}_i^{(1)} \exp\left(-\frac{U_*}{T_e}\right)\,,$$
(2.59b)

with

$$\mathring{\nu}_i^{(1)} = N_0 \Gamma_1 \Lambda_1 \frac{Ry^{3/2}}{T_e^{1/2} U_*} \,.$$

Here $Ry = 13.6$ eV is the Rydberg constant, $\Gamma_1 = 4\sqrt{2}\pi e^4/\sqrt{m}\, Ry^{3/2} = 1.73 \times 10^{-7}$, and $\Lambda_1 \simeq 10^{-2}$ to 10^{-1} for $T_e/U_* \in (0.1, 1)$; see Fig. 4.4 in [2.14].

- At comparatively low temperatures

$$T_{\mathrm{e}} < \frac{2}{3}(U_{\mathrm{i}} - U_*)\,, \tag{2.60a}$$

the ionization frequency is

$$\nu_{\mathrm{i}}^{(2)} = \mathring{\nu}_{\mathrm{i}}^{(2)} \exp\left(-\frac{U_{\mathrm{i}}}{T_{\mathrm{e}}}\right) \tag{2.60b}$$

with

$$\mathring{\nu}_{\mathrm{i}}^{(2)} = \frac{2\Gamma\bar{\Lambda}}{3\sqrt{\pi}}\,\frac{\Sigma_{\mathrm{i}}}{g_0}\,\frac{\mathrm{Ry}^{3/2}}{T_{\mathrm{e}}^3}\,,$$

where $\bar{\Lambda} = 0.2$, Σ_{i} is the partition function of the ions and g_0 is the statistical weight of the ground state.

However, it is usually important to retain the nature of the step ionization as a source of nonlinearity. In cases of step ionization from metastable states the general formulation – discussed in [2.6] – leads to the following expression [2.18] for the density changes due to step ionization

$$\left(\frac{\partial n_{\mathrm{e}}}{\partial t}\right)_{\mathrm{i}(j)} = n_{\mathrm{e}}^2 N_0 \tau_j \left\langle v s_{0j}\right\rangle_{[U_j,\infty]} \left\langle v s^{\mathrm{i}(j)}\right\rangle_{[U_{\mathrm{i}}-U_j,\infty]}\,, \tag{2.61}$$

where τ_j is the lifetime of a metastable state with energy U_j and cross-section for excitation s_{0j}, and $s^{\mathrm{i}(j)}$ is the cross-section for ionization from a metastable state. Averaging over the electron velocity (v) distribution function is over the energy interval noted. The rate coefficient ϱ_{si} for step ionization obtained from (2.61) is

$$\varrho_{\mathrm{si}} = \mathring{\varrho}_{\mathrm{si}} \exp\left(-\frac{U_{\mathrm{i}}}{T_{\mathrm{e}}}\right)\,. \tag{2.62a}$$

An alternative expression, which may hold at comparatively high electron temperatures, is

$$\varrho_{\mathrm{si}}' = \mathring{\varrho}_{\mathrm{si}}' \exp\left(-\frac{U_*}{T_{\mathrm{e}}}\right)\,. \tag{2.62b}$$

In (2.62), $\mathring{\varrho}_{\mathrm{si}}$ and $\mathring{\varrho}_{\mathrm{si}}'$ are slowly varying functions of temperature. However, owing to saturation in the metastable population densities caused by de-excitation via electron collisions, which may become more important than the destruction of metastables by diffusion or radiation, the effectiveness of the step ionization becomes limited with growing electron density [2.17] and the contribution of the step ionization to the particle balance transforms to

$$\varrho_{\mathrm{si}} n^2 \to \frac{\varrho_{\mathrm{si}} n^2}{1 + \tilde{\mu} n}\,. \tag{2.63}$$

An increase in $\tilde{\mu} n$ leads to the metastable densities becoming virtually independent of the electron density, and then the step ionization becomes linear in the electron density.

The way that the recombination losses involved in (2.56) vary with n_e and T_e depends on the type of recombination considered [2.6,7,15,19]. When dissociative recombination is the predominant loss mechanism, the recombination losses, which are given by $\varrho_r n_e n_i$ in general, reduce to $\varrho_r \tilde{\xi} n_e^2$, where $\tilde{\xi}$ is the proportion of the molecular ions. At comparatively low electron temperatures ϱ_r is a slowly varying function of T_e only [2.7] and it can be taken as a constant. In the case of associative ionization involving excited and ground-state atoms, $\tilde{\xi}$ is independent of n_e. When this association relates to atomic ions and neutrals,

$$\tilde{\xi} = 1 - \frac{\nu_i}{C_a N_0^2} ; \tag{2.64}$$

with an ionization frequency ν_i which includes also step ionization, this depends on n_e; C_a characterizes the associative ionization process. However, at comparatively high gas pressures, when, in general, volume recombination can be considered as the predominant loss mechanism, the contribution of the second term in (2.64) is small, $\tilde{\xi} \approx 1$ (e.g. in argon gas, this applies at $N_0 > 10^{17}\,\mathrm{cm}^{-3}$) and therefore ϱ_r in (2.56) can also be considered as a constant.

The solutions of (2.56) represent the diffusion- and recombination-controlled discharge regimes. The diffusion-controlled regime is usually considered at a constant temperature across the discharge cross-section, and longitudinal diffusion and nonlinear terms are neglected. The stationary form of (2.56) is

$$-\Delta_\perp (D_A n) = \nu_i n \tag{2.65a}$$

and determines, in discharges with a planar geometry (a plasma slab: $-d \leq x \leq d$) and with a cylindrical geometry (a plasma column: $r \leq R$), the well-known profiles of the particle density given by a cosine function and a Bessel function, respectively:

$$n = n_0(x{=}0) \, \cos\left(\frac{x}{L_D}\right), \tag{2.65b}$$

$$n = n_0(r{=}0) \, J_0\left(\frac{r}{L_D}\right). \tag{2.65c}$$

Here

$$L_D = \sqrt{\frac{D_A}{\nu_i}} \tag{2.66a}$$

is the diffusion length, which – with the boundary conditions $n(x = \pm d) = 0$ and $n(r = R) = 0$ corresponding to the geometry under consideration – is fixed by the size (d or R) of the discharge as

$$L_D = \frac{2d}{\pi}, \tag{2.66b}$$

$$L_{\mathrm{D}} = \frac{R}{2.405}. \tag{2.66c}$$

Therefore, when the nonlinear terms in (2.56) are neglected, the density at the discharge axis is not related to the electron temperature and remains undefined from the point of view of the particle balance equation. The condition for discharge maintenance is given by the Schottky relation [2.20]

$$\frac{\pi}{2d} = \sqrt{\frac{\nu_{\mathrm{i}}}{D_{\mathrm{A}}}}, \tag{2.67a}$$

or

$$\frac{2.405}{R} = \sqrt{\frac{\nu_{\mathrm{i}}}{D_{\mathrm{A}}}}. \tag{2.67b}$$

Later studies [2.21,22] take into account the nonlinearity in the particle balance equation through the process of step ionization.

In the recombination-controlled regime

$$\varrho_{\mathrm{r}} n \gg D_{\mathrm{A}}/L_n^2, \tag{2.68}$$

the particle losses are due to volume recombination. With direct ionization as the predominant mechanism of charged-particle gain, the particle balance equation (2.56) reduces to

$$\nu_{\mathrm{i}} = \varrho_{\mathrm{r}} n \tag{2.69a}$$

and directly relates the plasma density to the electron temperature:

$$n = \frac{\nu_{\mathrm{i}}}{\varrho_{\mathrm{r}}}. \tag{2.69b}$$

The different regimes of plasma heating stem from the electron energy balance equation. Later in this book, it is used in the much simplified form

$$-\frac{5}{2}\,\mathrm{div}(D_{\mathrm{e}} n\,\mathrm{grad}\,T_{\mathrm{e}}) = \frac{\delta(n_{\mathrm{e}}\bar{u})}{\delta t} + \frac{\sigma_{\mathrm{p(r)}}}{2}\,|E|^2. \tag{2.70}$$

Electron energy losses due to thermal conduction (the term on the left-hand side) and through collisions (the first term on the right-hand side) are included. The second term on the right-hand side gives the Joule heating by the field E sustaining the discharge. The heat flux vector associated with the left-hand side of (2.70),

$$q_{\mathrm{e}} = -\frac{5}{2}\frac{nT_{\mathrm{e}}}{m\nu_{\mathrm{en}}}\,\mathrm{grad}\,T_{\mathrm{e}}, \tag{2.71a}$$

introduces the thermal-conductivity coefficient and the temperature conductivity coefficient

$$\mathcal{X} = \frac{5}{2}\frac{nT_{\mathrm{e}}}{m\nu_{\mathrm{en}}} \equiv \frac{5}{2}nD_{\mathrm{e}}, \tag{2.71b}$$

$$\mathcal{K} \equiv \frac{\mathcal{X}}{n} = \frac{5}{2}\frac{T_{\mathrm{e}}}{m\nu_{\mathrm{en}}} \equiv \frac{5}{2}D_{\mathrm{e}}. \tag{2.71c}$$

The first term on the right-hand side of (2.70), originating from the collisional integral in Boltzmann's equation, accounts for both elastic and inelastic collisions by

$$\frac{\delta(n\bar{u}_e)}{\delta t} = -\frac{3}{2}\overline{\delta^s_{en}\nu^s_{en}}n(T_e - T_n)$$
$$\equiv -\frac{3}{2}\left[\delta_{en}\nu_{en}(T_e - T_n) + \delta^l_{en}\nu^l_{en}T_e + \nu^h_{en}T_e\right]n \; . \tag{2.72a}$$

In the first line of (2.72a) an effective frequency for the energy transfer through collisions is introduced. In the second line, the energy transfers in elastic collisions and in inelastic collisions with small (of order δ^l_{en}) and large (in principal, the total) energy loss are taken separately; $\nu^{l,h}_{en} = \sum_j \nu^j_{en}$ are summary frequencies of inelastic collisions. Assuming that the electrons with energies in the vicinity of the excitation energy of each excited state are those taking part in the processes of excitation of the given level, the term describing the energy losses by collisions can also be written as

$$\frac{\delta(n\bar{u}_e)}{\delta t} = -\left[\frac{3}{2}\delta_{en}\nu_{en}(T_e - T_n) + \sum_j \nu^{(j)}_{en}U_{0j}\right]n. \tag{2.72b}$$

In (2.72b), direct excitation from the ground state is the only process included. Generalization for stepwise excitation does not introduce basic complications.

Due to the small value of δ ($\delta \approx 10^{-4}$ to 10^{-5}), the contribution of the elastic collisions to the energy balance is small. The ionization frequency is also small compared with the excitation frequency. The main channel of electron energy loss by collisions is through inelastic collisions with excitation. Later, in the fluid-model theory of discharges (2.72b) is used in the simplified form

$$\frac{\delta(n\bar{u})}{\delta t} = -\frac{3}{2}n\nu_*U_* \; , \tag{2.73}$$

where ν_* is the frequency of excitation from the ground state to the dominant excited level, U_* the excitation energy of this level and the factor $(3/2)$ gives some weight for excitation to the other levels. Equation (2.73) can also be used in a form without any numerical factor on the right-hand side, with ν_* defined accordingly [2.12]. For example, the excitation frequency ν_* estimated for discharges in argon gas [2.11] with the assumption of a Maxwellian distribution (2.52c) and the following approximation for the total cross-section [2.12],

$$\sigma_* = \alpha_{c\;(exc)}\left(\frac{U}{U_*}\right)^{-1/2}\left(\frac{U}{U_*} - 1\right) \; , \tag{2.74a}$$

where $\alpha_{c\;(exc)} = 1.56 \times 10^{-16}\;\mathrm{cm}^{-2}$, is

$$\nu_* = 2N_0\alpha_{\mathrm{c\ (exc)}}\left(\frac{2}{\pi m U_*}\right)^{1/2} T_{\mathrm{e}}\left[\Gamma\left(\frac{5}{2},\frac{U_*}{T_{\mathrm{e}}}\right) - \frac{U_*}{T_{\mathrm{e}}}\Gamma\left(\frac{3}{2},\frac{U_*}{T_{\mathrm{e}}}\right)\right],$$

$$(2.74\mathrm{b})$$

yielding the number of collisions (per sec and electron) with energy loss U_* effectively taking into account all excited states.

In general, the excitation frequency – similarly to the ionization frequency – can be represented as

$$\nu_* = \mathring{\nu}_* \exp\left(-\frac{U_*}{T_{\mathrm{e}}}\right), \tag{2.75}$$

in which $\mathring{\nu}_*$ is a slowly varying function of the temperature. In the case of argon discharges $\mathring{\nu}_* = 1.04 \times 10^{-8} T_{\mathrm{e}}^{1/2} N_0$ with $\mathring{\nu}_*$ in s^{-1}, T_{e} in eV and N_0 in cm^{-3}, and, at pressures $p \approx 10$ to 40 Pa, for example, the excitation frequency is about $\nu_* \approx 5 \times 10^5\,\mathrm{s}^{-1}$.

The competition of the nonlocal energy losses through thermal conduction and the local energy losses through collisions determines the mechanisms of heating by the electric field: local and nonlocal [2.23,24]. Of course, the scale length of the field inhomogeneity is also involved in the definition of the heating mechanism. The relative contributions of the two processes of energy loss are characterized by the scale length $L_\chi = (\chi/n\nu_\varepsilon)^{1/2}$, i.e. the characteristic length of thermal conduction. In cases when the energy losses occur mainly through elastic collisions (the first term on the right-hand side of (2.72a), L_χ is defined by

$$L_\chi = \frac{\lambda_{\mathrm{f.p.}}}{\sqrt{\delta}}, \tag{2.76a}$$

where $\lambda_{\mathrm{f.p.}}$ is the electron mean free path. In cases of predominant energy loss through inelastic collisions with excitation, the corresponding length is

$$L_\chi = \lambda_{\mathrm{f.p.}}\sqrt{\frac{\nu}{\nu_*}}. \tag{2.76b}$$

The electron temperature T_{e} and its spatial distribution are the final results of the electron energy balance equation (2.70). The spatial profile of T_{e} is characterized by the scale length $L_T = (|\nabla T_{\mathrm{e}}|/T_{\mathrm{e}})^{-1}$ of its variation.

In cases where $L_T \gg L_\chi$, the length scale of the temperature variation is much larger than the characteristic length of thermal conduction, the effects of the latter are unimportant and the heating is local. The electron temperature, being uncoupled from the plasma density, is locally determined by the heating field. The scale length $L_E = (|\nabla|E|^2|/|E|^2)^{-1}$ of the heating-field intensity determines the scale length L_T of the temperature variation ($L_T \simeq L_E$). The criterion for local heating ($L_T \gg L_\chi$) becomes $L_E \gg L_\chi$ and the approach is applicable to plasmas of large size and sufficiently high neutral-gas density ($R > L_\chi$ in the case of a discharge column of radius

R) which are heated by a field with not too strong a space variation of its amplitude.

In the other limit ($L_T \lesssim L_\chi$) the spatial distribution of T_e is essentially determined by the effect of the thermal conductivity of the electrons. Different cases of this nonlocal heating, specified by the boundary conditions for electron energy relaxation at the walls and the heat flux to the walls as well as by the penetration of the heating field into the plasma, are possible. However, the general procedure for obtaining a solution for the spatial distribution of the temperature in all these cases is based on a perturbation method and an asymptotic series expansion with respect to small parameters that are specified according to the formulation of the problem. Such procedures are justified by the expectation of an almost homogeneous temperature distribution that is ensured by the thermal conductivity even with a strong spatial variation of the heating-field amplitude. Moreover, strongly inhomogeneous high-amplitude fields leading to strong local increases of T_e immediately call for switching on of the effect of thermal conduction. This results in spatial equalization of the electron temperature.

Depending on the mechanisms of particle loss and electron energy loss, different combinations of effects of locality and nonlocality can appear [2.25]: nonlocal heating with local particle losses through recombination, nonlocal heating with nonlocal particle losses through diffusion and local heating with local or nonlocal charged-particle losses. These situations are determined by the ratio of the characteristic lengths of diffusion L_D and thermal conduction L_χ, and their relation to the scale of the energy input $Q \equiv (\sigma_{p(r)}/2)|E|^2$.

3. Surface-Wave Propagation in Homogeneous Plasmas

The research on SWs in gas-discharge plasmas performed in the 1960s and 1970s was so intensive and successful that it established plasma waveguide systems as a new branch of research on waves in plasmas. After the first indications about conditions for the existence of EM SWs [3.1] and the pioneering work by Trivelpiece and Gould [3.2] on their behaviour, a lot of information was accumulated, especially on the properties and propagation characteristics of SWs in waveguides filled with homogeneous plasmas [3.3–75]. The wealth of results – both theoretical and experimental – led to, at the beginning of the 1980s, an opinion that the linear approach to wave propagation in homogeneous plasma waveguides was clearly understood and even that this area of activity had settled down. The definition and basic features of SWs entered textbooks on plasma physics [3.76,77]. Experimental and theoretical results on the properties of SWs were summarized in monographs [3.78,79] and review papers [3.80–82]. During the period up to the beginning of the 1980s, SWs were considered in terms of low-amplitude SW propagation in plasmas with given parameters. On this basis of thinking about plasma waveguides realized by DC discharges at comparatively low pressures, wave propagation under conditions of weak collisions, weak space-dispersion effects and weak electron drift was considered. During all these years, studies on SWs in plasma waveguide systems were strongly supported by very active research on EM surface modes in solid-state waveguide structures [3.83–85], which also provided a lot of experience and results.

However, the growing activity – starting at the beginning of the 1980s – on discharges produced by propagating SWs [3.86–88] and, in general, on nonlinear SWs ([3.89–92] and references therein) created the necessity for additional investigations of the dispersion behaviour of the waves. It appeared that even knowledge about wave dispersion in homogeneous plasmas was not complete or sufficient. For instance, more detailed studies on the effect of collisions extended understanding of the behaviour of the waves. However, the previously known main defining features of SWs in waveguides with homogeneous plasmas retained their validity:

(1) the SWs are proper modes of bounded systems
(2) they propagate along interfaces between two media with different permittivities

(3) the wave field amplitudes – which have maximum values at the interface
    – decrease with distance from the interface into the two media.

3.1 Remarks on Classification

In the following presentation, the guiding structures for SW propagation
are plasma waveguides, in which at least one of the media constituting the
waveguide is a plasma. Weakly ionized low-temperature plasmas, i.e. gas-
discharge plasmas, are considered. Discharges in an external magnetic field
are not included.

The configurations of the systems considered in both this chapter and the
next, (i) are a single planar interface between two media with permittivities
different in sign, i.e. plasmas semi-bounded by a dielectric or vacuum; (ii)
planar waveguides with plasma slabs; and (iii) cylindrical waveguides consisting of a plasma column surrounded by a dielectric and/or vacuum. In this
chapter the considerations are limited to the case of homogeneous plasmas
with sharp boundaries.

The SWs considered here are TM modes, i.e. waves with a magnetic-
field component transverse to the direction of the wave vector. The notation
"E-modes", also often used in the literature, emphasizes the existence of an
electric-field component along the direction of propagation. SWs in semi-
bounded plasmas and plasma slabs and the azimuthally symmetric mode in
cylindrical plasma waveguides are such modes.

Moreover, only *high-frequency* SWs, for which the electron motion (2.8a)
is the only motion of importance, are considered. As distinguished from bulk
HF waves (Sect. 2.1), which in isotropic plasmas split into pure longitudinal
and pure transverse waves, HF SWs combine EM-wave properties and the
behaviour of space-charge (longitudinal) waves in a unified way. Such a nature
of SWs is directly related to their field configuration: the electric field of
the wave has components both along and perpendicular to the interface.
In the approach of a *cold homogeneous plasma model* the plasma density
perturbations caused by the wave field appear as a periodic surface charge
at the interface and determine the longitudinal character of the waves. The
bulk plasma density is not perturbed by the SW field. In the framework of
this model – cold homogeneous *collisionless* plasmas – the propagation of the
SWs is ensured *only* by waveguide systems which include an interface between
two media with permittivities of *opposite* sign, and the waves travel along
the interface as *slow* waves. The phase velocity (v_{ph}) of the SWs is smaller
than the speed of light $c/\sqrt{\varepsilon_{\mathrm{d}}}$ in the dielectric which bounds the plasma (ε_{d}
being the permittivity of the surrounding dielectric). The frequency range
($\omega < \omega_{\mathrm{p}}$) of the waves determined by the condition of a negative value of the
plasma permittivity ($\varepsilon < 0$) is below the range of the bulk waves (2.24). SWs
propagate at frequencies up to the frequency of their quasi-static resonance

$$\omega = \frac{\omega_p}{\sqrt{1 + \varepsilon_d}}, \tag{3.1a}$$

or

$$\omega = \frac{\omega_p}{\sqrt{2}} \tag{3.1b}$$

in the case of a plasma–vacuum interface.

When *collisions* are introduced as the next step in moving the modelling closer to real situations, their effect (even in the case of weak collisions $\omega > \nu$) is not only to cause damping but also to extend the frequency range of the wave by introducing new branches into the SW dispersion behaviour (as is indicated in [3.93]). Owing to collisions, *another* region of weakly damped SWs appears for

$$\varepsilon_r > 0, \tag{3.2}$$

which overlaps the frequency range of the bulk EM waves. In that region the SWs are *fast* (their phase velocity is larger than the speed of light in the surrounding dielectric) and they are bound to the interface only because of collisions. The two branches of *weakly damped* waves for

$$\omega \leq \frac{\omega_p}{\sqrt{1 + \varepsilon_d}} \tag{3.3}$$

and

$$\omega > \omega_p \tag{3.4}$$

are connected through a region of *strongly damped* (evanescent) waves, which are also surface modes.

Conditions of *strong collisions* ($\nu > \omega$) also allow propagation of weakly damped SWs. However, in this case, the SWs of plasma waveguides change their behaviour and appear as EM Sommerfeld-type waves [3.94], which are known to propagate along a conducting surface (a metal–air interface).

Another effect that removes the SW resonance in cold homogeneous plasmas is space dispersion due to *thermal electron motion*. The wave propagation is extended above the SW resonance frequency (3.1). However, because of the additional damping – collisionless Landau damping – this extension is not too significant.

Changes in the *waveguide configuration* also introduce changes in the wave dispersion behaviour. For example, when the waveguide configuration is plasma/thin dielectric layer/vacuum (a configuration which matches more closely the gas-discharge plasma experiments), changes in the SW dispersion occur also in the vicinity of the SW resonance of a cold collisionless plasma bounded by a dielectric (with permittivity ε_d). The formation of a maximum in the dispersion curve shows up in the appearance of a backward wave.

In general, the dispersion relation of SWs is of the form

$$\mathcal{D}(\omega, k, \varepsilon) = 0. \tag{3.5}$$

In fact, the dispersion relation can contain some other quantities [3.95] associated with the geometry, the structure of the waveguide and the plasma properties (such as sizes, the permittivities of the surrounding media, the electron–neutral collision frequency and the electron thermal velocity). However, these quantities are fixed and they, or combinations of them, appear as parameters when the wave dispersion behaviour is represented.

Since the plasma permittivity ε ((2.8) and (2.9)) is a function of the ratio $\omega/\omega_{\mathrm{p}}$, the obvious way of representing the dispersion behaviour of the waves in normalized quantities is to give the dependence of $\omega/\omega_{\mathrm{p}}$ on a normalized wave number. However, different physical meanings related to different situations, both in the formulation of the problem and in conducting the experiment, can be involved in such a normalization. For example, when studying SW dispersion behaviour in DC discharges or in any discharges except those created by SWs themselves, the dispersion behaviour of the wave can be obtained either by changing the wave frequency and keeping the discharge conditions, i.e. the plasma density and therefore ω_{p}, fixed, or by changing the plasma frequency (through changing the plasma density by varying, for example, the discharge current) and keeping the wave frequency fixed. In the accepted terminology [3.81,87,96], the representation of the wave dispersion behaviour in the first case ($\omega_{\mathrm{p}} = $ const.) is in terms of a *dispersion diagram or dispersion curve*, whereas in the second case ($\omega = $ const.) it is in terms of a *phase diagram*.

In the case of waveguided discharges, when an SW at a given frequency ω produces a plasma with a density varying in the longitudinal direction it is the *phase diagram* which describes the wave behaviour along the plasma column. Such situations address cases of wave propagation in inhomogeneous media – a problem which will be treated in detail in the next chapter. In such cases the phase diagrams should be considered as giving information about the wave behaviour in a local approach (zero-order approximation in terms of the geometrical-optics [WKB] approach).

Since the dependence of the dispersion function $\mathcal{D}$ (3.5) on ω and ω_{p} occurs not only through their ratio $\omega/\omega_{\mathrm{p}}$, dispersion and phase diagrams look different. Moreover, at the same $\omega/\omega_{\mathrm{p}}$ values they may present different behaviours of the SW field configuration.

In both cases – dispersion and phase diagrams – a quantity related to the size of the plasma waveguide (e.g. the thickness of the slab in the case of a planar waveguide or the radius of the column in the case of a cylindrical waveguide) can be used for normalizing the wave number. When electrodynamic quantities are used for the normalization, these are the skin depth ((2.26a) in the case of weak collisions) and the vacuum wave number

$$k_{\mathrm{v}} = \omega/c, \tag{3.6}$$

which may appear as proper normalization quantities for the dispersion and phase diagrams, respectively.

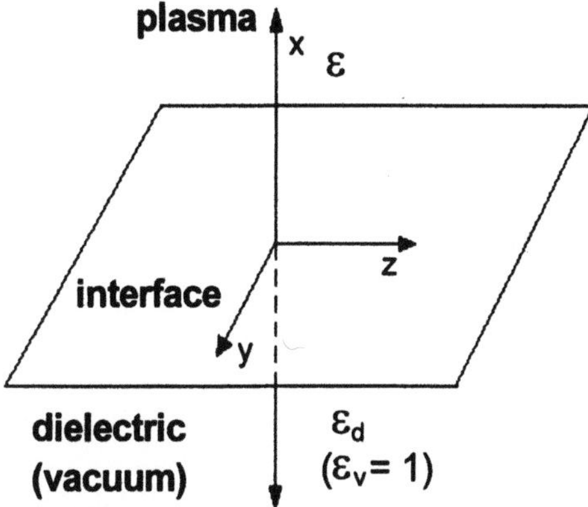

Figure 3.1. Waveguide configuration: single interface

HF SWs in different configurations of the waveguide system are considered below. To describe the physical features of the problem, the simplest configuration – a single interface – is treated first. The next steps proceed to slab and cylindrical configurations. With view to a further description of SW-sustained discharges, the dispersion behaviour is explained in terms of phase diagrams.

3.2 Single Plane Interface

The waveguide structure is depicted schematically in Fig. 3.1. A plasma (with permittivity ε) occupying the semi-space $x > 0$ is bounded by a dielectric (or vacuum) of permittivity ε_d (or $\varepsilon_v = 1$). The wave propagation is in the z direction. The field variation, taken in the form (2.5), reduces to

$$\propto \exp(-\mathrm{i}\,\omega t + \mathrm{i}\,kz)\,, \tag{3.7}$$

where $k = \beta + \mathrm{i}\,\alpha$ is the complex wave number (β and α are the propagation constant and the space damping rate, respectively).

The behaviour of the SW in the *cold plasma* approximation is considered, i.e. the wave phase velocity v_{ph} is much larger than the electron thermal velocity v_{th}. The field amplitudes are

$$\boldsymbol{E}(x) = \left(\mathrm{i}\,\frac{k}{\varkappa_{\mathrm{p}}}, 0, 1\right) A\,\mathrm{e}^{-\varkappa_{\mathrm{p}} x}\,,$$

$$\boldsymbol{B}(x) = \left(0, \mathrm{i}\,\frac{\omega\varepsilon}{c^2 \varkappa_{\mathrm{p}}}, 0\right) A\,\mathrm{e}^{-\varkappa_{\mathrm{p}} x} \tag{3.8a}$$

at $x > 0$, and

$$E(x) = \left(-i\,\frac{k}{\varkappa_d}, 0, 1 \right) A\,e^{\varkappa_d x} ,$$

$$B(x) = \left(0, -i\,\frac{\omega\varepsilon_d}{c^2\varkappa_d}, 0 \right) A\,e^{\varkappa_d x} \tag{3.8b}$$

at $x < 0$, where $A = E_z(x = 0)$ and the quantities

$$\varkappa_p = \sqrt{k^2 - \frac{\omega^2}{c^2}\varepsilon} , \tag{3.9a}$$

$$\varkappa_d = \sqrt{k^2 - \frac{\omega^2}{c^2}\varepsilon_d} \tag{3.9b}$$

characterize the field distribution in the transverse direction; in the case of a plasma semi-bounded by a vacuum (3.9b) reduces to

$$\varkappa_v = \sqrt{k^2 - \frac{\omega^2}{c^2}} . \tag{3.9c}$$

Equality of the constants in (3.8a) and (3.8b) accounts for the boundary condition (2.17a) applied at the interface $x = 0$.

The boundary condition (2.17b) leads to the SW dispersion law [3.1,77,79,80]

$$\mathcal{D}(\omega, k) \equiv \frac{\varkappa_p}{\varepsilon} + \frac{\varkappa_d}{\varepsilon_d} = 0 \tag{3.10a}$$

or, equivalently, to

$$k^2 = \frac{\omega^2}{c^2}\,\tilde{\varepsilon} , \tag{3.10b}$$

where

$$\tilde{\varepsilon} = \frac{\varepsilon_d \varepsilon}{\varepsilon_d + \varepsilon} . \tag{3.10c}$$

Written in the form (3.10b), the dispersion relation of SWs looks like that of EM bulk waves (2.24b) with ε replaced by $\tilde{\varepsilon}$. The latter can be considered as an "effective" permittivity of the waveguide, represented in terms of both plasma and dielectric permittivities. When dissipation of the wave energy through collisions in the plasma is taken into account ($\varepsilon = \varepsilon_r + i\varepsilon_i$ as given by (2.8a) with real (ε_r) and imaginary (ε_i) parts presented by (2.9)), (3.10b) gives

$$\beta = \frac{1}{\sqrt{2}}\frac{\omega}{c}\left(\sqrt{\tilde{\varepsilon}_r^2 + \tilde{\varepsilon}_i^2} + \tilde{\varepsilon}_r \right)^{1/2} , \tag{3.11a}$$

$$\alpha = \frac{1}{\sqrt{2}}\frac{\omega}{c}\left(\sqrt{\tilde{\varepsilon}_r^2 + \tilde{\varepsilon}_i^2} - \tilde{\varepsilon}_r \right)^{1/2} , \tag{3.11b}$$

where

$$\tilde{\varepsilon}_{\mathrm{r}} = \frac{\left(\varepsilon_{\mathrm{r}}^2 + \varepsilon_{\mathrm{i}}^2 + \varepsilon_{\mathrm{r}}\varepsilon_{\mathrm{d}}\right)\varepsilon_{\mathrm{d}}}{(\varepsilon_{\mathrm{r}} + \varepsilon_{\mathrm{d}})^2 + \varepsilon_{\mathrm{i}}^2}, \tag{3.12a}$$

$$\tilde{\varepsilon}_{\mathrm{i}} = \frac{\varepsilon_{\mathrm{i}}\varepsilon_{\mathrm{d}}^2}{(\varepsilon_{\mathrm{r}} + \varepsilon_{\mathrm{d}})^2 + \varepsilon_{\mathrm{i}}^2}. \tag{3.12b}$$

The quantities $\varkappa_{\mathrm{p,d}}$ of (3.9) take the forms

$$\varkappa_{\mathrm{p}} = \frac{\omega}{c}\frac{|\varepsilon|}{\sqrt{-(\varepsilon + \varepsilon_{\mathrm{d}})}}, \tag{3.13a}$$

$$\varkappa_{\mathrm{d}} = \frac{\omega}{c}\frac{\varepsilon_{\mathrm{d}}}{\sqrt{-(\varepsilon + \varepsilon_{\mathrm{d}})}}. \tag{3.13b}$$

As (3.11) shows, weak dissipation of the waveguide with respect to SW propagation means

$$\tilde{\varepsilon}_{\mathrm{r}} \gg \tilde{\varepsilon}_{\mathrm{i}}, \tag{3.14}$$

and this is the case considered below.

According to (3.11), the waveguided system is transparent ($\beta > \alpha$) to SW propagation if

$$\tilde{\varepsilon}_{\mathrm{r}} > 0, \tag{3.15a}$$

which leads to two regions of values of the plasma permittivity:

$$\varepsilon_{\mathrm{r}} < -\frac{1}{2}\left(\varepsilon_{\mathrm{d}} + \sqrt{\varepsilon_{\mathrm{d}}^2 - 4\varepsilon_{\mathrm{i}}^2}\right) \tag{3.15b}$$

and

$$1 > \varepsilon_{\mathrm{r}} > -\frac{1}{2}\left(\varepsilon_{\mathrm{d}} - \sqrt{\varepsilon_{\mathrm{d}}^2 - 4\varepsilon_{\mathrm{i}}^2}\right). \tag{3.15c}$$

SWs in these regions ((3.15b,c); compare to (3.3) and (3.4), respectively) are referred to as *surface* and *radiative* modes.

The opposite case of

$$\tilde{\varepsilon}_{\mathrm{r}} < 0, \tag{3.16a}$$

which gives strongly damped waves ($\alpha > \beta$), occurs in the region situated between the bands (3.15b) and (3.15c), i.e. at plasma permittivity values specified by

$$-\frac{1}{2}\left(\varepsilon_{\mathrm{d}} + \sqrt{\varepsilon_{\mathrm{d}}^2 - 4\varepsilon_{\mathrm{i}}^2}\right) < \varepsilon_{\mathrm{r}} < -\frac{1}{2}\left(\varepsilon_{\mathrm{d}} - \sqrt{\varepsilon_{\mathrm{d}}^2 - 4\varepsilon_{\mathrm{i}}^2}\right). \tag{3.16b}$$

These waves are referred to as *evanescent* mode [3.93]. As (3.15b,c) and (3.16b) show, these three regions represent the existence of SWs in plasmas with weak enough dissipation ($\varepsilon_{\mathrm{d}} > 2\varepsilon_{\mathrm{i}}$). In the opposite case of strongly dissipative plasmas ($2\varepsilon_{\mathrm{i}} > \varepsilon_{\mathrm{d}}$), the waveguide is transparent ($\tilde{\varepsilon}_{\mathrm{r}} > 0$) over the complete range of ε_{r} values.

The wave number β and the space damping rate α, obtained from (3.11), are, in the case (3.15) of a transparent waveguide,

$$\beta = \frac{\omega}{c} \sqrt{\tilde{\varepsilon}_r} \left(1 + \frac{1}{8} \frac{\tilde{\varepsilon}_i^2}{\tilde{\varepsilon}_r^2} \right) , \tag{3.17a}$$

$$\alpha = \frac{1}{2} \frac{\omega}{c} \frac{\tilde{\varepsilon}_i}{\sqrt{\tilde{\varepsilon}_r}} . \tag{3.17b}$$

In the case (3.16) of the evanescent mode, the corresponding results are

$$\beta = \frac{1}{2} \frac{\omega}{c} \frac{\tilde{\varepsilon}_i}{\sqrt{-\tilde{\varepsilon}_r}} , \tag{3.18a}$$

$$\alpha = \frac{\omega}{c} \sqrt{-\tilde{\varepsilon}_r} \left(1 + \frac{1}{8} \frac{\tilde{\varepsilon}_i^2}{\tilde{\varepsilon}_r^2} \right) . \tag{3.18b}$$

Equations (3.17) and (3.18) are considered below for both weak ($\omega \gg \nu$) and strong ($\nu \gg \omega$) collisions.

3.2.1 Case of Weak Collisions ($\omega \gg \nu$)

With $\nu \ll \omega$, $|\varepsilon_r| \gg \varepsilon_i$ (see (2.9)), and (3.12) takes the form

$$\tilde{\varepsilon}_r = \frac{\varepsilon_r \varepsilon_d}{\varepsilon_r + \varepsilon_d} \left(1 + \frac{\varepsilon_i^2 \varepsilon_d}{\varepsilon_r (\varepsilon_r + \varepsilon_d)^2} \right) , \tag{3.19a}$$

$$\tilde{\varepsilon}_i = \frac{\varepsilon_i \varepsilon_d^2}{(\varepsilon_r + \varepsilon_d)^2} \left(1 - \frac{\varepsilon_i^2}{(\varepsilon_r + \varepsilon_d)^2} \right) . \tag{3.19b}$$

The second term in each of the large brackets is a small correction.

The wave number and the space damping rate obtained (in terms of ε and ε_d) from (3.17) in the case of a transparent waveguide are

$$\beta = \frac{\omega}{c} \sqrt{\frac{\varepsilon_r \varepsilon_d}{\varepsilon_r + \varepsilon_d}} \left[1 + \frac{\varepsilon_i^2 \varepsilon_d}{2\varepsilon_r (\varepsilon_r + \varepsilon_d)^2} \left(1 + \frac{\varepsilon_d}{4\varepsilon_r} \right) \right] , \tag{3.20a}$$

$$\alpha = \frac{\omega}{2c} \frac{\varepsilon_i \varepsilon_d^{3/2}}{\sqrt{\varepsilon_r (\varepsilon_r + \varepsilon_d)^3}} \left[1 - \frac{\varepsilon_i^2}{(\varepsilon_r + \varepsilon_d)^2} \left(1 + \frac{\varepsilon_d}{4\varepsilon_r} \right) \right] . \tag{3.20b}$$

As comparison of (3.17) and (3.18) shows, the propagation constant of the evanescent mode has the form of the space damping rate of the surface mode and its space damping rate is in the form of the propagation constant of the surface mode:

$$\beta = \frac{\omega}{2c} \frac{\varepsilon_i \varepsilon_d^{3/2}}{\sqrt{-\varepsilon_r (\varepsilon_r + \varepsilon_d)^3}} \left[1 - \frac{\varepsilon_i^2}{(\varepsilon_r + \varepsilon_d)^2} \left(1 + \frac{\varepsilon_d}{2\varepsilon_r} \right) \right] , \tag{3.21a}$$

$$\alpha = \frac{\omega}{c} \sqrt{\frac{\varepsilon_r \varepsilon_d}{-(\varepsilon_r + \varepsilon_d)}} \left[1 + \frac{\varepsilon_i^2 \varepsilon_d}{2\varepsilon_r (\varepsilon_r + \varepsilon_d)^2} \left(1 + \frac{\varepsilon_d}{4\varepsilon_r} \right) \right] . \tag{3.21b}$$

The behaviour of the complete solution (3.11) is displayed in Fig. 3.2. The regions of the surface mode (a) and of the radiative mode (c) are those

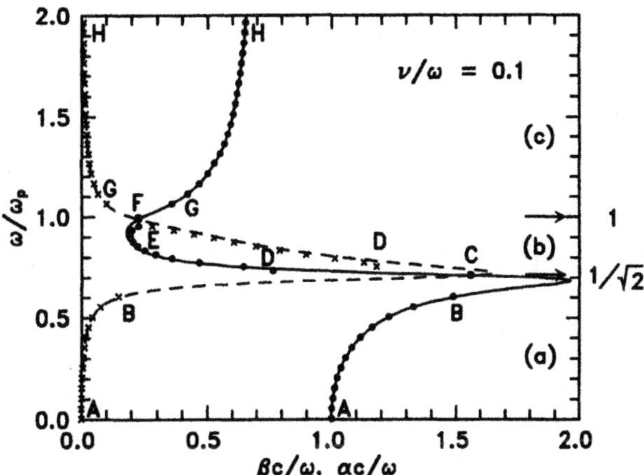

Figure 3.2. Wave number β (*solid curve*) and space damping rate α (*dashed curve*) of SW propagation along a single interface between cold weakly collisional plasma and vacuum: plot of (3.11) in terms of a phase diagram at $\nu/\omega = 0.1$. Approximate solutions for β and α: in region (a), surface mode, AB, (3.20) with $\varepsilon_d = 1$; in region (b), evanescent mode, DE, (3.21) with $\varepsilon_d = 1$; and in region (c), radiative mode, GH, (3.20) with $\varepsilon_d = 1$. At points C and F the values of β and α are equal

where the waves are weakly damped. The solutions in these regions are connected through the solution in region (b), i.e. the evanescent mode, which is a strongly damped wave. At the points $\omega/\omega_p = \sqrt{2}$ and $\omega/\omega_p = 1$, which correspond to $\varepsilon = -1$ (since the case of a plasma–vacuum interface is depicted in Fig. 3.2) and to $\varepsilon = 0$, the real and imaginary parts of the wave number are equal.

Apart from causing damping, the collisions influence the transverse structure of the waves, introducing an oscillatory (imaginary) part $\varkappa_{p,d}^{(i)}$ in the quantities $\varkappa_p = \varkappa_p^{(r)} + i\,\varkappa_p^{(i)}$ and $\varkappa_d = \varkappa_d^{(r)} + i\,\varkappa_d^{(i)}$; the real parts $\varkappa_{p,d}^{(r)}$ give an exponential decay of the field amplitudes. For surface modes, the transverse variation of the field is

$$\propto e^{-\varkappa_p x} = \exp\left(-\frac{\omega}{c}\frac{\varepsilon_r}{\sqrt{-(\varepsilon_r + \varepsilon_d)}}x\right)\exp\left(i\frac{\omega}{2c}\varepsilon_i\sqrt{\frac{-(\varepsilon_r + \varepsilon_d)}{\varepsilon_r^2}}\,x\right)$$

$$\text{for } x > 0\,, \tag{3.22a}$$

$$\propto e^{\varkappa_d x} = \exp\left(\frac{\omega}{c}\frac{\varepsilon_d}{\sqrt{-(\varepsilon_r + \varepsilon_d)}}x\right)\exp\left(i\frac{\omega}{2c}\frac{\varepsilon_d\varepsilon_i}{\sqrt{-(\varepsilon_r + \varepsilon_d)^3}}\,x\right)$$

$$\text{for } x < 0 \tag{3.22b}$$

in the plasma ($x > 0$) and dielectric ($x < 0$) semi-spaces, respectively. The oscillatory parts of $\varkappa_{p,d}$ appear as transverse components of the wave vector and lead to a power flux in the transverse direction related to the dissipation of the EM energy into the plasma region. The reciprocity in the behaviour of the real and imaginary parts of the axial wave numbers of the evanescent and surface modes causes a reciprocity in the behaviour of the real and imaginary parts of $\varkappa_{\mathrm{p,d}}$. The oscillatory part of the field distribution of the evanescent mode in the transverse direction is larger than that which ensures the decay of the field ($\varkappa_{\mathrm{p}}^{(i)} > \varkappa_{\mathrm{p}}^{(r)}$, $\varkappa_{\mathrm{d}}^{(i)} > \varkappa_{\mathrm{d}}^{(r)}$). In fact, the transverse components of the wave number for the evanescent mode have the same form as the real parts of $\varkappa_{\mathrm{p}}$ and $\varkappa_{\mathrm{d}}$ of the surface mode (see (3.22a,b)), with the minus sign dropped under the square roots. The transverse wave number is larger than the longitudinal one. The wave field is bound to the interface because of the collisions.

The field distribution of the radiative mode in the transverse direction can be analysed more easily if (3.20) is simplified by subdividing the region $0 \le \varepsilon_{\mathrm{r}} \le 1$ into two parts: (i) $\varepsilon_{\mathrm{r}} \ll \varepsilon_{\mathrm{d}}$ (or $\varepsilon_{\mathrm{r}} \ll 1$), i.e. a region which is above but close to $\varepsilon_{\mathrm{r}} = 0$, and (ii) $\varepsilon_{\mathrm{r}} \to 1$.

In the first case ($\varepsilon_{\mathrm{r}} \ll 1$) the expressions (3.20) take the form

$$\beta = \frac{\omega}{c}\sqrt{\varepsilon_{\mathrm{r}}}\left(1 + \frac{\varepsilon_{\mathrm{i}}^2}{8\varepsilon_{\mathrm{r}}^2}\right), \tag{3.23a}$$

$$\alpha = \frac{\omega}{2c}\frac{\varepsilon_{\mathrm{i}}}{\sqrt{\varepsilon_{\mathrm{r}}}}\left(1 - \frac{\varepsilon_{\mathrm{i}}^2}{2\varepsilon_{\mathrm{r}}\varepsilon_{\mathrm{d}}}\right) \tag{3.23b}$$

and the field distributions in the plasma and dielectric semi-spaces are

$$\propto \exp\left(-\frac{\omega}{2c}\frac{\varepsilon_{\mathrm{i}}^{3/2}}{\varepsilon_{\mathrm{r}}}x\right)\exp\left(-\mathrm{i}\frac{\omega}{2c}\frac{\varepsilon_{\mathrm{i}}^{3/2}}{\varepsilon_{\mathrm{r}}}x\right), \tag{3.24a}$$

$$\propto \exp\left(\frac{\omega}{4c}\frac{\varepsilon_{\mathrm{i}}}{\sqrt{\varepsilon_{\mathrm{d}}}}x\right)\exp\left(\mathrm{i}\frac{\omega}{c}\sqrt{\varepsilon_{\mathrm{d}}}x\right), \tag{3.24b}$$

respectively. The collisions bind the wave to the interface: the field decay in the transverse direction is determined by ε_{i}. In the second case ($\varepsilon_{\mathrm{r}} \to 1$), the longitudinal component of the wave vector and the space damping rate obtained from (3.20) are

$$\beta \approx \frac{\omega}{c\sqrt{2}}, \tag{3.25a}$$

$$\alpha = \frac{\omega}{2^{5/2}c}\varepsilon_{\mathrm{i}} \tag{3.25b}$$

and the field distribution in the transverse direction for the plasma and vacuum semi-spaces,

$$\propto \exp\left(-\frac{3}{4\sqrt{2}}\frac{\omega}{c}\varepsilon_{\mathrm{i}}x\right)\exp\left(\mathrm{i}\frac{\omega}{\sqrt{2}c}x\right), \tag{3.26a}$$

$$\propto \exp\left(\frac{1}{4\sqrt{2}}\frac{\omega}{c}\varepsilon_{\mathrm{i}}x\right)\exp\left(\mathrm{i}\frac{\omega}{\sqrt{2}c}(2\varepsilon_{\mathrm{d}} - 1)x\right), \tag{3.26b}$$

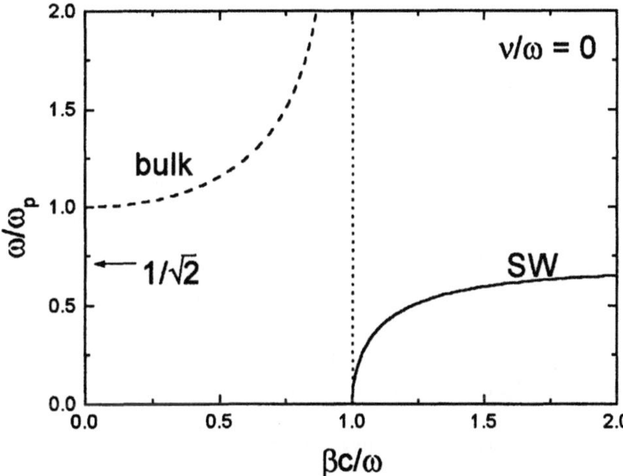

Figure 3.3. Phase diagram for SW propagation (*solid curve*) in a cold collision-less plasma semi-bounded by a vacuum. For comparison, the bulk transverse wave (*dashed curve*) is also shown. The *dotted line* represents the vacuum light speed

respectively, shows that again the collisions ensure the decay of the field.

When the plasma is nondissipative ($\varepsilon_i = 0$), the waveguide configuration of a single plasma–dielectric interface supports SW propagation ($\varkappa_p, \varkappa_d > 0$) only for ($\varepsilon_r + \varepsilon_d$) < 0, i.e. in the frequency range (3.3) indicated by (a) in Fig. 3.2. The wave propagation occurs up to the cut-off frequency (3.1) also called the frequency of the quasi-static resonance of the SWs, since in this case $\varkappa_{p,d} \to \infty$ and the magnetic-field component of the wave (and consequently the SW energy flux) vanishes. Expressions (3.11), (3.17) and (3.20) result in

$$\beta = \frac{\omega}{c}\sqrt{\frac{\varepsilon_r\varepsilon_d}{\varepsilon_r + \varepsilon_d}}, \qquad \alpha = 0, \tag{3.27}$$

describing the dispersion behaviour (Fig. 3.3) of SW propagation in collision-less plasmas (the plasma permittivity ε_r is given by (2.9a) with $\nu = 0$).

At frequencies above the resonance frequency (3.1) or, equivalently, at plasma densities n below n_c^*, where

$$n_c^* \equiv n_c(1 + \varepsilon_d), \tag{3.28}$$

which could be called the SW cut-off density, the waveguide cannot support SW propagation. For $-\varepsilon_d < \varepsilon_r < 0$, or $n_c < n < n_c^*$ (i.e. in the region occupied by the evanescent surface mode in the case of weak collisions), (3.11) and (3.18) result in

$$\beta = 0, \qquad \alpha = \frac{\omega}{c}\sqrt{-\tilde{\varepsilon}_r} \tag{3.29}$$

and $\varkappa_{p,d}$ (3.9) are purely imaginary. Such a situation corresponds to wave propagation in the x direction and damping in the z direction. For $0 < \varepsilon_r < 1$,

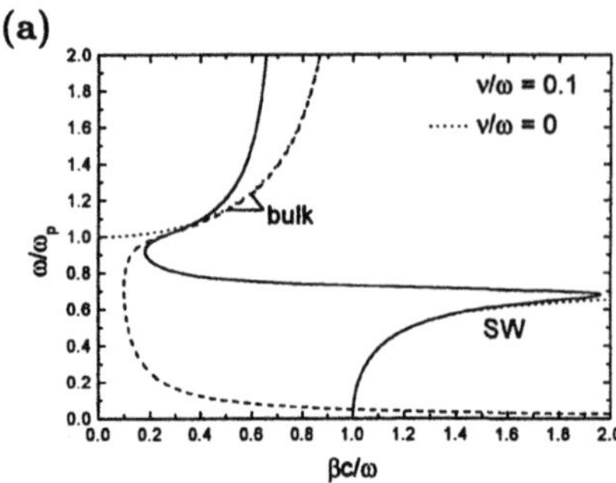
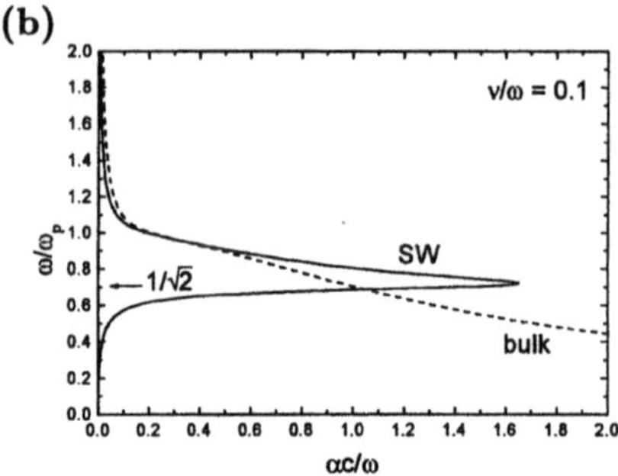

Figure 3.4. Phase diagram of bulk (*dashed curve*) and surface (*solid curve*) transverse waves in weakly collisional plasmas semi-bounded by vacuum: **(a)** real part β and **(b)** imaginary part α of the wave number; $\nu/\omega = 0.1$. In **(a)**, the dispersion behaviour at zero collision frequency (*dotted curves*) is also shown for comparison

i.e. at plasma densities n below the critical density n_{c}, which is the region occupied by the radiative surface mode when collisions are present, (3.11) and (3.17) give

$$\beta = \frac{\omega}{c}\sqrt{\tilde{\varepsilon}_{\mathrm{r}}}, \qquad \alpha = 0. \tag{3.30}$$

However, $\varkappa_{\mathrm{p,d}}$ (3.9) are purely imaginary now and this corresponds to refraction into the plasma at the angle $\Theta_{\mathrm{B}}^{(\mathrm{p})} = \arctan(\varepsilon_{\mathrm{d}}/\varepsilon_{\mathrm{r}})^{1/2}$ of a plane EM wave falling on the plasma–vacuum interface from the dielectric semi-space at the Brewster angle $\Theta_{\mathrm{B}}^{(\mathrm{d})} = \arctan(\varepsilon_{\mathrm{r}}/\varepsilon_{\mathrm{d}})^{1/2}$. Therefore, it is only because of collisions that the modes existing for $n < n_{\mathrm{c}}^{*}$ transform into SWs (Fig. 3.2). Moreover, the collisions cause overlapping of the frequency ranges of the surface and bulk waves. In Fig. 3.4, the propagation constants and space damping rates of the two types of waves are compared. The characteristics of the bulk transverse waves are given according to (2.24b).

For $\omega/\omega_{\mathrm{p}} > 1$ (i.e. for $\varepsilon_{\mathrm{r}} > 0$) the radiative surface mode has a smaller wave number, i.e. a larger phase velocity, than the transverse bulk wave, and its damping is weaker. In the range below $\omega/\omega_{\mathrm{p}} = 1$ the bulk waves are "skinned" and their damping strongly increases (the damping rate being larger than the real part of the propagation constant). Nevertheless their damping is weaker than that of the evanescent mode. The crossing of both the real and the imaginary parts of the wave numbers of the bulk and surface waves occurs at $\varepsilon_{\mathrm{r}} \simeq 0$, and this is also the point where the evanescent mode smoothly transforms into the radiative mode. In the range of the surface mode, the damping rate of the SWs is, of course, much smaller than that of the bulk waves.

Detailed discussions of the effects of collisions make the understanding of numerical solutions of the wave dispersion behaviour in waveguides with different geometries easier [3.97–101]. The results obtained for the SW characteristics (wave number, space damping rate, field distribution in the trans-

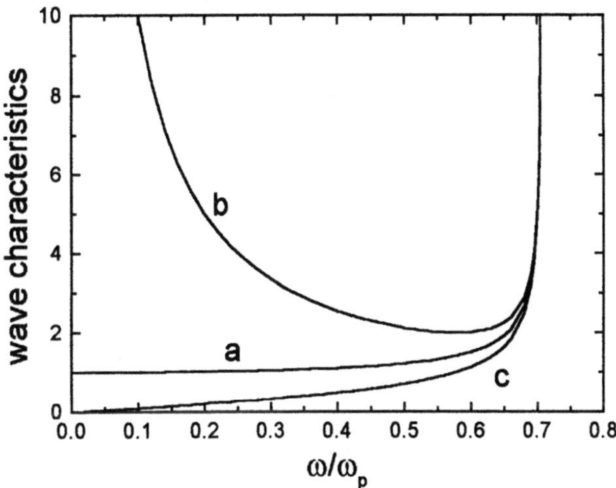

Figure 3.5. ω/ω_p dependence of the SW characteristics (plasma–vacuum interface): $\beta c/\omega$ (a), $\varkappa_\mathrm{p} c/\omega$ (b) and $\varkappa_\mathrm{v} c/\omega$ (c)

verse direction and its decaying and oscillating parts) justify the names used for the different branches of SWs. However, the branch of SWs called the "surface mode", which is below the SW resonance in cold homogeneous collisionless plasmas, is the branch that appears as proper SWs: they are weakly damped in the longitudinal direction and have a strongly decaying amplitude in the transverse direction. This is the branch which sustains the SW discharges to be described below. The propagation characteristics of this mode will be listed in more detail.

Surface Mode in Collisionless Plasmas (3.27). For $|\varepsilon_\mathrm{r}| \gg \varepsilon_\mathrm{d}$, i.e. at small ω/ω_p the surface mode is a comparatively fast wave with phase and group velocities less than but close to the light speed in the dielectric

$$\beta = \frac{\omega}{c}\,\sqrt{\varepsilon_\mathrm{d}}\,. \tag{3.31a}$$

In the quasi-static limit ($\omega/\beta \ll c$), i.e. for $\omega/\omega_\mathrm{p} \lesssim 1/\sqrt{1+\varepsilon_\mathrm{d}}$, the waves are much slower, with a frequency spectrum given by

$$\omega = \frac{\omega_\mathrm{p}}{\sqrt{1+\varepsilon_\mathrm{d}}}\left(1 - \frac{\varepsilon_\mathrm{d}^2\omega_\mathrm{p}^2}{2(1+\varepsilon_\mathrm{d})^2\beta^2 c^2}\right)\,. \tag{3.31b}$$

The variation of the group velocity with ε in the complete ω/ω_p region of existence of the wave is

$$v_\mathrm{gr} = c\,\frac{\sqrt{\varepsilon_\mathrm{r}(1+\varepsilon_\mathrm{r})^3}}{1+\varepsilon_\mathrm{r}^2}\,. \tag{3.31c}$$

The variation of $\varkappa_\mathrm{p,d}$ (Fig. 3.5) over the phase diagram shows that $\varkappa_\mathrm{p} c/\omega$ is always larger than unity, which corresponds to a strong field decay into the

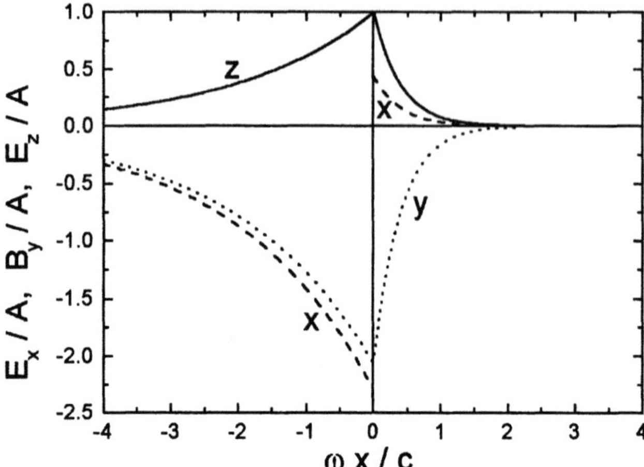

Figure 3.6. Transverse distribution of the normalized amplitudes of the field components E_z (z, *solid curve*), H_y (y, *dotted curve*) and E_x (x, *dashed curve*) of an SW at a plasma–vacuum interface for $\omega/\omega_\mathrm{p} = 0.4$. A is the amplitude of the E_z field component at the interface

plasma region. The amplitudes of the field components (as given by (3.8)) are presented in Fig. 3.6. The transverse electric-field component has opposite signs in the two media. As a consequence the longitudinal component P_z of the wave power flux (2.32b) has opposite signs in the plasma and dielectric semi-spaces. In the dielectric semi-space, the power flux is directed forward (with respect to the direction of propagation of the wave), whereas in the plasma semi-space it is directed backward. The larger value of the power flux in the surrounding dielectric determines the positive value of the total flux

$$P_z = \frac{|A|^2}{4\mu_0\omega} \frac{(\varepsilon_\mathrm{d} - \varepsilon_\mathrm{r})(\varepsilon_\mathrm{d} + \varepsilon_\mathrm{r})^2}{\varepsilon_\mathrm{d}^{3/2}(-\varepsilon_\mathrm{r})^{3/2}} \tag{3.32}$$

and the energy transport by the waves (Fig. 3.7).

In general, the SW behaviour reveals the following properties. For lower ω/ω_p values the waves are comparatively fast: their phase velocity and group velocity are smaller than the light speed in the surrounding dielectric, but are close to it. The EM character of the waves is pronounced. The scale $\varkappa_\mathrm{p}^{-1}$ of the field penetration into the plasma is much smaller than that in the dielectric, which gives a stronger wave field decay in the plasma semi-space. The EM character of the waves, together with the extension of the wave field into the dielectric, allows a larger power flux to be guided in this region. With increasing ω/ω_p values, the waves slow down, their longitudinal character becomes more pronounced and the scale of the field penetration – tending to the value of the wavelength – is of the same order of magnitude both in the plasma and in the dielectric. The portion of the wave power flux which is guided decreases in the dielectric semi-space, whereas that in

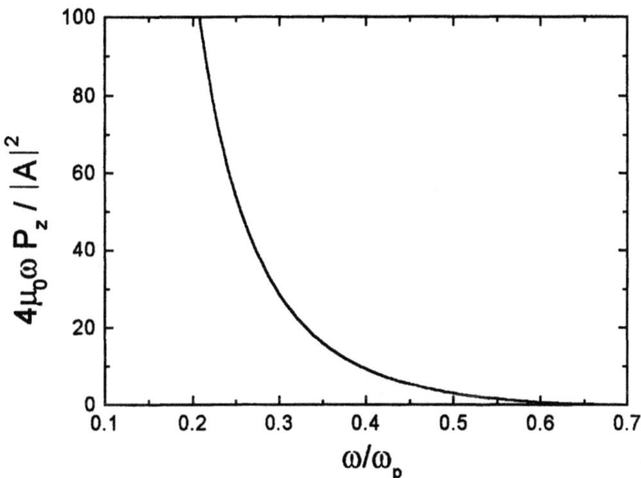

Figure 3.7. Dependence of the normalized SW power flux in a plasma–vacuum waveguide on ω/ω_p

the plasma increases (in absolute value) and this leads to a decrease of the total power flux. In the quasi-static region, for $\omega/\omega_\mathrm{p} \to (1/\sqrt{1+\varepsilon_\mathrm{d}})$, i.e. approaching the SW resonance, which in cold collisionless plasmas is the upper limit of the frequency band of wave propagation, the magnetic field of the wave becomes negligible, the electric-field components are bound by the interface, the waveguide channel collapses in the transverse direction and the guided-wave power flux tends to zero.

Surface Mode in Weakly Collisional Plasmas. The collisional damping rates are as follows:

- space damping rate

$$\alpha = \frac{\nu}{2c}\frac{1-\varepsilon_\mathrm{r}}{\sqrt{\varepsilon_\mathrm{r}(\varepsilon_\mathrm{r}+1)^3}} \qquad (3.33a)$$

- time damping rate

$$\gamma = -\frac{\nu}{2}\frac{1-\varepsilon_\mathrm{r}}{1+\varepsilon_\mathrm{r}^2}\ . \qquad (3.33b)$$

In the region where SWs are comparatively fast ($v_\mathrm{ph} \lesssim c/\sqrt{\varepsilon_\mathrm{d}}$, as given by (3.31a)), the group velocity is constant and the ω/ω_p dependence of the space damping rate

$$\alpha = \frac{\nu\omega^2}{2c^3}\lambda_\mathrm{sk}^2 \equiv \frac{\nu}{2c}\frac{\omega^2}{\omega_\mathrm{p}^2}\ , \qquad (3.33c)$$

where λ_sk is the skin depth (2.26a), is completely due to the dependence of γ on ε_r. With retardation of the wave ($\omega \to \omega_\mathrm{p}/\sqrt{2}$), the time damping rate tends to a constant value

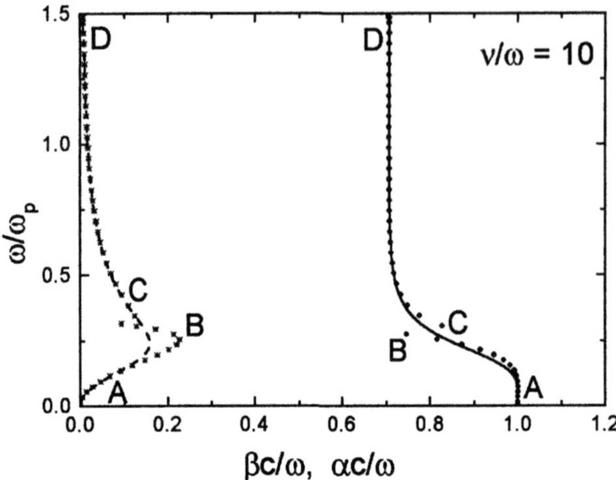

Figure 3.8. Wave number β (*solid curve*) and space damping rate α (*dashed curve*) of SW propagation along a single interface between a cold strongly collisional plasma and vacuum: plot of (3.11) in terms of a phase diagram for $\nu/\omega = 10$. Approximate solutions for β and α: AB, (3.35) with $\varepsilon_d = 1$; CD, (3.20) with $\varepsilon_d = 1$

$$\gamma = -\frac{\nu}{2} \tag{3.33d}$$

and the variation of the space damping rate

$$\alpha = \frac{\nu}{c}\left(\frac{\omega_p^2}{\omega^2} - 2\right)^{-3/2} \tag{3.33e}$$

with ε_r is due to the changes of the group velocity of the wave.

When dissipation is taken into account, the singularity of the refractive index kc/ω connected with the plasma density going through the cut-off density n_c^* disappears. Moreover, as has been commented before, the losses of EM energy in the plasma region give rise to the possibility of solutions with zero power flux at $x = \pm\infty$ for $n < n_c$ also.

3.2.2 Case of Strong Collisions ($\nu \gg \omega$)

The wave characteristics come out directly from the solution (3.11) of the dispersion relation (3.10) in the case $\nu > \omega$. For the case of a plasma–vacuum interface, they are shown in Fig. 3.8. The wave is weakly damped in the whole range of its existence. To clarify the nature of the wave dispersion, some analysis of (3.11) and (3.17) is given below.

Equation (2.9) shows that in the dense plasma range ($\omega/\omega_p < \sqrt{\omega/\nu}$) the imaginary part of the plasma permittivity is larger than its real part ($\varepsilon_i \gg |\varepsilon_r|$). Such a condition leads to SWs in plasmas similar to waves propagating along a conducting surface. The wave appears as a Sommerfeld-type wave

[3.94]. (In the literature on solid-state waveguide systems [3.93] this type of wave is known as the "Zenneck mode".) With decreasing densities, which corresponds to going up along the $\omega/\omega_{\rm p}$ axis, the real part of the plasma permittivity tends to 1 ($\varepsilon_{\rm r} \to 1$) and its imaginary part can become smaller than the real part ($\varepsilon_{\rm i} < \varepsilon_{\rm r}$). Such a situation corresponds to conditions which, in the case of weak collisions considered above, ensure the existence of a radiative surface mode.

Therefore, two regions of different types of waves can appear in the case of strong collisions: (i) $\varepsilon_{\rm i} \gg |\varepsilon_{\rm r}|$, i.e. a strongly dissipative dense plasma ($\omega/\omega_{\rm p} < \sqrt{\omega/\nu}$), and (ii) $\varepsilon_{\rm r} \gg \varepsilon_{\rm i}$, i.e. a weakly dissipative rarefied plasma ($\omega/\omega_{\rm p} > \sqrt{\omega/\nu}$). However, in both cases the waveguide is weakly dissipative, i.e. the imaginary part of its effective permittivity is smaller than the real part (3.14), ensuring weak damping of the waves in their complete frequency range. As has been mentioned, when the plasma is strongly dissipative ($\varepsilon_{\rm i} \gg |\varepsilon_{\rm r}|$), the waveguide structure is weakly dissipative if $\varepsilon_{\rm i} \gg \varepsilon_{\rm d}$. In this case (3.12) reduces to

$$\tilde{\varepsilon}_{\rm r} = \varepsilon_{\rm d} \left(1 - \frac{\varepsilon_{\rm d}(\varepsilon_{\rm r} + \varepsilon_{\rm d})}{\varepsilon_{\rm i}^2}\right) , \tag{3.34a}$$

$$\tilde{\varepsilon}_{\rm i} = \frac{\varepsilon_{\rm d}^2}{\varepsilon_{\rm i}} \left(1 - \frac{(\varepsilon_{\rm r} + \varepsilon_{\rm d})^2}{\varepsilon_{\rm i}^2}\right) , \tag{3.34b}$$

and (3.11) and (3.17) result in the following expressions for the wave dispersion characteristics:

$$\beta = \frac{\omega}{c} \sqrt{\varepsilon_{\rm d}} \left(1 - \frac{\varepsilon_{\rm r}\varepsilon_{\rm d}}{2\varepsilon_{\rm i}^2}\right) , \tag{3.35a}$$

$$\alpha = \frac{\omega \varepsilon_{\rm d}^{3/2}}{2c\varepsilon_{\rm i}} \left(1 - \frac{\varepsilon_{\rm r}^2}{\varepsilon_{\rm i}^2}\right) . \tag{3.35b}$$

The second terms in each of the large brackets is a small correction. In Fig. 3.8 the results obtained for the longitudinal wave number and the space damping rate from (3.35) are labelled as the curves AB. With $\alpha/\beta = (2\sqrt{\varepsilon_{\rm d}}\,\varepsilon_{\rm i})^{-1}$ the waves are weakly damped ($\alpha/\beta \ll 1$).

In the region of dense plasmas ($1 \ll \nu/\omega \ll (\omega_{\rm p}/\omega)^2$), the wave phase velocity is very close to the light speed c in the surrounding dielectric but a little smaller than c, since

$$\beta = \frac{\omega}{c} \sqrt{\varepsilon_{\rm d}} \left[1 + \frac{\varepsilon_{\rm d}}{2} \left(\frac{\omega}{\omega_{\rm p}}\right)^2\right] . \tag{3.36a}$$

With a decrease of the density ($1, (\omega_{\rm p}/\omega) \ll (\nu/\omega) \ll (\omega_{\rm p}/\omega)^2$) the wave transforms into a fast one. Its phase velocity is larger, although only slightly so, than the light speed in the dielectric and

$$\beta = \frac{\omega}{c} \sqrt{\varepsilon_{\rm d}} \left(1 - \frac{\varepsilon_{\rm d}}{2} \frac{\omega^2 \nu^2}{\omega_{\rm p}^4}\right) . \tag{3.36b}$$

In the region of weakly dissipative plasmas ($\varepsilon_r > \varepsilon_i$, curves CD in Fig. 3.8), (3.11) takes the form (3.20), presented earlier as giving the behaviour of the radiative mode in weakly collisional plasmas. Here, in strongly collisional plasmas, this corresponds to a region of ω/ω_p values which satisfy the inequalities

$$\frac{\nu}{\omega} \gg 1, \ \frac{\omega_p}{\omega}, \ \left(\frac{\omega_p}{\omega}\right)^2 . \tag{3.37}$$

With $\varepsilon_r \approx 1$ and $\varepsilon_i = \omega_p^2/\omega\nu$, (3.20) reduces to

$$\beta = \frac{\omega}{c} \sqrt{\frac{\varepsilon_d}{1+\varepsilon_d}} \left(1 + \frac{\varepsilon_d(4+\varepsilon_d)}{8(1+\varepsilon_d)^2} \frac{\omega_p^4}{\omega^2\nu^2}\right), \tag{3.38a}$$

$$\alpha = \frac{\omega}{2c} \left(\frac{\varepsilon_d}{1+\varepsilon_d}\right)^{3/2} \frac{\omega_p^2}{\omega\nu} \left(1 - \frac{\omega_p^2}{\omega\nu} \frac{1}{(1+\varepsilon_d)^2}\right), \tag{3.38b}$$

where the second term in each of the brackets is a correction. The phase velocity tends to a constant value (for a plasma–vacuum interface, equal to $c\sqrt{2}$).

In the case of strong collisions ($\nu > \omega$) a quasi-static resonance does not show up and SWs can propagate in the total region of plasma densities as fast, weakly damped waves except for the region $\omega/\omega_p \approx (\omega/\nu)^{1/2}$ (Fig. 3.8), where the real and imaginary parts of the plasma permittivity become comparable and of the order of unity.

The solutions obtained for the case of radiative modes in strongly collisional plasmas show in essence that SWs can be formed not only at interfaces between two media with different permittivities but also at interfaces with a jump in the conductivity of the media. In fact, the case of a plasma–vacuum interface corresponds to a case with the same values for the real parts of the permittivities of the two media ($\varepsilon_r = 1$, $\varepsilon_d \equiv \varepsilon_v = 1$), and it is the plasma conductivity ($\varepsilon_i \neq 0$) which makes the difference between the two media.

3.2.3 Semi-Bounded Plasmas Overlain by a Dielectric or a Plasma

The presence of dielectric or plasma layers between the spaces occupied by a plasma and a vacuum modifies the dispersion properties of the SWs. This can be demonstrated with the configuration of the waveguide presented in Fig. 3.9. A semi-space ($0 < x < \infty$) occupied by a cold plasma with permittivity ε (2.8) is separated from the space ($-\infty < x < -d$) occupied by the vacuum ($\varepsilon_v = 1$) by a layer ($-d < x < 0$) with thickness d and permittivity ε_s. The media in the three space regions are homogeneous. Propagation of a TM mode ($E_x, B_y, E_z \neq 0$) guided by the interface $x = 0$ (and $x = -d$) and decaying as $x \rightarrow \pm\infty$ is considered.

The dispersion relation describing SW propagation in the z direction is [3.102–107]

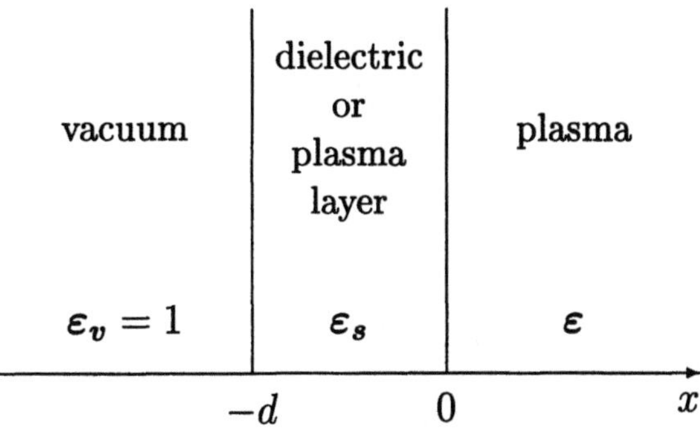

Figure 3.9. Waveguide configuration

$$\mathcal{D}(\omega, k) = \left(\frac{\varepsilon}{\varkappa_p} + \frac{\varepsilon_s}{\varkappa_s} \right) \left(\frac{\varepsilon_s}{\varkappa_s} + \frac{1}{\varkappa_v} \right)$$
$$- \left(\frac{\varepsilon_s}{\varkappa_s} - \frac{\varepsilon}{\varkappa_p} \right) \left(\frac{\varepsilon_s}{\varkappa_s} - \frac{1}{\varkappa_v} \right) e^{-2\varkappa_s d} = 0 \,. \tag{3.39}$$

Here, ε and $\varkappa_p$, as given by (2.8a) and (3.9a), respectively, characterize the plasma medium $(x > 0)$; $\varkappa_v$ given by (3.9c) is related to the vacuum region $(x < -d)$; and

$$\varkappa_s = \sqrt{k^2 - \frac{\omega^2}{c^2} \varepsilon_s} \tag{3.40}$$

characterizes the field variation in the layer $(-d < x < 0)$. In the case of a dielectric layer, $\varepsilon_s \equiv \varepsilon_d = $ const. and (3.40) takes the form of (3.9b). With a layer consisting of a weakly collisional plasma with plasma frequency ω_s, the permittivity ε_s is as given by (2.8a) with ω_p replaced by ω_s.

As is known [3.81,82], the presence of a thin dielectric layer $(\varepsilon_s \equiv \varepsilon_d, \varkappa_s d \equiv \varkappa_d d \ll 1)$ between the space regions occupied by a plasma and a vacuum influences the SW dispersion properties mainly in the region of the SW resonance in cold collisionless plasmas. Depending on the layer thickness, a region of variation of the phase velocity between c and $c/\sqrt{\varepsilon_d}$ appears, the ω/ω_p range of wave propagation extends beyond the SW resonance (3.1a), the dispersion curve turns (Sect. 3.3) and, through a region of a backward wave, reaches the SW resonance (3.1a) with ω/ω_p values higher than the resonance value. The dispersion relation (3.39) reduces to

$$\varkappa_v + \frac{\varkappa_p}{\varepsilon} + \frac{\varkappa_d^2 d}{\varepsilon_d} \frac{\varkappa_v \varkappa_p}{\varepsilon} \left(\frac{\varepsilon_d}{\varkappa_d} - \frac{\varepsilon}{\varkappa_p} \right) \left(\frac{\varepsilon_d}{\varkappa_d} - \frac{1}{\varkappa_v} \right) = 0 \,. \tag{3.41a}$$

The first two terms represent the dispersion law (3.10a) of SW propagation along a single plasma–vacuum interface and the third term

$$\mathcal{D}_1(\omega, k) = -\frac{k^2 d}{\varepsilon_d} \left(\varepsilon_d^2 - 1 \right) \left(1 - \frac{\omega^2}{k^2 c^2} \frac{\varepsilon_d}{\varepsilon_d + 1} \right) \approx -\frac{k^2 d}{\varepsilon_d} \left(\varepsilon_d^2 - 1 \right) \tag{3.41b}$$

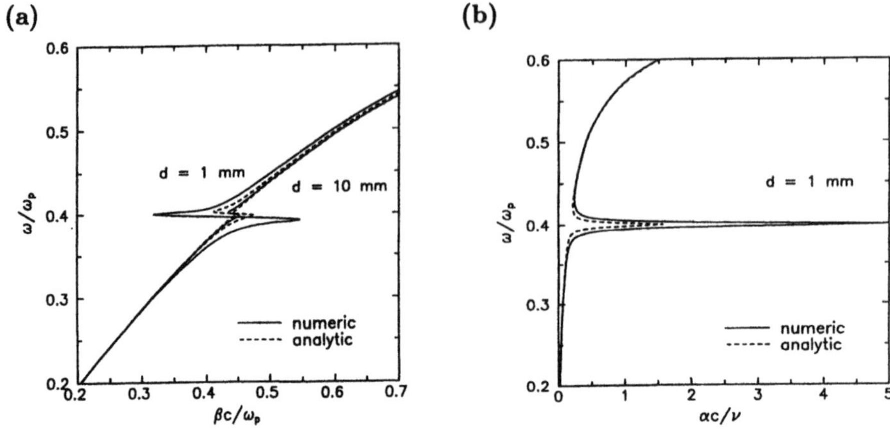

Figure 3.10. Dispersion (**a**) and space damping rate (**b**) of the SW mode influenced by the effect of linear mode conversion due to its interaction with the quasi-static mode of the plasma layer, given by analytical solutions (*dashed curves*) of (3.42) and numerical solutions (*solid curves*) of (3.39); $\omega_s/\omega_p = 0.4$, $\nu/\omega_p = 5 \times 10^{-3}$, $d = 0.1\,\mathrm{cm}$ and $d = 1\,\mathrm{cm}$ (the stronger effect in (**a**) is with $d = 1\,\mathrm{cm}$) and $\omega_p = 10^9\,\mathrm{s}^{-1}$

describes the influence of the dielectric layer on the wave dispersion behaviour.

In the presence of a plasma layer the solution of (3.39) accounts for a possible linear mode interaction of the SW of a semi-infinite weakly collisional plasma (wave number β and space damping rate α, which are given by (3.20) with ε_d replaced by 1) with the bulk mode ($\varepsilon_s = 0$) of the layer. This is described by the approximation

$$\varepsilon_s \left(\varkappa_v + \frac{\varkappa_p}{\varepsilon} \right) = -k^2 d \tag{3.42}$$

to (3.39) for $\varkappa_s d \ll 1$. The solution of (3.42) is depicted in Fig. 3.10. The effect of the mode transformation is stronger for larger thickness d of the plasma layer.

Coupling of the modes of the structure can also be demonstrated by starting from a thick plasma layer ($\varkappa_s d \gg 1$). The solution of (3.39) for $\varkappa_s d \to \infty$ gives two surface modes, which are attached to the vacuum–layer interface and the layer–plasma interface. A finite value of the layer thickness d together with dissipation in the two regions occupied by plasmas leads to closing of the gap between the characteristics of these modes (Fig. 3.11) through formation of "anomalous dispersion" bends.

3.2.4 Influence of the Thermal Motion of the Electrons

Thermal effects influence the properties of SWs in the range of their existence as slow waves. This is the case when the phase velocity v_{ph} is comparable

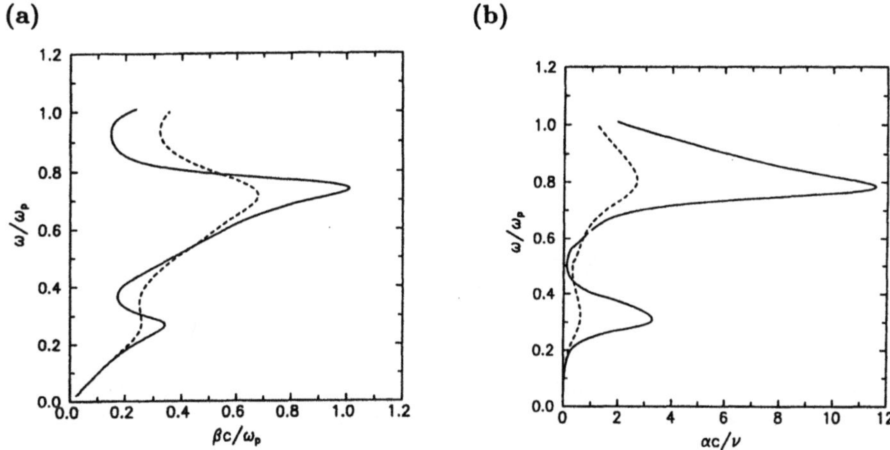

(a)

(b)

Figure 3.11. Numerical solutions of (3.39) for wave number β **(a)** and damping rate α **(b)** for $\omega_\mathrm{s}/\omega_\mathrm{p} = 0.4$, $d\omega_\mathrm{p}/c = 3$ and $\nu/\omega_\mathrm{p} = 8 \times 10^{-2}$ (*solid curves*) and $\nu/\omega_\mathrm{p} = 2 \times 10^{-1}$ (*dashed curves*)

to the thermal velocity $v_\mathrm{th} = \sqrt{T_\mathrm{e}/m}$, and the wave–particle interactions are more efficient than the usual particle–particle collisions. The solution of the problem requires a boundary condition for the electron velocity at the plasma–dielectric interface $x = 0$ (Fig. 3.1). The condition of mirror reflection of the electrons from the boundary has commonly been used.

A description within the kinetic plasma model ([3.77,80] and references therein) completes the results by introducing the effect of Landau damping of the SW. As has been mentioned, owing to the thermal electron motion the resonance of SWs in cold collisionless plasmas disappears, and the ω/ω_p range extends smoothly above the resonance value (3.1) with propagation of an increasingly damped wave (via the Landau mechanism).

The dispersion relation of SWs is

$$\frac{1}{\pi}\left(\int_{-\infty}^{+\infty}\frac{k^2\,\mathrm{d}k_x}{\varkappa^2\varepsilon^\mathrm{l}(\omega,\varkappa)} + \int_{-\infty}^{+\infty}\frac{\mathrm{d}k_x}{\varkappa^2}\frac{k_x^2}{\varepsilon^\mathrm{tr}(\omega,\varkappa) - (c^2k^2/\omega^2)}\right) = -\frac{\varkappa_d}{\varepsilon_\mathrm{d}},$$

$$(3.43)$$

where $\varkappa^2 = k_x^2 + k^2$, and $\varepsilon^\mathrm{l}(\omega,\varkappa)$ and $\varepsilon^\mathrm{tr}(\omega,\varkappa)$ are the longitudinal and transverse plasma dielectric functions of an infinite plasma, well known from textbooks [3.77]. With plasma spatial dispersion neglected, (3.43) reduces to dispersion relation (3.10a) of SWs in semi-infinite cold collisionless plasmas. In the quasi-static limit

$$\frac{\omega^2}{c^2}|\varepsilon| \ll k^2,$$

$$(3.44)$$

(3.43) simplified to

$$\frac{k}{\pi} \int_{-\infty}^{+\infty} \frac{\mathrm{d}k_x}{\varkappa^2 \varepsilon^{\mathrm{l}}(\omega, \varkappa)} = -\frac{1}{\varepsilon_{\mathrm{d}}} , \tag{3.45}$$

results in a spectrum of SW propagation in the case of a plasma–vacuum ($\varepsilon_{\mathrm{d}} = 1$) interface given by

$$\omega = \frac{\omega_{\mathrm{p}}}{\sqrt{2}}(1 + 1.22\,\beta r_{\mathrm{D}}) \tag{3.46a}$$

and a time damping rate given by

$$\gamma = -\frac{\omega_{\mathrm{p}}}{\sqrt{2}} 0.176\,\beta r_{\mathrm{D}} \tag{3.46b}$$

[3.56,108]. Owing to the thermal motion the quasi-static surface-charge oscillations at the frequency (3.1b) are transformed into a wave with space dispersion which propagates at frequencies slightly above the resonance frequency (3.1b) for cold collisionless plasmas. In contrast to the bulk longitudinal waves [3.77], the SW frequency is related to the first power of the wave number β. While the Landau damping of the bulk waves is exponentially small, the damping of the SWs due to the thermal electron motion is pronounced. This is due to the fact that the SW field decaying exponentially into the plasma depth contains Fourier harmonics with $k_x \in (0, \infty)$, i.e. not only weakly damped oscillations $\varkappa r_{\mathrm{D}} < 1$, but also relatively highly damped short-wavelength oscillations $\varkappa r_{\mathrm{D}} > 1$. These short-wavelength components in the Fourier expansion of the SW field interact more efficiently with the slow electrons, causing a stronger Cherenkov dissipation of the SW field energy.

3.3 Plasma Slabs and Plasma Columns

SW propagation along planar and cylindrical plasma waveguides – plasma slabs and plasma columns surrounded by a dielectric and/or vacuum – is presented here in the form of a survey for the purpose of the discharge modelling described in Chaps. 5 and 6. Results for the SW mode, which, because of its weak damping and weak radiation, can efficiently produce SW-sustained discharges, are mainly discussed. The terminology of "fast" and "slow" SWs here relate to the EM and quasi-static ranges of the SW mode (region (a) in Fig. 3.2) and should not be confused with the "radiative mode" and the "surface mode" discussed in Sect. 3.2. Although the cylindrical geometry is commonly used in experiments on SW-produced discharges, the slab configuration is also considered, since the slab model – being simpler – provides a more transparent understanding of the physical basis of the discharge maintenance. Moreover, planar discharges have recently started appearing in experiments and applications too.

The main physical features of SWs discussed above were shown by the model of a single plasma–dielectric interface. However, the geometry of the waveguide – plasma slab or plasma column – introduces some important pecularities into the wave behaviours related to the ratios of the SW field penetration depths $\varkappa_{\mathrm{p,d}}^{-1}$ (3.9) into the media and the transverse dimension of the guide. The thickness $2d$ of the slab or the radius R of the cylinder appears as a natural quantity for normalization of the wave number β in the presentation of the phase diagrams: $(\omega/\omega_{\mathrm{p}})$ versus (βd) and $(\omega/\omega_{\mathrm{p}})$ versus (βR) in the cases of planar and cylindrical waveguides, respectively. With an additional dimension of the guide present, the phase diagram transforms into a family of curves with a parameter $\sigma = \omega d/c$ or $\sigma = \omega R/c$. In a way, the size of the waveguide is compared with the SW wavelength $(\lambda = 2\pi/\beta)$ and the vacuum wavelength $(\lambda_{\mathrm{v}} = 2\pi/k_{\mathrm{v}})$.

For $\varkappa_{\mathrm{p}}d$, $\varkappa_{\mathrm{p}}R \ll 1$ the SW field penetrates deeply into the plasma and its decay across the slab/cylinder is weak. In the opposite case, for $\varkappa_{\mathrm{p}}d$, $\varkappa_{\mathrm{p}}R \gg 1$, the field variation in the transverse direction is strong and the field is concentrated close to the boundary. The scaling of the field penetration depth in the dielectric/vacuum can also be different: $\varkappa_{\mathrm{d}}d$, $\varkappa_{\mathrm{d}}R \ll 1$ or $\varkappa_{\mathrm{d}}d$, $\varkappa_{\mathrm{d}}R \gg 1$. In the former case the field is greatly extended in the dielectric/vacuum region whereas in the latter it is strongly attached to the boundary. Since $\varkappa_{\mathrm{p}} > \varkappa_{\mathrm{d,v}}$ (see (3.9)), the following cases are possible:

$$\varkappa_{\mathrm{p}}d,\ \varkappa_{\mathrm{p}}R \gg 1;\quad \varkappa_{\mathrm{d}}d,\ \varkappa_{\mathrm{d}}R \ll 1, \tag{3.47a}$$

$$\varkappa_{\mathrm{p}}d,\ \varkappa_{\mathrm{p}}R \ll 1;\quad \varkappa_{\mathrm{d}}d,\ \varkappa_{\mathrm{d}}R \ll 1, \tag{3.47b}$$

$$\varkappa_{\mathrm{p}}d,\ \varkappa_{\mathrm{p}}R \gg 1;\quad \varkappa_{\mathrm{d}}d,\ \varkappa_{\mathrm{d}}R \gg 1. \tag{3.47c}$$

Independently of the value of σ, the phase diagrams end in the range of high $\omega/\omega_{\mathrm{p}}$ values in the region (3.47c), in which the SW resonance (3.1) in cold collisionless plasmas is positioned. Changes of the waveguide configuration – semi-bounded plasmas, slabs, cylindrical waveguides – do not influence the wave properties in this region. Since small $\omega/\omega_{\mathrm{p}}$ means large ω_{p}, the inequalities $\varkappa_{\mathrm{p}}d$, $\varkappa_{\mathrm{p}}R \gg 1$ determine the wave behaviour at the "bottom" of the phase diagrams and the phase diagrams always, i.e. independently of the value of σ, begin with region (3.47a). The limit $\omega_{\mathrm{p}} \to \infty$, $\omega/\omega_{\mathrm{p}} \to 0$ fixes $\beta R = \omega R/c \equiv \sigma$ (or $\beta d = \omega d/c \equiv \sigma$) as a value of β there. For a large σ this starting value of β is large enough and leads to large values of $\varkappa_{\mathrm{p}}d$ or $\varkappa_{\mathrm{p}}R$ even with a decrease of ω_{p} for larger $\omega/\omega_{\mathrm{p}}$. Therefore, for large σ, the conditions $\varkappa_{\mathrm{p}}d$, $\varkappa_{\mathrm{p}}R \gg 1$ hold over the complete phase diagram, which in this case is formed by the consecutive locations of the regions (3.47a,c). This case gives, in general, a behaviour similar to that in semi-bounded plasmas. For small σ, the starting value of βd or $\beta R = \sigma$ for $\omega_{\mathrm{p}} \to \infty$ $(\omega/\omega_{\mathrm{p}} \to 0)$ is too small, and in spite of the increase of β across the phase diagram, the decrease of ω_{p} makes possible the appearance of a region where $\varkappa_{\mathrm{p}}d$, $\varkappa_{\mathrm{d}}R \ll 1$. Therefore, for small σ the waves start as fast SWs with a field which is strongly attached to the boundary in the plasma region and greatly extended in the dielectric region, case (3.47a). Keeping their behaviour of fast waves, they enter the region of a thin slab/cylinder defined by (3.47b), in which the field variation in the transverse direction is weak both in the plasma and in the dielectric

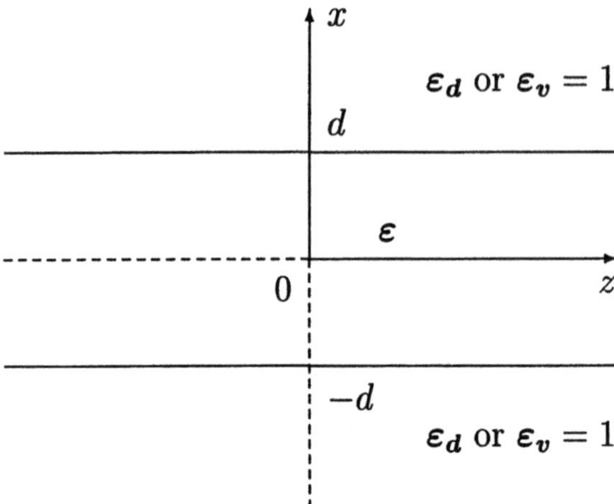

Figure 3.12. Planar waveguide: a plasma slab bounded by a dielectric/vacuum

regions. Here the waves transform into slow SWs. In fact for small σ values the region (3.47b) of a thin slab/cylinder is quite extended and covers a large range of variation of ω/ω_p and, correspondingly, of ω_p. Since experiments on SW-produced discharges are usually performed at small σ ($\sigma < 0.2$), this is the region which covers almost the complete length of the discharge column. The phase diagram ends with an approach to the SW resonance within region (3.47c), which gives slow SWs in the thick-slab/cylinder case, characterized by strong attachment of the field to the boundary from both the plasma and the dielectric sides. Therefore, for small σ values the phase diagrams contain the three regions (3.47) consecutively, given by (3.47a), (3.47b) and (3.47c) starting from the bottom to the top. The limiting case of $\sigma \to 0$ gives the quasi-static approach to the description of SWs.

3.3.1 Plasma Slabs: Planar Waveguides

Planar plasma waveguides – plasma slabs of thickness $2d$ bounded by a dielectric/vacuum – support propagation of two modes [3.79]: symmetric and antisymmetric waves, specified with respect to the transverse distribution of their E_z field component (which is symmetric or antisymmetric with respect to the slab axis). The waveguide configuration is shown in Fig. 3.12. The wave propagation is in the z direction ($\propto \exp(-i\omega t + i kz)$); the transverse structure of the wave field amplitude is in the x direction. The two SWs are TM modes: E_x, B_y, $E_z \neq 0$. In the case of the symmetric mode, $E_z(x = d) = E_z(x = -d)$, $B_y(x = d) = -B_y(x = -d)$, whereas for the antisymmetric mode $E_z(x = d) = -E_z(x = -d)$, $B_y(x = d) = B_y(x = -d)$.

For example, the solutions of the wave field equation (2.16) for H_y in the spaces ($x > d$, $x < -d$) occupied by the dielectric are

$$B(x > d) = B(x = d)e^{-\varkappa_{\mathrm{d}}(x-d)}, \tag{3.48a}$$

$$B(x < -d) = B(x = -d)e^{\varkappa_{\mathrm{d}}(x+d)}, \tag{3.48b}$$

whereas the solutions for the field distribution of the symmetric and anti-symmetric mode in the plasma region $(-d \le x \le d)$ are

$$B(-d \le x \le d) = B(x = d)\frac{\sinh(\varkappa_{\mathrm{p}}x)}{\sinh(\varkappa_{\mathrm{p}}d)}, \tag{3.49a}$$

$$B(-d \le x \le d) = B(x = d)\frac{\cosh(\varkappa_{\mathrm{p}}x)}{\cosh(\varkappa_{\mathrm{p}}d)}, \tag{3.49b}$$

respectively. The boundary condition (2.17a) leads to the dispersion relations of the symmetric and antisymmetric surface modes:

$$\mathcal{D}(\omega, k) = \frac{\varkappa_{\mathrm{p}}}{\varepsilon}\coth(\varkappa_{\mathrm{p}}d) + \frac{\varkappa_{\mathrm{d}}}{\varepsilon_{\mathrm{d}}} = 0, \tag{3.50a}$$

$$\mathcal{D}(\omega, k) = \frac{\varkappa_{\mathrm{p}}}{\varepsilon}\tanh(\varkappa_{\mathrm{p}}d) + \frac{\varkappa_{\mathrm{d}}}{\varepsilon_{\mathrm{d}}} = 0, \tag{3.50b}$$

respectively.

Their solutions are given in Fig. 3.13a. The field distribution in the transverse direction is shown in Fig. 3.13b. The symmetric mode propagates as a forward wave, and its phase diagrams are similar in shape to that of SWs in semi-infinite plasmas. The antisymmetric mode behaves differently, showing a backward-wave propagation at small σ values in collisionless plasmas. In the quasi-static limit $((kc/\omega)^2 \gg \varepsilon^2)$ (3.50) reduces to

$$\frac{\varepsilon}{\varepsilon_{\mathrm{d}}} = -\coth(kd), \tag{3.51a}$$

$$\frac{\varepsilon}{\varepsilon_{\mathrm{d}}} = -\tanh(kd). \tag{3.51b}$$

The quasi-static resonance of the two modes is the same as that given by (3.1). However, the antisymmetric mode tends to the resonance from above (as a backward wave).

In the thick-slab limit $(\varkappa_{\mathrm{p}}d \gg 1)$ the two dispersion relations (3.50) degenerate to that for SW propagation along a single interface (3.10a).

In the thin-slab limit $(\varkappa_{\mathrm{p}}d \ll 1)$ the behaviour of the symmetric and antisymmetric modes is quite different. In the case of the symmetric mode [3.109] (3.50a) reduces to

$$\varkappa_{\mathrm{d}}^2 = \frac{\varepsilon_{\mathrm{d}}^2}{\varepsilon^2 d^2}, \tag{3.52a}$$

with the solution for the wave number given by

$$\beta^2 = \frac{\omega^2}{c^2}\varepsilon_{\mathrm{d}}\left(1 + \frac{\varepsilon_{\mathrm{d}}}{\sigma^2\varepsilon^2}\right). \tag{3.52b}$$

In the case of an antisymmetric mode [3.110], the EM region of its existence as a fast SW $(|\varepsilon| \gg 1)$ is described by

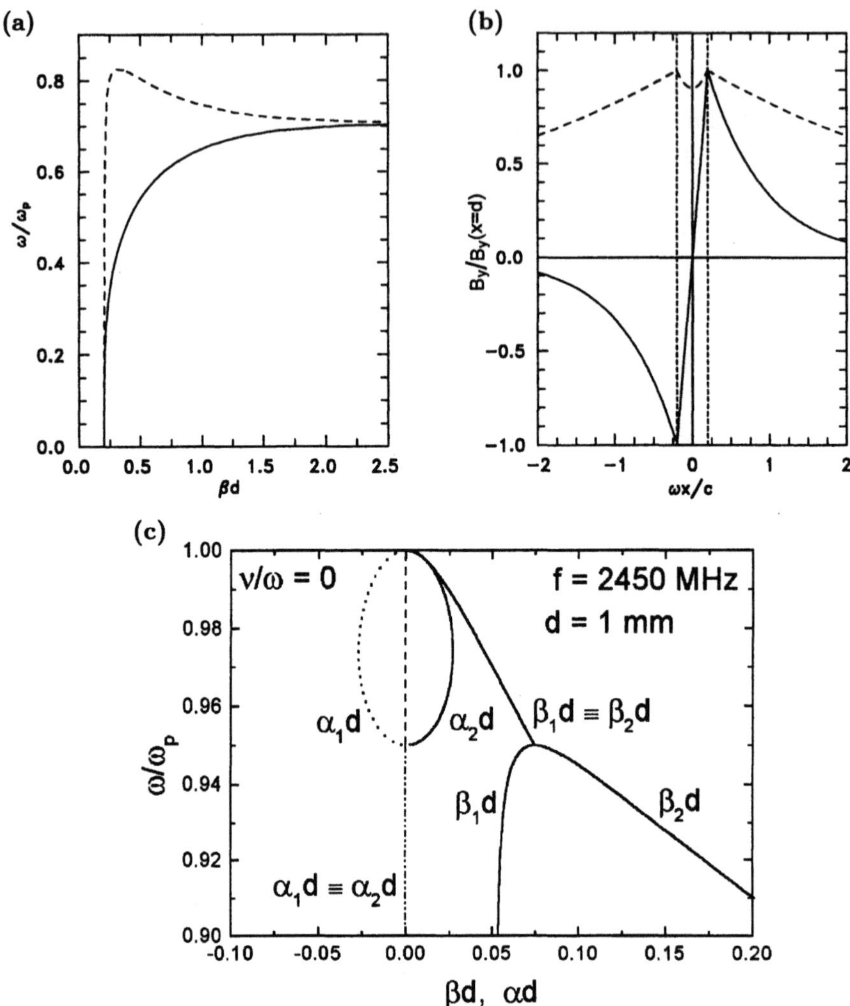

Figure 3.13. (a) Phase diagrams with $\sigma = 0.2$ and (b) transverse distribution of the normalized B_y field component for $\omega/\omega_\mathrm{p} = 0.45$, of the symmetric (*solid line*) and antisymmetric (*dashed line*) surface modes in a planar plasma waveguide: plasma slab of thickness $2d$ bounded by vacuum; (c) phase diagram of the anti-symmetric mode with its complex solutions ([3.101], Fig. 2)

$$\frac{\varkappa_\mathrm{p}^2 d}{\varepsilon} = -\frac{\varkappa_\mathrm{d}}{\varepsilon_\mathrm{d}}\,, \tag{3.53a}$$

obtained from (3.50b), and the wave number is

$$\beta^2 = \frac{\omega^2}{c^2}\,\varepsilon_\mathrm{d}(1 + \sigma^2\varepsilon_\mathrm{d})\,. \tag{3.53b}$$

In the limit of weak retardation ($\beta c/\omega \approx \sqrt{\varepsilon_\mathrm{d}}$) the dispersion behaviour of the two modes results in

$$\beta d \approx \sigma \,. \tag{3.54}$$

The phase velocities of the long-wavelength EM waves tend, as $\omega/\omega_\mathrm{p} \to 0$, like SWs along a single interface, to the light speed in the dielectric $v_\mathrm{ph} = \omega\sqrt{\varepsilon_\mathrm{d}}/c$.

The dispersion characteristics of the symmetric mode in the different regions of its phase diagram are as given below [3.109]. A plasma slab bounded by a vacuum is considered. As has been mentioned, when the parameter $\sigma = \omega d/c$ is comparatively small, the three regions (3.47) appear consecutively with increasing ω/ω_p value. In the case (3.47a) of $\varkappa_\mathrm{p}d \gg 1$ with $|\varepsilon| \gg 1$, which first takes place at small ω/ω_p values, (3.50a) takes the form of (3.10a) with the solutions for the real (β) and imaginary (α) parts of the wave number as given by (3.31a) and (3.33c), respectively. The value of $\varkappa_\mathrm{p}d = 1$ corresponds to a transition from a thick slab ($\varkappa_\mathrm{p}d > 1$) to a thin slab ($\varkappa_\mathrm{p}d < 1$). As (3.52b) shows, the thin-slab case (3.47b) appears on the phase diagrams if

$$\sigma^2 + \frac{1}{\varepsilon^2} + \sigma^2|\varepsilon| \ll 1 \,, \tag{3.55}$$

i.e. only for $\sigma \ll 1$ and for $1 \ll |\varepsilon| \ll 1/\sigma^2$. Under such a condition $\varkappa_\mathrm{v}d = |\varepsilon|^{-1}$, i.e. $\varkappa_\mathrm{v}d \ll 1$. In terms of plasma-density values, the transition from a thick ($\varkappa_\mathrm{p}d \gg 1$) to a thin ($\varkappa_\mathrm{p}d \ll 1$) slab occurs at $n/n_\mathrm{c} \lesssim (1-\sigma^2)/\sigma^2$ (with $\sigma < 1$). In the thin-slab region (3.47b) the wave continues its propagation as a fast wave at small ω/ω_p. According to (3.52) and (3.55), the region of fast waves in the thin-slab approach is

$$\frac{1}{\sigma} \ll |\varepsilon| \ll \frac{1}{\sigma^2} \,. \tag{3.56a}$$

The dispersion relation obtained from (3.50a) for $\varkappa_\mathrm{p}d \ll 1$ is

$$\mathcal{D}(\omega, k) = \frac{1}{\varepsilon d} + \varkappa_\mathrm{v} = 0 \,. \tag{3.56b}$$

The real part of the wave number β is given by (3.52b), whereas the space damping rate due to the collisions is

$$\alpha = \frac{\omega^3}{c^3} \frac{\lambda_\mathrm{sk}^4}{d^2} \frac{\nu}{\omega} \equiv \frac{\nu c \omega^2}{\omega_\mathrm{p}^4 d^2} \,. \tag{3.56c}$$

With increasing ω/ω_p values the phase velocity of the wave decreases, and the wave in a thin plasma slab with a comparatively high plasma densitiy

$$1 \ll |\varepsilon| < \frac{1}{\sigma} \tag{3.57a}$$

becomes slow, and is described by the quasi-static approach of (3.52a) and (3.56b):

$$\mathcal{D}(\omega, k) = \frac{1}{\varepsilon d} + k = 0 \,. \tag{3.57b}$$

With a wave spectrum given by

$$\omega = \omega_{\mathrm{p}}(\beta d)^{1/2} \,, \tag{3.57c}$$

the group velocity of the waves is:

$$v_{\mathrm{gr}} = d\frac{\omega_{\mathrm{p}}^2}{2\omega} \,. \tag{3.57d}$$

In the quasi-static limit the wave has a constant value of the time damping rate ($\gamma = -\nu/2$), and the variation with $\omega/\omega_{\mathrm{p}}$ of its space damping rate

$$\alpha = \frac{\omega^2}{c^2 d}\,\lambda_{\mathrm{sk}}^2\,\frac{\nu}{\omega} \equiv \frac{\nu\omega}{\omega_{\mathrm{p}}^2 d} \tag{3.57e}$$

is due to changes of the group velocity (3.57d). With increasing $\omega/\omega_{\mathrm{p}}$ values, the transverse variation of the wave field amplitude becomes stronger and the wave behaviour extends again into the region of the thick-slab limit $\varkappa_{\mathrm{p}}d \gg 1$ (3.47c). The wave spectrum approaches its resonance value (3.1b) and the time and space damping rates are those – (3.33d,e), respectively – of slow waves in semi-bounded plasmas. Both the fastest and the slowest waves in plasma slabs, which appear at the bottom ($\omega/\omega_{\mathrm{p}} \to 0$) and the top ($\omega/\omega_{\mathrm{p}} \to 1/\sqrt{2}$), respectively, of the phase diagrams, are described in the limit of a thick-slab approach ($\varkappa_{\mathrm{p}}d \gg 1$) and behave similarly to SWs in semi-bounded plasmas. In both cases the SW field in the plasma region is strongly bound to the interface at $x = \pm d$. The field decay in the vacuum is different: whereas in the case of fast waves, $\varkappa_{\mathrm{v}}d \ll 1$ and the field is greatly extended into the vacuum, in the case of slow waves, $\varkappa_{\mathrm{v}}d \gg 1$ and the field is bound to the interfaces. At large σ values only the two regions (3.47a,c) exist in the phase diagrams. In fact the case of small σ values, which corresponds to the thin-slab case (3.47b), is the only case that traces out the specific details of the slab configuration of waveguide systems.

The behaviour of the antisymmetric mode is discussed here with respect to its specific feature of the existence of a maximum in its phase diagram (($\omega/\omega_{\mathrm{p}}$) versus ($\beta d$)). The solution (for $\varepsilon_{\mathrm{d}} = 1$) of the dispersion relation (3.51b) of the wave in the quasi-static region $\varkappa_{\mathrm{p}} \approx k$ in the thin-slab approximation ($\sigma^2 \ll 1$, $|\varepsilon|/2$) in collisionless plasmas is

$$(kd)_{1,2}^2 = \frac{\varepsilon^2}{2}\left(1 \mp \sqrt{1 - 4\frac{\sigma^2}{\varepsilon^2}}\right) \,. \tag{3.58a}$$

The sign "+" corresponds to the backward wave and the sign "−" gives the forward wave. The transition from the first case to the second is at $|\varepsilon| = 2\sigma$, the value at which the expression under the square root in (3.58a) is equal to zero. It forms a maximum in the phase diagram at

$$(\beta d)_{\mathrm{m}} = \sqrt{2}\,\sigma \,. \tag{3.58b}$$

Above this maximum, i.e. for $4\sigma^2/\varepsilon^2 > 1$, the two solutions (3.58a) transform – although the plasma is collisionless – into complex solutions (in fact, complex conjugates) $k_{1,2} = \beta + \mathrm{i}\,\alpha_{1,2}$:

$$(\beta d) = \frac{1}{2}\sqrt{2}\,|\varepsilon|\sqrt{1 + 2\frac{\sigma}{\varepsilon}}\,, \tag{3.58c}$$

$$(\alpha d)_{1,2} = \pm\frac{\varepsilon^2}{4\beta d}\sqrt{4\frac{\sigma^2}{\varepsilon^2} - 1}\,. \tag{3.58d}$$

The numerical solution of (3.50b) confirms the behaviour commented on above (Fig. 3.13c). The forward and backward waves existing below the maximum of the $(\omega/\omega_{\mathrm{p}})$ versus (βd) plot continue as a backward wave above the maximum. The wave with positive α is a continuation of the forward wave, whereas that with negative α is a continuation of the backward wave. For decreasing $|\varepsilon|$ towards $\varepsilon \to 0$, βd approaches zero. If weak collisions $(\nu/\omega \ll 1)$ are taken into account, a small separation of the solutions at $\varepsilon_{\mathrm{r}} = -2\sigma$ appears:

$$(kd)_1^2 - (kd)_2^2 = -2\varepsilon_{\mathrm{r}}\sqrt{\mathrm{i}\,\frac{\nu}{\omega}(1 + 2\sigma)\sigma}\,. \tag{3.58e}$$

In fact, the maximum with the complex conjugate solutions branching off is a transformation at the complex turning point, as was discussed in [3.101].

The case of $\nu = 0$, $\alpha \neq 0$, i.e. the range of $\omega/\omega_{\mathrm{p}}$ values above the maximum, where the wave is damped in spite of the absence of absorption, corresponds to a kind of reactive behaviour. The power flux along the z direction in the vacuum region is completely turned over into the plasma region, where the flux is directed in the opposite z direction; the net flux integrated transversely is zero at each z position. With $\nu \neq 0$ the energy flux in the vacuum is larger than the flux in the opposite direction in the plasma.

The numerical solution of (3.50b) for thicker slabs demonstrates that the qualitative behaviour depicted in Fig. 3.13c is unchanged. The maximum is less pronounced. It should also be mentioned that (3.50b) possesses additional solutions with larger $|\alpha|$ (i.e. strongly damped).

Similar situations, i.e. the appearance of a maximum in the phase diagram, occur also in cases of azimuthally symmetric SW propagation along a plasma column surrounded by a dielectric and vacuum (see Sect. 3.3.2) and of higher-mode propagation along a plasma column surrounded by a vacuum. An asymmetry of the field configuration is the common feature of all these cases of different geometries or different modes. With increasing axial wave number β the transverse field penetration diminishes, and the electric-field intensity becomes concentrated closer to the interface between the plasma and the outside. Therefore, the properties of regions further away from the interface become less and less important. These include the opposite signs of the electric-field strength in the case of an asymmetric field structure, the vacuum beyond a thin dielectric layer, and probably also an electron density that is higher inside the plasma than in the region near the interface in the case of transverse density nonuniformity, to be considered in the next chapter. All three influences favour the formation of a maximum in the collisionless case and the appearance of new branches at this maximum.

Situations similar to those described above can occur elsewhere [3.111], for example, in the cases of cyclotron waves [3.112] and of lower-hybrid heating (conversion of "cold" lower-hybrid backward waves into "warm" ion Bernstein waves [3.113]), though here the type of turning point encountered is different.

3.3.2 Plasma Columns: Cylindrical Waveguides

The dispersion characteristics of the azimuthally symmetric ($\partial/\partial\varphi = 0$) surface modes (with the field components E_r, H_φ, $E_z \neq 0$) in cylindrical waveguides (Fig. 3.14), the plasma columns of radius R surrounded by a dielectric and/or vacuum, are discussed in this subsection. The wave propagation is in the z direction, $\propto \exp(-i\omega t + i kz)$.

(a)

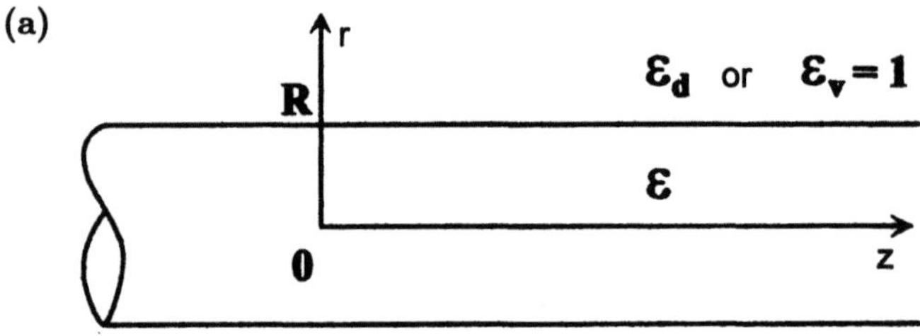

(b)

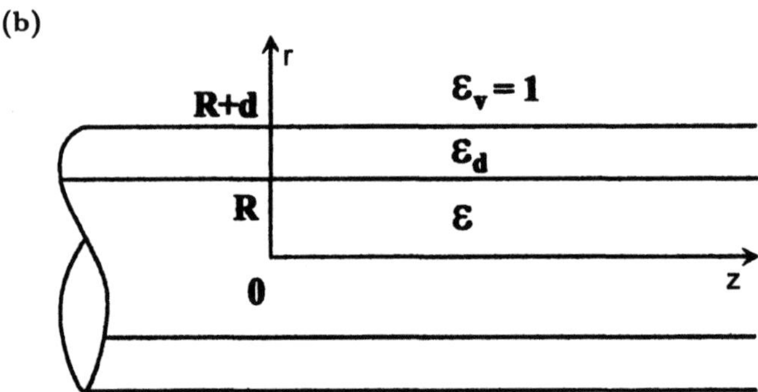

Figure 3.14. Cylindrical plasma waveguide: plasma column of radius R surrounded by a dielectric or vacuum (**a**) and by a dielectric of thickness d *and* a vacuum (**b**)

The solutions of the wave field equation (2.16) together with Maxwell's equations (2.1) and (2.2), for the waveguide in Fig. 3.14a with a cold homogeneous weakly collisional plasma inside, are

$$B_\varphi(r) = B(r = R)\frac{I_1(\varkappa_p r)}{I_1(\varkappa_p R)}, \qquad (3.59a)$$

$$E_r(r) = \frac{kc^2}{\omega\varepsilon} B(r = R)\frac{I_1(\varkappa_{\mathrm{p}}r)}{I_1(\varkappa_{\mathrm{p}}R)}, \tag{3.59b}$$

$$E_z(r) = \frac{\mathrm{i}c^2\varkappa_{\mathrm{p}}}{\omega\varepsilon} B(r = R)\frac{I_0(\varkappa_{\mathrm{p}}r)}{I_1(\varkappa_{\mathrm{p}}R)} \tag{3.59c}$$

for $r < R$ and $\varkappa_{\mathrm{p}}$ as given by (3.9a), and

$$B_\varphi(r) = B(r = R)\frac{K_1(\varkappa_{\mathrm{d}}r)}{K_1(\varkappa_{\mathrm{d}}R)}, \tag{3.60a}$$

$$E_r(r) = \frac{kc^2}{\omega\varepsilon_{\mathrm{d}}} B(r = R)\frac{K_1(\varkappa_{\mathrm{d}}r)}{K_1(\varkappa_{\mathrm{d}}R)}, \tag{3.60b}$$

$$E_z(r) = -\frac{\mathrm{i}c^2}{\omega\varepsilon_{\mathrm{d}}} B(r = R)\frac{K_0(\varkappa_{\mathrm{d}}r)}{K_1(\varkappa_{\mathrm{d}}R)} \tag{3.60c}$$

for $r > R$ and $\varkappa_{\mathrm{d}}$ as given by (3.9b). In (3.59a) and (3.60a) the boundary condition (2.17b) is taken into account. The condition (2.17a) for continuity of the E_z field component at $r = R$ gives the dispersion relation [3.79,81,82] of the waves

$$\mathcal{D}(\omega, k) = \frac{\varkappa_{\mathrm{p}}}{\varepsilon} \frac{I_0(\varkappa_{\mathrm{p}}R)}{I_1(\varkappa_{\mathrm{p}}R)} + \frac{\varkappa_{\mathrm{d}}}{\varepsilon_{\mathrm{d}}} \frac{K_0(\varkappa_{\mathrm{d}}R)}{K_1(\varkappa_{\mathrm{d}}R)} = 0, \tag{3.61a}$$

or

$$\mathcal{D}(\omega, k) = \frac{\varkappa_{\mathrm{p}}}{\varepsilon} \frac{I_0(\varkappa_{\mathrm{p}}R)}{I_1(\varkappa_{\mathrm{p}}R)} + \varkappa_{\mathrm{v}} \frac{K_0(\varkappa_{\mathrm{v}}R)}{K_1(\varkappa_{\mathrm{v}}R)} = 0, \tag{3.61b}$$

for a plasma column surrounded by a vacuum.

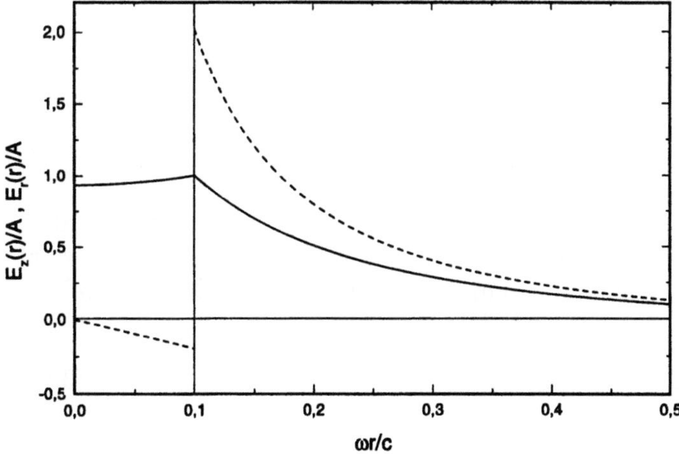

Figure 3.15. Variation of the wave field components E_z (*solid curve*) and E_r (*dashed curve*) in the transverse direction for $\sigma = 0.1$ and $\omega/\omega_{\mathrm{p}} = 0.3$; waveguide configuration of Fig. 3.14a; normalization to $A = |E_z(r = 0)|$

First, the dispersion behaviour for weak collisions ($\nu \ll \omega$) is discussed. Figure 3.15 illustrates the transverse variation of the electric-field components.

In the limit of the thick-cylinder approximation $\varkappa_\mathrm{p} R$, $\varkappa_\mathrm{d} R \gg 1$ (3.47c), (3.61a) reduces to the dispersion relation for SW propagation along a single interface (3.10a) and the corresponding solutions (3.31a) and (3.33d,e) for the wave number and space and time damping rates. The influence of the plasma-column radius is given by the correction terms [3.68]:

$$\beta = \frac{\omega}{c}\left(\frac{\varepsilon_\mathrm{d}\varepsilon_\mathrm{r}}{\varepsilon_\mathrm{r}+\varepsilon_\mathrm{d}}\right)^{1/2}\left(1+\frac{1}{2\sigma\sqrt{-(\varepsilon_\mathrm{r}+\varepsilon_\mathrm{d})}}\right),\tag{3.62a}$$

$$\alpha = \frac{1}{2}\frac{\omega}{c}\frac{\nu}{\omega}\frac{\varepsilon_\mathrm{d}^{3/2}(1-\varepsilon_\mathrm{r})}{\sqrt{\varepsilon_\mathrm{r}}(\varepsilon_\mathrm{r}+\varepsilon_\mathrm{d})^3}\left(1+\frac{\varepsilon-\varepsilon_\mathrm{r}}{2\sigma\varepsilon_\mathrm{r}\sqrt{-(\varepsilon_\mathrm{r}+\varepsilon_\mathrm{d})}}\right).\tag{3.62b}$$

In the limiting case of a thin cylinder $\varkappa_\mathrm{p} R$, $\varkappa_\mathrm{d} R \ll 1$ (3.47b), which occurs on phase diagrams with comparatively small σ values ($\sigma < 0.3$ [3.114]), (3.61a) results in

$$\varkappa_\mathrm{d}^2 = -\frac{4\varepsilon_\mathrm{d}}{R^2\varepsilon}\left[\ln\left(-\frac{\varepsilon}{4\varepsilon_\mathrm{d}}\right)\right]^{-1},\tag{3.63a}$$

giving the following expression for the wave number:

$$k^2 = \varepsilon_\mathrm{d}\left\{\frac{\omega^2}{c^2}-\frac{4}{R^2\varepsilon}\left[\ln\left(-\frac{\varepsilon}{4\varepsilon_\mathrm{d}}\right)\right]^{-1}\right\}.\tag{3.63b}$$

For $\sigma^2 < (4/|\varepsilon_\mathrm{r}|)\ln(-\varepsilon_\mathrm{r}/4\varepsilon_\mathrm{d})$ the SW propagation characteristics are

$$\beta = \frac{2\sqrt{\varepsilon_\mathrm{d}}}{R\sqrt{-\varepsilon_\mathrm{r}}}\left[\ln\left(-\frac{\varepsilon_\mathrm{r}}{4\varepsilon_\mathrm{d}}\right)\right]^{-1/2},\tag{3.63c}$$

$$\alpha = \frac{\varepsilon_\mathrm{i}\sqrt{\varepsilon_\mathrm{d}}}{R(-\varepsilon_\mathrm{r})^{3/2}}\left[\ln\left(-\frac{\varepsilon_\mathrm{r}}{4\varepsilon_\mathrm{d}}\right)\right]^{-1/2}.\tag{3.63d}$$

In the third limiting case (3.47a) – $\varkappa_\mathrm{p} R \gg 1$, $\varkappa_\mathrm{d} R \ll 1$ – which is at the very bottom of the phase diagram when the σ values is comparatively small, and is greatly extended on it at comparatively large σ, the dispersion relation (3.61a) reduces to

$$\varkappa_\mathrm{p} R = -(\varkappa_\mathrm{d} R)^2\frac{\varepsilon}{\varepsilon_\mathrm{d}}\,|\ln \varkappa_\mathrm{d} R|\tag{3.64a}$$

and gives

$$k^2 = \frac{\omega^2}{c^2}\varepsilon_\mathrm{d}\left\{1+\frac{2c}{R\omega\sqrt{-\varepsilon}}\left[\ln\left(\frac{c}{2\omega R}\frac{\sqrt{-\varepsilon}}{\varepsilon_\mathrm{d}}\right)\right]^{-1}\right\}.\tag{3.64b}$$

The phase curve starts from $\beta R = \sigma\sqrt{\varepsilon_\mathrm{d}}$ at $\omega/\omega_\mathrm{p} \to 0$.

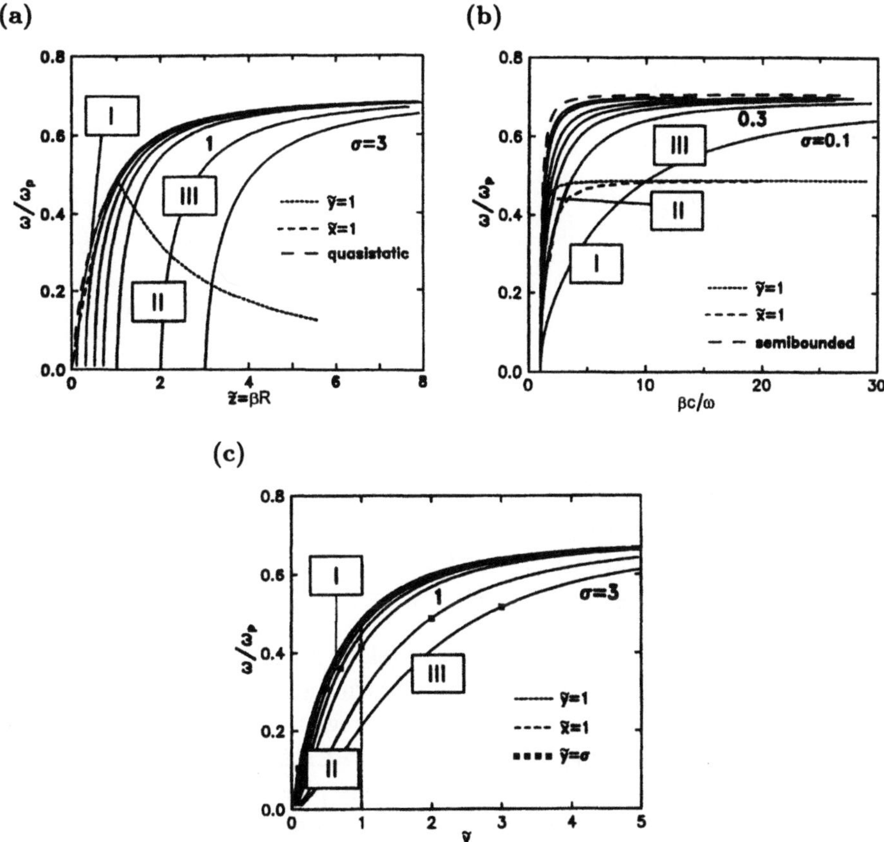

Figure 3.16. Phase diagrams (σ = const.) of SW propagation in a plasma column surrounded by vacuum. The curves $\tilde{x} \equiv \varkappa_{\mathrm{p}} R = 1$, $\tilde{y} = \varkappa_{\mathrm{v}} R = 1$ mark the boundaries between the three regions I ($\tilde{x}, \tilde{y} < 1$), II ($\tilde{x} > 1$, $\tilde{y} < 1$) and III ($\tilde{x}, \tilde{y} > 1$). The parameter σ takes values of 0.1, 0.3, 0.5, 0.7, 1, 2 and 3 in sequence. (a) Phase diagrams presented as ($\omega/\omega_{\mathrm{p}}$) versus $\tilde{z} \equiv \beta R$ and compared with the dispersion curve in the quasi-static approximation ($c \to \infty$, $\sigma \to 0$, *broken curve*). (b) Phase diagrams presented as $\omega/\omega_{\mathrm{p}}$ versus $\beta c/\omega$ and compared with the phase curve in a semi-bounded plasma ($R \to \infty$, $\sigma \to \infty$, *broken curve*). (c) Phase diagrams presented as ($\omega/\omega_{\mathrm{p}}$) versus $\tilde{y} \equiv \varkappa_{\mathrm{v}} R$. Points (■) of the $\tilde{y} = \sigma$ curve split regions II and III into two parts ([3.114], Figs. 1–3)

Phase diagrams (σ = const.) representing numerical solutions of (3.61b) are given in Fig. 3.16. The dependences of $\omega/\omega_{\mathrm{p}}$ on $\tilde{z} \equiv \beta R$, on $\beta c/\omega$ and on $\tilde{y} \equiv \varkappa_{\mathrm{v}} R = (\tilde{z}^2 - \sigma^2)^{1/2}$ are shown; $\tilde{x} \equiv \varkappa_{\mathrm{p}} R$ is related to $\tilde{z}$ according to $\tilde{x} = (\tilde{z}^2 - \varepsilon\sigma^2)^{1/2}$. The curves $\tilde{x} = 1$ and $\tilde{y} = 1$ separate the planes (($\omega_{\mathrm{p}}/\omega$), $\tilde{z}$ or $\beta c/\omega$ or $\tilde{y}$) into the three regions (3.47b,a,c) of different possible values of $\tilde{x}$ and $\tilde{y}$, denoted in Fig. 3.16 by I ($\tilde{x}, \tilde{y} < 1$), II ($\tilde{x} > 1$, $\tilde{y} < 1$) and III ($\tilde{x}$, $\tilde{y} > 1$), respectively.

The presentation of the phase diagrams in terms of (ω/ω_p) versus $\tilde{z} = \beta R$ makes them comparable with the quasi-static approximation $(c \to \infty, \sigma \to 0)$ to the dispersion behaviour of the waves:

- dispersion law

$$|\varepsilon_r| = \frac{I_0(\tilde{z})K_1(\tilde{z})}{I_1(\tilde{z})K_0(\tilde{z})} \qquad (3.65a)$$

- time damping rate

$$\gamma = -\frac{\nu}{2} \qquad (3.65b)$$

- group velocity

$$v_{gr} = \frac{1}{2}\,\omega R\left(\frac{I_0(\tilde{z})}{I_1(\tilde{z})} - \frac{K_1(\tilde{z})}{K_0(\tilde{z})}\right) \qquad (3.65c)$$

- space damping rate

$$\alpha = \frac{1}{R}\,\frac{\nu}{\omega}\left(\frac{I_0(\tilde{z})}{I_1(\tilde{z})} - \frac{K_1(\tilde{z})}{K_0(\tilde{z})}\right)^{-1}. \qquad (3.65d)$$

The dispersion curve of the quasistatic (slow) SWs (3.65), positioned completely in the regions I and III, is the limit which the phase diagrams approach as $\sigma \to 0$. Values of $\sigma \neq 0$ give evidence of region II. However, for small σ values ($\sigma < 0.2$–0.3), although the phase diagram starts from region II, region I still constitutes its main part. The end of the phase diagram lies in region III. With further increase of σ ($\sigma > 0.3$), region I disappears from the phase diagram and region II enlarges. In this case the phase diagrams are positioned in regions II and III. Since these are the regions which complete the dispersion behaviour of SW propagation in semi-bounded plasmas, large values of σ make the phase diagram of an SW in a cylindrical waveguides (Fig. 3.16b) comparable to that of the wave at a single interface ($R \to \infty, \sigma \to \infty$). The characteristics of the latter are given by (3.27) for the wave number, (3.31c) for the group velocity and (3.33a,b) for the space and time damping rates, respectively.

Thus, the dispersion curves of the quasi-static SWs ($c \to \infty, \sigma = 0$) along a plasma column and of the EM SWs in a semi-bounded plasma ($R \to \infty$, $\sigma \to \infty$) can be considered as the two limiting cases of the phase diagrams of the propagation of EM SWs in SW-produced plasmas. A comparison of these two limiting cases yields the following. (i) A convenient way to solve the problem for quasi-static SWs in a plasma column is to treat it as an initial-value problem $\omega(\beta)$, whereas a solution of a boundary value problem $\beta(\omega)$ is the proper representation of the SW characteristics in a semi-bounded plasma. (ii) The proper normalizing parameters for the first and second cases are R and $k_v = \omega/c$, respectively. (iii) The variation of the space damping rate with the electron density is determined by changes of the group velocity in the first case, and by changes of the time damping rate in the second one.

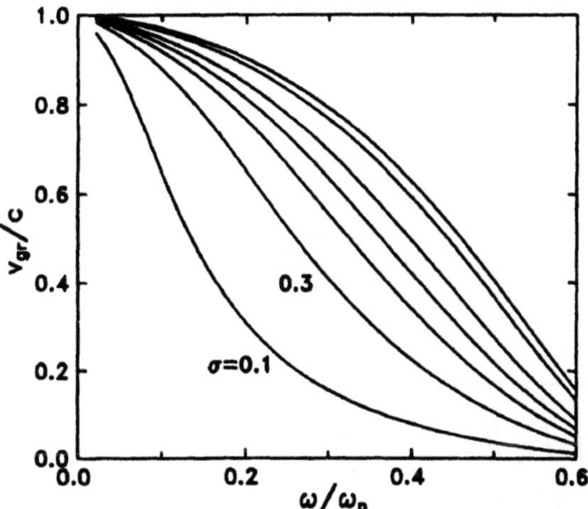

Figure 3.17. Wave group velocity v_{gr} normalized to the vacuum light speed c against $\omega/\omega_{\mathrm{p}}$ at different σ values (as given in Fig. 3.16) ([3.114], Fig. 4)

The parameter $\sigma \equiv \omega R/c$ of the phase diagrams of the EM SWs is formed as a combination of the normalizing parameters of the two limiting cases considered.

Analytical solutions (in explicit form) for SW behaviour are possible not only in the limiting case of quasi-static waves in a cylindrical waveguide and EM waves in a semi-bounded plasma, but also for the phase diagrams at arbitrary values of σ [3.114]. The results are as follows:

- dispersion law

$$\frac{|\varepsilon_{\mathrm{r}}|}{\tilde{x}} \frac{\mathrm{I}_1(\tilde{x})}{\mathrm{I}_0(\tilde{x})} = \frac{1}{\tilde{y}} \frac{\mathrm{K}_1(\tilde{y})}{\mathrm{K}_0(\tilde{y})} \tag{3.66a}$$

- time damping rate

$$\gamma = -\frac{\nu}{2} \left(1 - \sigma^2 |\varepsilon_{\mathrm{r}}| \frac{\tilde{M}}{\tilde{x}^2}\right) \left(1 + \sigma^2 \frac{|\varepsilon_{\mathrm{r}}|}{|\varepsilon_{\mathrm{r}}| + 1} \tilde{T}\right)^{-1} \tag{3.66b}$$

- normalized group velocity (Fig. 3.17)

$$\frac{v_{\mathrm{gr}}}{c} = \sigma \sqrt{\tilde{y}^2 + \sigma^2} \frac{\varepsilon_{\mathrm{r}}}{|\varepsilon_{\mathrm{r}}| + 1} \tilde{T} \left(1 + \sigma^2 \frac{|\varepsilon_{\mathrm{r}}|}{|\varepsilon_{\mathrm{r}}| + 1} \tilde{T}\right)^{-1} \tag{3.66c}$$

- space damping rate

$$\alpha = \frac{\nu}{\omega} \frac{\omega}{c} \frac{1}{2\sigma\sqrt{\tilde{y}^2 + \sigma^2}} \cdot \frac{|\varepsilon_{\mathrm{r}}| + 1}{|\varepsilon_{\mathrm{r}}|} \frac{1 - \sigma^2 |\varepsilon_{\mathrm{r}}|(\tilde{M}/\tilde{x}^2)}{\tilde{T}} . \tag{3.66d}$$

The latter is displayed in Figs. 3.18 and 19a,b and compared there with the space damping rates of SWs along a single interface and of quasi-static SWs along a plasma column. Here

$$\tilde{T} = \frac{\tilde{L}}{\tilde{y}^2} - \frac{\tilde{M}}{\tilde{x}^2} \, ,$$

$$\tilde{M} = 1 - \frac{\tilde{x}}{2}\left(\frac{I_0(\tilde{x})}{I_1(\tilde{x})} - \frac{I_1(\tilde{x})}{I_0(\tilde{x})}\right) \, ,$$

$$\tilde{L} = 1 + \frac{\tilde{x}}{2|\varepsilon_r|}\frac{I_0(\tilde{x})}{I_1(\tilde{x})} - \frac{\tilde{y}^2|\varepsilon_r|}{2\tilde{x}}\frac{I_1(\tilde{x})}{I_0(\tilde{x})} = 1 + \frac{\tilde{y}}{2}\left(\frac{K_0(\tilde{y})}{K_1(\tilde{y})} - \frac{K_1(\tilde{y})}{K_0(\tilde{y})}\right) \, .$$

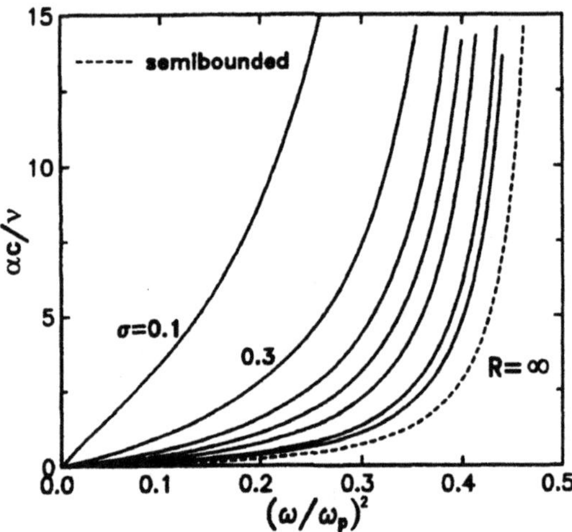

Figure 3.18. Variation of the space damping rate with the electron density at different σ values. Comparison with semi-bounded plasma (*broken curve*); σ values as in Fig. 3.16 ([3.114], Fig. 5)

In connection with the use of the results to obtain the wave characteristics in models of discharges sustained by SWs, the phase diagrams with $\sigma < 2\text{--}3$ and the regions on them where the waves are weakly damped are of interest. These are the results applicable to region I ($\tilde{x} < 1$) and to the bottom of the region $\tilde{x} > 1$ (i.e. region II and the beginning of region III).

In region I ($\tilde{x} < 1$) the wave characteristics can be approximated by the following expressions:

- dispersion law

$$|\varepsilon_r| = \frac{2K_1(\tilde{y})}{\tilde{y}K_0(\tilde{y})} \tag{3.67a}$$

- time damping rate

$$\gamma = -\frac{\nu}{2}\left(1 + \frac{\sigma^2}{\tilde{y}^2}\frac{|\varepsilon_r|}{|\varepsilon_r| + 1}\tilde{A}\right)^{-1} \tag{3.67b}$$

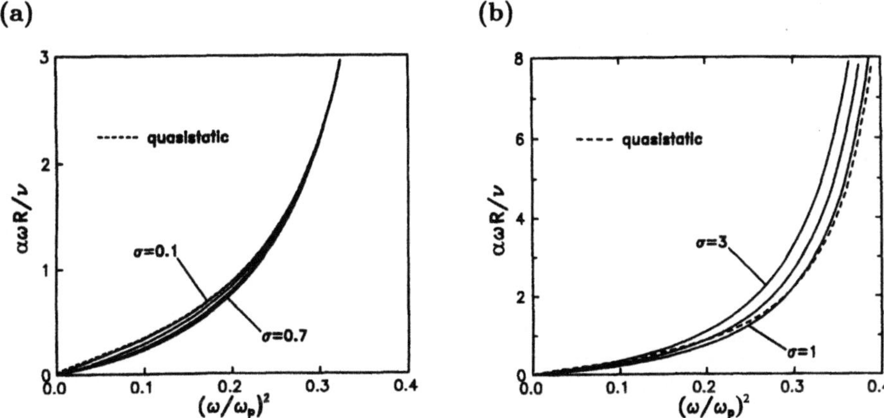

Figure 3.19. Variation of the space damping rate with the electron density compared with the quasi-static approximation (*broken curve*) for **(a)** $\sigma \leq 0.7$, i.e. $\sigma = 0.1,\ 0.3,\ 0.5,\ 0.7$ in sequence, and **(b)** $\sigma \geq 1$ (in sequence $\sigma = 1,\ 2,\ 3$) ([3.114], Fig. 6)

- normalized group velocity

$$\frac{v_{\text{gr}}}{c} = \sigma(|\varepsilon_{\text{r}}| + 1)\bar{F}_1 \left(1 + \frac{\sigma^2}{\tilde{y}^2}\frac{|\varepsilon_{\text{r}}|}{|\varepsilon_{\text{r}}| + 1}\tilde{A}\right)^{-1} \tag{3.67c}$$

- space damping rate

$$\alpha = \frac{\nu}{\omega}\frac{\omega}{c}\frac{1}{2\sigma(|\varepsilon_{\text{r}}| + 1)}\frac{1}{\bar{F}_1}\ . \tag{3.67d}$$

Here

$$\tilde{A} = 1 - \frac{\tilde{y}^2}{4}|\varepsilon_{\text{r}}| + \frac{1}{|\varepsilon_{\text{r}}|}\ ,$$

$$\frac{1}{\bar{F}_1} = \frac{\tilde{y}(|\varepsilon_{\text{r}}| + 1)^2}{|\varepsilon_{\text{r}}|\tilde{A}\sqrt{1 + (\sigma^2/\tilde{y}^2)}}\ .$$

In regions II and III, i.e. for $\tilde{x} > 0$, and with $\tilde{y}/\sigma < 1$, the corresponding results are as given below:

- dispersion law

$$\sqrt{|\varepsilon_{\text{r}}|} = \frac{\sigma}{\tilde{y}}\frac{\text{K}_1(\tilde{y})}{\text{K}_0(\tilde{y})}\ , \tag{3.68a}$$

or, for $0.3 < \sigma < 1$,

$$\sqrt{|\varepsilon_{\text{r}}|} = \frac{\sigma}{\tilde{y}^2\left[\ln(\tilde{y}/2) + \Gamma\right]}\ , \tag{3.68a'}$$

with $\Gamma = 0.577\ldots$

- time damping rate

$$\gamma = -\frac{\nu}{4}\frac{\tilde{y}^2}{\sigma^2}\frac{1}{\tilde{B}} \tag{3.68b}$$

- normalized group velocity

$$\frac{v_{\mathrm{gr}}}{c} = \left(1 + \frac{\tilde{y}^2}{\sigma^2}\frac{1}{c}\right)^{-1}\sqrt{1 + \frac{\tilde{y}^2}{\sigma^2}} \tag{3.68c}$$

- space damping rate

$$\alpha = \frac{\nu}{\omega}\frac{\omega}{c}\frac{1}{2\sigma(|\varepsilon_{\mathrm{r}}| + 1)}\frac{1}{\bar{F}_2} \ . \tag{3.68d}$$

Here

$$\tilde{B} = \tilde{C} + \frac{\tilde{y}^2}{\sigma^2} \ ,$$

$$\frac{1}{\bar{F}_2} = \frac{\tilde{y}^2(|\varepsilon_{\mathrm{r}}| + 1)}{2\tilde{C}\sigma\sqrt{1 + (\tilde{y}^2/\sigma^2)}} \ ,$$

$$\tilde{C} = \tilde{L} = 1 + \frac{\sigma}{2\sqrt{|\varepsilon_{\mathrm{r}}|}} - \frac{\tilde{y}^2\sqrt{|\varepsilon_{\mathrm{r}}|}}{2\sigma} \ .$$

The different behaviour of the SW properties in the two cases given above is due to the difference in the penetration of the wave field into the plasma determined by the ratio of the skin depth λ_{sk} to the plasma radius R, $\lambda_{\mathrm{sk}}/R > 1$ in the first case, $\lambda_{\mathrm{sk}}/R < 1$ in the second one. Although in both cases the space damping rate can be represented as

$$\alpha \propto \frac{\nu}{(|\varepsilon_{\mathrm{r}}| + 1)\bar{F}} \ , \tag{3.69a}$$

for $\tilde{x} < 1$ the group velocity determines the variation of α with the electron density (here $\gamma \approx -\nu/2$), whereas for $\tilde{x} > 1$ $v_{\mathrm{gr}} \approx c$ and the dependence of α on n is determined by the variation of the time damping rate with the electron density. The analogy of these two cases with the behaviour of quasi-static SWs in cylindrical waveguides and of SW propagation along a single interface is evident. The representation of the type (3.69a) has also been discussed in [3.115]. In both cases ($\tilde{x} < 1$ and $\tilde{x} > 1$) the function $\bar{F}$ in (3.69a) is given by [3.114]

$$\frac{1}{\bar{F}} = \frac{\tilde{y}^2|\varepsilon_{\mathrm{r}}|}{\tilde{G}\sqrt{\tilde{y}^2 + \sigma^2}} \ . \tag{3.69b}$$

For $\tilde{x} < 1$, i.e. in region I in Fig. 3.16, $\tilde{G} \equiv \tilde{A}$ and $\bar{F} \equiv \bar{F}_1$. For $\tilde{x} > 1$, with $\tilde{y} < \sigma < 1$, i.e. in regions II and III in Fig. 3.16, $\tilde{G} \equiv 2\tilde{C}$ and $\bar{F} \equiv \bar{F}_2$. In relation to SW-sustained discharges, the cases of SW behaviour for $\sigma < 3$ are of interest and they are covered by the regions of small ($\sigma < 0.3$) and intermediate

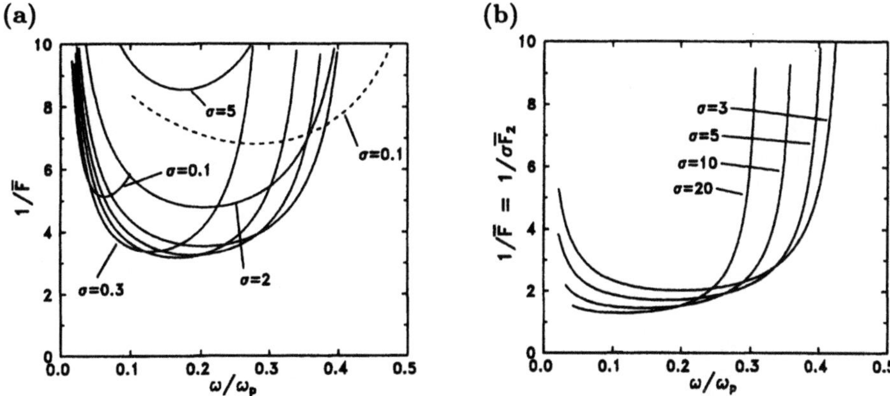

Figure 3.20. Behaviour of the function $\bar{F}$: (a) $\bar{F} \equiv \bar{F}_1$ at $\sigma = 0.1$ (*broken curve*), $\bar{F} \equiv \bar{F}_2$ at $\sigma = 0.1$, 0.3, 0.5, 0.7, 1, 2 (*solid curves*) and $\sigma = 5$, which is out of the range of validity of the conclusions drawn on the basis of this figure; (b) $\bar{F} = \sigma \bar{F}_2$, $\sigma = 3$, 5, 10, 20 ([3.114], Fig. 10)

$(0.3 \leq \sigma \leq 3)$ σ values. Moreover, the region of $\omega/\omega_\mathrm{p} = 0.1$–$0.4$ is the one that usually appears over the axial density profile. It should be recalled that the phase curves for small σ start from region II (Fig. 3.16), i.e. with $\bar{F} \equiv \bar{F}_2$. However, at small σ the values of ω/ω_p belonging to region II are too low to be achieved in experiments and thus it is usually region I ($\bar{F} \equiv \bar{F}_1$) which describes the discharge behaviour in this case. In region I ($\sigma < 0.3$) $\bar{F}_1$ varies slowly over the phase diagram (Fig. 3.20a) and an averaged value of $\bar{F}$ can be used for describing the SW behaviour. At intermediate values of σ ($0.3 \leq \sigma \leq 3$) $\bar{F}_2$ – although its variation over the phase diagrams is stronger ($1/\bar{F}_2$ starts from large values at small ω/ω_p, goes through a minimum at some ω/ω_p value and increases again with increasing ω/ω_p) – can also be considered as a slowly varying function over the ω/ω_p interval appearing in experiments with SW-sustained discharges. Thus an averaged value for $\bar{F}_2$ can be used in this case, too. A common behaviour of the function $\bar{F}$, which generalizes the cases of small ($\sigma < 0.3$) and intermediate ($0.3 \leq \sigma \leq 3$) σ values, is the fact that its variation with ω/ω_p at different values of σ converges in a comparatively narrow interval (Fig. 3.20a). With comparatively large σ values ($\sigma > 3$) it is the function $\bar{F} \equiv \sigma \bar{F}_2$ which, being a slowly varying function over any given phase diagram with $\sigma > 3$, converges at different σ values in a comparatively narrow interval (Fig. 3.20b).

These slowly varying functions improve the accuracy of the analytical results by avoiding the use of an asymptotic approximation to the modified Bessel functions. For instance, the applicability of the approximations to the $K_\nu(\tilde{y})$ modified Bessel functions for small values of their argument is limited to extremely small values of the argument. This does not allow their use in the description of the electrodynamics of SW-sustained discharges in the region of plasma densities which are of interest from an experimental point of view.

The validity of the approximations to the $I_\nu(\tilde{x})$ modified Bessel functions for small values of their arguments holds even at values of $\tilde{x}$ close to 1 and their use in obtaining the results (3.67), (3.68) does not influence the accuracy of the calculations. Similar procedures which introduce approximations of the dispersion behaviour were also used in [3.116,117].

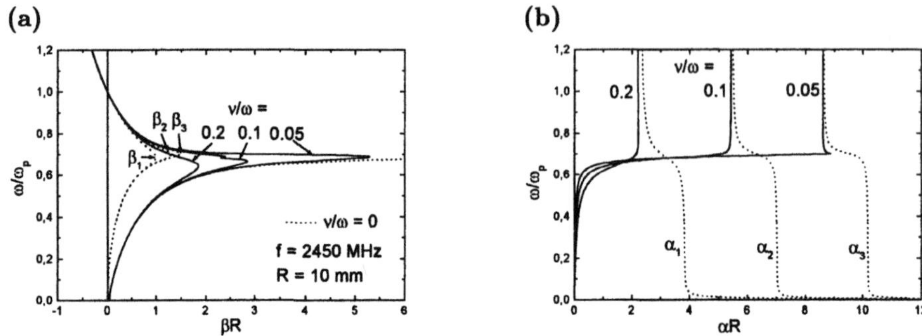

Figure 3.21. (a) Influence of collisions on the phase diagrams of SWs in a plasma column (*solid curves*); the corresponding space damping rates are shown in (b). The highly damped modes which exist as complex modes even in collisionless plasmas are also shown (*dashed curves*)

The mode discussed up to here is the "surface mode" according to the classification given in Sect. 3.2. Weak collisions influence the phase diagrams by extending them above the SW resonance in cold collisionless plasmas. In Fig. 3.21 phase diagrams (with a relative low value of σ) comparing solutions at different collision frequencies with that for $\nu = 0$ are illustrated. The turn-back of the phase diagrams in the case of cylindrical waveguides (Fig. 3.21a) reminds one of the discussions on the effects of collisions on SW behaviour in a semi-infinite plasma (Sect. 3.2, Fig. 3.2). Figure 3.21 also reveals the existence of complex solutions in the collisionless case (see Sect. 3.3.1). The small kink that can be recognized in the β curve for $\nu/\omega = 0.1$ ($\beta R \approx 2.3$) is connected to the asymptotic behaviour of αR. In the case of collisions the tendency of $\beta \to 0$ for $\omega/\omega_\mathrm{p} \approx 1$ and the vertical asymptotes of α are similar to the character of the series of strongly damped solutions that exist as complex modes for $\nu = 0$.

The power flux of an SW in a cylindrical waveguide [3.89,118,119] defined according to (2.28b) and (2.32b) is

$$P_z = \frac{1}{2\mu_0} \operatorname{Re} \left\{ \int_0^{2\pi} \mathrm{d}\varphi \int_0^\infty r E_r(r) B_\varphi^*(r) \, \mathrm{d}r \right\} . \tag{3.70a}$$

After using the solutions for the wave field (3.59) and (3.60), (3.70a) transforms into

$$P_z = \frac{\pi\varepsilon_0}{2} \frac{\beta\omega}{8} R^2 \Delta \left(\frac{\mathcal{A}}{\varepsilon} + \mathcal{B} \right) |E(r = R)|^2 , \tag{3.70b}$$

where

$$\Delta = \frac{\varepsilon^2}{\varkappa_p^2} \frac{I_1^2(\varkappa_p R)}{I_0^2(\varkappa_p R)} = \frac{1}{\varkappa_v^2} \frac{K_1^2(\varkappa_v R)}{K_0^2(\varkappa_v R)} \tag{3.70c}$$

stems directly from the dispersion law (3.61b), and

$$\mathcal{A} = 1 - \frac{I_0(\varkappa_p R)I_2(\varkappa_p R)}{I_1^2(\varkappa_p R)}, \tag{3.71a}$$

$$\mathcal{B} = \frac{K_0(\varkappa_v R)K_2(\varkappa_v R)}{K_1^2(\varkappa_v r)} - 1 \tag{3.71b}$$

account for the power flux in the regions occupied by the plasma and the vacuum, respectively. These two contributions to the power flux are of opposite sign: positive, i.e. in the direction of the wave propagation, in the vacuum and negative in the plasma. For $\varkappa_v R \ll 1$ (regions I and II in the phase diagrams in Fig. 3.16) the power flux is larger in the vacuum than in the plasma ($|P_{z_p}| \ll |P_{z_v}|$), whereas in the region close to the SW resonance in a collisionless plasma (region III in Fig. 3.16) they are of the same order of magnitude, and wave propagation stops.

The Joule loss of SW power in a cylindrical waveguide, given by (2.32b) and (2.29), is

$$Q = \frac{1}{2} \int_0^{2\pi} d\varphi \int_0^R r\sigma_{p(r)}|E(r)|^2 \, dr, \tag{3.72a}$$

with $|E(r)|^2 = |E_r(r)|^2 + |E_z(r)|^2$ and $\sigma_{p(r)}$ determined by (2.13). By using (3.59), (3.72a) reduces to

$$Q = \frac{\pi\varepsilon_0}{2} \nu R^2 \frac{\omega_p^2}{\omega^2} (\mathcal{C} + \mathcal{D})|E(r = R)|^2, \tag{3.72b}$$

where

$$\mathcal{C} = 1 - \frac{I_1^2(\varkappa_p R)}{I_0^2(\varkappa_p R)}, \tag{3.72c}$$

and

$$\mathcal{D} = \frac{\beta^2}{\varkappa_p^2} \left[\frac{I_1^2(\varkappa_p R)}{I_0^2(\varkappa_p R)} - \frac{I_2(\varkappa_p R)}{I_0(\varkappa_p R)} \right] \tag{3.72d}$$

account for the contributions of the E_z and E_r field components, respectively, to the absorbed power. In the case where the thin-cylinder approximation ($\tilde{x} \ll 1$) is valid, $|E_z|^2 \gg |E_r|^2$ and the main part of the loss is related to the E_z field component.

For describing the variation of the power flux P_z of an SW in a cylindrical waveguide in region I of the phase diagram at small σ values (Fig. 3.16, case $\varkappa_p R < 1$, $\varkappa_v R < 1$, in the region of slow waves with a behaviour similar to quasi-static waves), the function

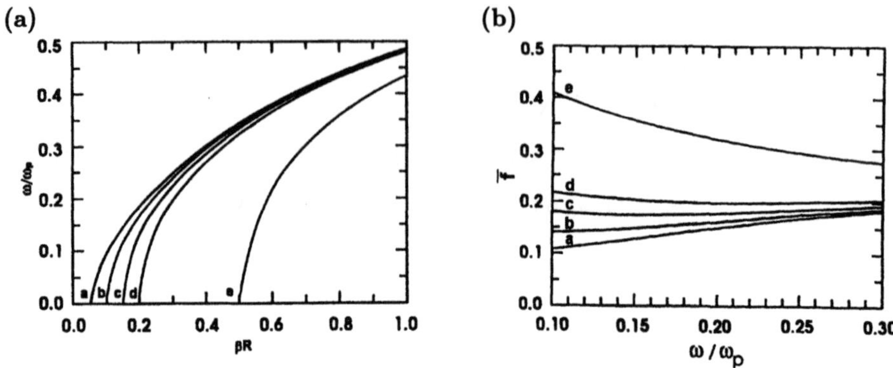

Figure 3.22. (a) Calculated phase diagrams of SW propagation along a plasma column bounded by vacuum for different values of $\sigma = 0.05$ *(curve a)*, 0.1 *(curve b)*, 0.15 *(curve c)*, 0.2 *(curve d)* and 0.5 *(curve e)*; (b) calculated variation of the function $\bar{f}$ (according to (3.73)) over the corresponding phase diagrams in (a). The case of $\sigma = 0.5$ shows that $\bar{f} \approx$ const. is not valid anymore when a region of a thin-cylinder approximation ($\varkappa_{\mathrm{p}} R \ll 1$) does not exist on the phase diagrams ([3.119], Figs. 1 and 2)

$$\bar{f} = \frac{\beta R \mathcal{B}}{4}, \tag{3.73}$$

which is a slowly varying function (Fig. 3.22) over the phase diagram, can be introduced. This function can be taken as a constant $\bar{f} \equiv \langle \bar{f} \rangle$ with a value equal to the mean value $\langle \bar{f} \rangle$ of $\bar{f}$ over the $\omega/\omega_{\mathrm{p}}$ interval considered, for the given phase diagram with a specified value (and even a range) of σ (Fig. 3.22b). The group velocity and the space damping rate of the wave are related to this function according to

$$v_{\mathrm{gr}} = \frac{\omega_{\mathrm{p}}^2 R}{\omega} \bar{f}, \tag{3.74a}$$

$$\alpha = \frac{\nu\omega}{2\omega_{\mathrm{p}}^2 R \bar{f}}; \tag{3.74b}$$

the time damping rate is $\gamma = -\nu/2$. The slowly varying function $\bar{f}$ is included (with a precision of the order of $\sigma^2/\tilde{y}^2$) in the slowly varying function $\bar{F}_1$ defined by (3.67).

As has been discussed in Sect. 3.2.2, weakly damped SWs can be supported not only in weakly collisional but also in strongly collisional plasma waveguides. Possessing comparatively high phase velocities, these belong to region II (Fig. 3.16) of the phase diagram, i.e. with $\varkappa_{\mathrm{v}} R \ll 1$, $\varkappa_{\mathrm{p}} R \gg 1$. Equation (3.61b) reduces to the following approximate form:

$$\frac{\varkappa_{\mathrm{p}} R}{\varepsilon} + (\varkappa_{\mathrm{v}} R)^2 \ln\left(\frac{2}{\Gamma' \varkappa_{\mathrm{v}} R}\right) = 0, \tag{3.75a}$$

where $\Gamma' = 1.781$ and the solution at $\nu > \omega$ for the space damping rate is

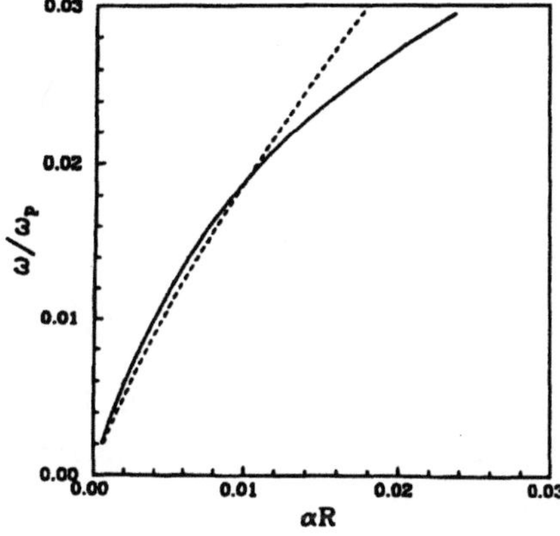

Figure 3.23. Space damping rate of fast SWs in a plasma column bounded by vacuum for $\sigma = 0.15$ and $\nu/\omega = 10$. Numerical solution of (3.61b) shown as *solid curve* and approximate solution (3.75b) as *dashed curve* ([3.120], Fig. 2)

$$\alpha = R^{-1} \frac{\omega}{\omega_{\mathrm{p}}} \sqrt{\frac{\nu}{2\omega}} \left| \ln \left(\frac{\omega R}{c} \frac{\Gamma'^2}{2} \frac{\omega}{\omega_{\mathrm{p}}} \sqrt{\frac{\nu}{\omega}} \right) \right|^{-1} . \tag{3.75b}$$

The applicability of (3.75b) is illustrated in Fig. 3.23, where it is compared with the numerical solution of (3.61b).

The presence of a dielectric ($R < r < R + d$) of thickness d and permittivity ε_{d} overlying the plasma column of radius R and separating the plasma region ($r < R$) from the vacuum ($r > R + d$) transforms the dispersion relation (3.61b) into [3.121]

$$-\frac{\varkappa_{\mathrm{d}}\varepsilon}{\varkappa_{\mathrm{p}}\varepsilon_{\mathrm{d}}} \frac{\mathrm{I}_1(\varkappa_{\mathrm{p}}R)}{\mathrm{I}_0(\varkappa_{\mathrm{p}}R)} = \frac{\delta_1 + (\varkappa_{\mathrm{d}}/\varepsilon_{\mathrm{d}}) Z_2 \delta_2}{\delta_3 - (\varkappa_{\mathrm{d}}/\varepsilon_{\mathrm{d}}) Z_2 \delta_4} . \tag{3.76a}$$

The expression

$$Z_2 = \frac{1}{\varkappa_{\mathrm{v}}} \frac{\mathrm{K}_1\left[\varkappa_{\mathrm{v}}(R+d)\right]}{\mathrm{K}_0\left[\varkappa_{\mathrm{v}}(R+d)\right]} \tag{3.76b}$$

is associated with the description of the wave field in the vacuum region. The more complicated right-hand part of (3.76a) is connected to the presence of the dielectric layer. The form of the expressions for δ_1, δ_2, δ_3 and δ_4, related to the wave field variation in the dielectric layer, depends on the phase velocity of the wave. In the region of $v_{\mathrm{ph}} < c/\sqrt{\varepsilon_{\mathrm{d}}}$ (i.e. for $\beta > \omega\sqrt{\varepsilon_{\mathrm{d}}}/c$), $\varkappa_{\mathrm{d}}^2 \equiv \beta^2 - (\omega^2\varepsilon_{\mathrm{d}}/c^2)$ is positive and

$$\delta_1 = \mathrm{K}_1(\bar{R})\mathrm{I}_1(\overline{R+d}) - \mathrm{I}_1(\bar{R})\mathrm{K}_1(\overline{R+d}) ,$$
$$\delta_2 = \mathrm{I}_1(\bar{R})\mathrm{K}_0(\overline{R+d}) + \mathrm{I}_0(\overline{R+d})\mathrm{K}_1(\bar{d}) ,$$
$$\delta_3 = \mathrm{I}_0(\bar{R})\mathrm{K}_1(\overline{R+d}) + \mathrm{K}_0(\bar{R})\mathrm{I}_1(\overline{R+d}) ,$$
$$\delta_4 = \mathrm{I}_0(\bar{R})\mathrm{K}_0(\overline{R+d}) - \mathrm{I}_0(\overline{R+d})\mathrm{K}_0(\bar{R}) \tag{3.76c}$$

include the modified Bessel functions. Here the notation $\bar{R} \equiv \varkappa_{\mathrm{d}}R$, $\overline{R+d} \equiv \varkappa_{\mathrm{d}}(R+d)$ is used. In the case of $c/\sqrt{\varepsilon_{\mathrm{d}}} < v_{\mathrm{ph}} < c$ (i.e. for $\omega/c < \beta < (\omega/c)\sqrt{\varepsilon_{\mathrm{d}}}$), which describes the very bottom of the phase diagram in the wave number interval $\sigma < \beta R < \sigma\sqrt{\varepsilon_d}$, $\varkappa_d^2$ is negative and the expressions for $\delta_1 \ldots \delta_4$ in (3.76a) have to be replaced by

$$\delta_1' = -N_1(\bar{R}')J_1(\overline{R+d}') + J_1(\bar{R}')N_1(\overline{R+d}'),$$

$$\delta_2' = J_1(\bar{R}')N_0(\overline{R+d}') - N_1(\bar{R}')J_0(\overline{R+d}'),$$

$$\delta_3' = -J_0(\bar{R}')N_1(\overline{R+d}') + N_0(\bar{R}')J_1(\overline{R+d}'),$$

$$\delta_4' = J_0(\bar{R}')N_0(\overline{R+d}') - N_0(\bar{R}')J_0(\overline{R+d}'), \qquad (3.76d)$$

where $\bar{R}' \equiv R\sqrt{-\varkappa_d^2}$, $\overline{R+d}' \equiv (R+d)\sqrt{-\varkappa_d^2}$ and J_ν, N_ν are Bessel functions. In this case the replacement $\varkappa_d \rightarrow \sqrt{-\varkappa_d^2}$ should also be made in (3.76a). Figure 3.24 represents a numerical solution of (3.76a).

In the case of a thin dielectric layer ($\varkappa_d R \ll 1$) the dispersion relation (3.76a) takes the form [3.120]

$$\frac{\varkappa_p}{\varepsilon}\frac{I_0(\varkappa_p R)}{I_1(\varkappa_p R)} + \varkappa_v \frac{R}{R+d}\frac{K_0[\varkappa_v(R+d)]}{K_1[\varkappa_v(R+d)]} = \frac{\varkappa_d^2 R}{\varepsilon_d}\ln\left(\frac{R+d}{R}\right), \qquad (3.77)$$

analogous to that of (3.61b). Its right-hand part gives the contribution of the thin dielectric (glass tube) surrounding the plasma.

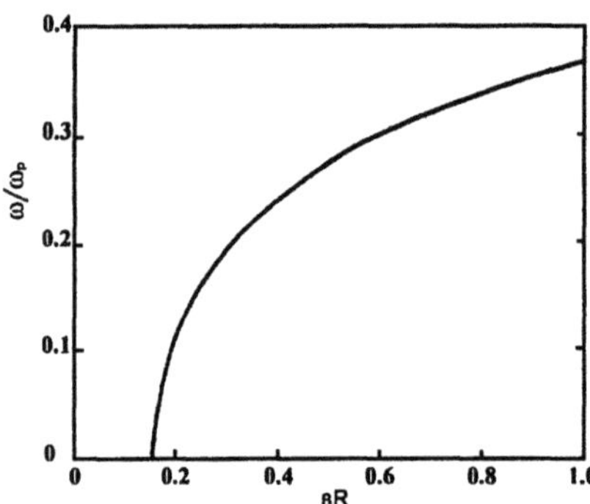

Figure 3.24. Phase diagram of SWs along a cold homogeneous plasma column surrounded by a dielectric and vacuum at $\sigma \equiv \omega R/c = 0.149$, $(R+d)/R = 1.33$ and $\varepsilon_d = 3.78$ ([3.122], Fig. 2)

Therefore, with a dielectric layer surrounding a plasma column which separates the plasma region from the vacuum, SWs start their propagation in the range of fast waves (i.e. at the bottom of the phase diagram), with a phase velocity tending to the vacuum light speed ($v_{ph} \rightarrow c$). With a smooth transition through the light speed in the dielectric $v_{ph} = c/\sqrt{\varepsilon_d}$, the waves enter the region of slow SWs (Fig. 3.25). In the region of the latter – the shorter-wavelength region – the presence of the dielectric layer causes the appearance of a maximum in the phase diagram owing to a transition from a curve close to that of the dispersion behaviour in a plasma–vacuum waveguide to one close to the dispersion behaviour in a plasma–dielectric waveguide. Related to this transition, a region of backward wave propagation appears (Fig. 3.25). As has been discussed before with respect to the antisymmetric mode in a

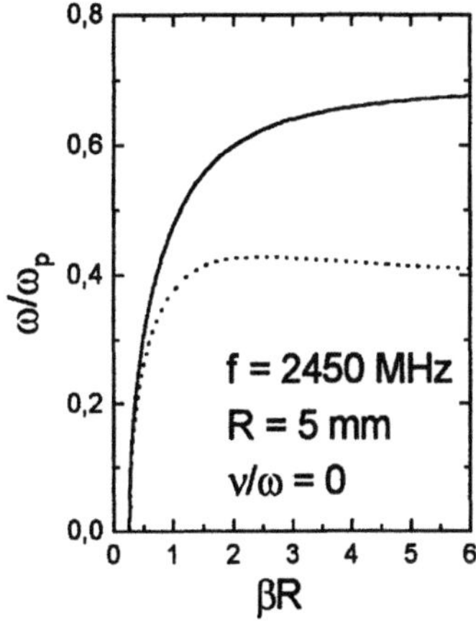

Figure 3.25. Phase diagram of azimuthally symmetric SWs, in a plasma column surrounded by a vacuum (*solid curve*) and by a dielectric ($\varepsilon_d = 4.7$ and thickness $d = 0.1\,cm$) and vacuum ([3.101], Fig. 1)

plasma slab, the existence of such a maximum leads to the appearance of complex solutions for k in the region above the maximum, regardless of the fact that the plasma is collisionless.

The inclusion of a metal cylinder which surrounds the plasma column–dielectric–free-space structure is the next step in introducing geometrical effects on SW behaviour. However, calculations [3.123–126] of the SW characteristics in such a waveguide configuration do not show big differences [3.101] when they are made for the situations of interest for experiments on SW-produced discharges: in these cases the metal screen used for avoiding perturbation of the wave field without changing the field configuration too much is not too close to the discharge tube.

Discussions of the effects of the electron thermal motion on SW behaviour were summarized in [3.81,82].

4. Surface-Wave Propagation in Inhomogeneous Plasmas

Compared with the numerous studies in the literature of the 1960s and 1970s on SWs in homogeneous plasmas, the number of publications during the same period on properties of waves in inhomogeneous plasmas was quite smaller [4.1–13]. This can be blamed on the complexity of the analytical treatment of the problem and on the lack of good computing capabilities at that time. Since the research on SWs in this period was concerned with DC discharge plasmas, the plasma-density inhomogeneity involved was in the transverse direction. In spite of not too much extended research, the main effects related to the plasma inhomogeneity were well understood. Weak inhomogeneity leads to changes in the wave dispersion which are quite pronounced in the range towards the SW resonance (3.1) in cold collisionless plasmas (Fig. 3.3). Like the effects of collisions (Fig. 3.2) and thermal motion, the inhomogeneity also removes the resonance. However, whereas the influence of collisons and thermal motion is in the direction of extending the range of the waves above the resonance, the inhomogeneity, forming a maximum of $\omega/\omega_{\mathrm{p}}$ in the phase curves moves the $\omega/\omega_{\mathrm{p}}$ range down. Excluding the range close to the resonance (3.1), the inhomogeneity only weakly affects the dispersion, and the wave behaviour in nonuniform plasmas is quite well described by the phase diagrams for homogeneous plasmas (Figs. 3.3, 3.13a, 3.16) with ω_{p}^2 replaced by $\overline{\omega_{\mathrm{p}}^2}$, i.e. by presenting the plasma frequency in terms of the density averaged over the cross-section. The early studies [4.2,3] also discovered resonance absorption of SWs in cold strongly inhomogeneous plasmas, a phenomenon which is now considered as the most important effect of plasma inhomogeneity. This new mechanism of dissipation of the SW energy – which appears in addition to the standard collisional and Landau damping – is due to a linear transformation of SWs into volume plasmons at the point r_{r} (of the density profile) at which the local plasma frequency $\omega_{\mathrm{p}}(x_{\mathrm{r}}|r_{\mathrm{r}})$ is equal to the wave frequency, i.e. $\omega = \omega_{\mathrm{p}}(x_{\mathrm{r}}|r_{\mathrm{r}})$, and the real part of the local permittivity becomes zero: $\varepsilon_{\mathrm{r}}(r_{\mathrm{r}}) = 0$.

Interest in the dispersion behaviour of SWs in inhomogeneous plasmas has recently been renewed by research on SW-sustained discharges which has stressed again the effects of transverse inhomogeneity [4.14]. The increased capability for numerical calculation at this later stage of studies has allowed the investigation of combined effects of collisions and inhomogeneity [4.15–17]. The results confirmed the importance of inhomogeneity as a factor with a basic physical meaning, and showed that inhomogeneity and collisions acting together make drastic changes in the dispersion behaviour of SWs. All the changes in the wave dispersion which in homogeneous plasmas are in the range of the SW resonance (3.1), i.e. the appearance of a maximum of $\beta d|\beta R$ in the phase curve and its turning back, are shifted now to a lower ω/ω_p value. The appearance of the point of resonance absorption at the plasma boundary, and its shift inside the plasma at higher ω/ω_p turns the phase diagram forwards again. Interpretation of the numerical results provoked interest in the analytical treatment of the problem [4.18].

In spite of interest in understanding the changes of the dispersion behaviour caused by plasma-density inhomogeneity, the current studies on the effects of transverse inhomogeneity are mainly motivated by interest in understanding the possible mechanisms of electron heating in the discharge. The Joule heating in the plasma volume (Chaps. 5 and 6), which, with respect to the wave, means collisional damping of the wave (considered in Chap. 3), is the main mechanism of heating. The axial (E_z) component of the SW electric field is mainly involved in this mechanism. A new mechanism of SW damping means, in general, a new mechanism for electron heating and discharge maintenance. The dissipation of the wave energy in regions of plasma resonance in an inhomogeneous plasma determines the mechanisms of Joule and collisionless heating in the resonance regions close to the discharge walls. As is shown in Chaps. 5 and 6, these mechanisms of heating may also be important for the maintenance of the discharge [4.16,19–24]. Because of the strong resonance enhancement of the radial (E_r) component of the SW electric field in the regions of resonance absorption, the E_r field component is the component mainly involved in these mechanisms of heating. The Joule heating in regions of resonance absorption is related to the collisional damping of the volume plasmons into which SWs transform. The collisionless electron heating in regimes of resonance absorption is related to the noncollisional Landau-type damping of the volume partial waves which result from the Fourier decomposition of the SW field in the transverse direction. This spatial decomposition contain large wave numbers, i.e. short-wavelength waves, which are strongly damped.

The research on waveguided discharges brought up another aspect of the problem of SW propagation in inhomogeneous plasmas. This is the wave behaviour in plasmas inhomogenous in the longitudinal direction, i.e. in the direction of the wave propagation. The importance of this problem is obvious since the varying density along the plasma column is a basic feature of these discharges. The axial plasma inhomogeneity shows up in both experimental and theoretical results [4.22–30] and is considered to be part of the physical basis of the creation and maintenance of the discharge. In spite of this, the dispersion law for SWs in a homogeneous plasma has always been used in the discharge models. In addition, the interferogram method and the field radial decay method [4.28,29], which are employed in almost all experiments on SW-produced plasmas as diagnostic tools, are essentially based on the applicability of the dispersion law for homogeneous plasmas. Since the plasma has obvious axial inhomogeneity, questions about the validity and degree of applicability of the dispersion relation used need to be answered. This requires consideration of the manner of application of the geometrical-optics approach [4.31] (i.e. the WKB approximation) to SW propagation [4.18,32–35]. If this approximation holds, the SW dispersion law for homogeneous plasmas is a result of the zero-order approach in terms of geometrical optics, and it may be used as a "local" law. Since SWs are waves with a two-dimensional variation of their field, a procedure for application of the geometrical-optics approach to such situations had to be developed. The deviation of the slowly varying amplitude in an inhomogeneous plasma from the constant amplitude in a homogeneous, collisionless plasma gives an indication of the degree of applicability of the "local" dispersion law. Obviously, in the range close to the SW resonance the wave field changes strongly and fast [4.18,22,23,36–40] in the collisionless limit ($\nu \to 0$) of cold plasmas, and questions about the manner of the dissipation of the wave energy arise.

In general, the problem of SW propagation characteristics in SW-produced discharges is a two-dimensional one with respect to both wave field configuration and plasma inhomogeneity. Codes based on integral formulations of Maxwell's equations allow full numerical solutions when both transverse and longitudinal nonuniformities are present simultaneously [4.39–41].

4.1 Main Aspects Treated in this Chapter

The presentation in this chapter of the behaviour of SWs in inhomogeneous plasmas is mainly based on considerations in [4.17,18,37–40]. The properties of SW propagation along plasma–plasma and plasma–dielectric interfaces are described here, as a rule, for the case of weak collisions ($\omega \gg \nu$), employing as models different configurations of plane and cylindrical geometries. Approaches to plasma nonuniformity either in the transverse or the longitudinal direction with respect to the wave propagation are developed; results addressing both nonuniformities considered together are presented are at the end.

Regions of weak inhomogeneity over which the wavelength variation is sufficiently slow and the geometrical-optics approach is applicable, as well as resonance regions, where the geometrical-optics approach is not valid, are covered. The latter include (i) in the case of transverse inhomogeneity, regions of arbitrary (including strong) inhomogeneity in which plasma resonances and mode conversion occur (here, specifically, transformation of SWs into plasmons, i.e. into localized volume plasma oscillations) and (ii) in the case of longitudinal inhomogeneity, the region of the quasi-static resonance (3.1) of the SWs, where the SW group velocity tends to zero.

The analysis and main results presented in this chapter are listed below.

(1) Case of inhomogeneity in the transverse direction.
   (i) The absorption of SW energy due to plasma resonances localized close to the plasma boundary is analysed in detail. Cases where plasmons excited by the SW are absorbed by either collisional damping or space dispersion effects are considered.
   (ii) The contributions to the dispersion laws of SWs determined from the geometrical-optics solutions for weak plasma inhomogeneity are given.
   (iii) Dispersion relations which govern SW propagation for arbitrary variation of the plasma density profile in a thin plasma layer and a thin plasma cylinder are discussed.
   (iv) Numerical results for phase diagrams in plasma slabs and plasma cylinders are presented for a variety of conditions. Details are given on the influence of collisions and transverse inhomogeneity, on complex modes and on backward waves. The effect of resonance absorption of SWs is emphasized.
(2) Case of inhomogeneity in the longitudinal direction.
   (i) A method for applying the geometrical-optics approach to two-dimensional problems of wave-field variation is discussed. The space dependences of the SW amplitude are given. It is shown that within the geometrical-optics approach SWs propagate without radiation and obey the local dispersion relations.
   (ii) A generalization of the geometrical-optics approach for obtaining the distribution of SW fields is presented for guiding structures of arbitrary geometry.
   (iii) The region of the quasi-static resonance of the SWs is considered specially. Analytical solutions of the equation which describes the wave behaviour around the point of the quasi-static resonance are given. It is shown that in the limit (defined below) of high enough dissipation, the SW energy is totally absorbed in the vicinity of the resonance point. The opposite limit of weak dissipation, which goes beyond the scope of the geometrical-optics approach, is analysed qualitatively.
   (iv) Numerical results are presented for exemplary cases, and comparisons of different approaches are given for semi-space and cylindrical plasmas.
(3) The essential results of some numerical solutions obtained from integral codes for cases containing both transverse and longitudinal nonuniformity are presented for plasma cylinders, together with some comparisons with one-dimensional solutions.

As has been discussed in Chap. 3, the SW studied is a TM mode with field components E_x, B_y and $E_z \neq 0$ in rectangular coordinates and E_r, B_φ, $E_z \neq 0$ in cylindrical coordinates; the field variation is of the form

$\propto f(x, z) \exp(-\mathrm{i}\omega t)$ and $\propto f(r, z) \exp(-\mathrm{i}\omega t)$, respectively. In this chapter, the wave equations are solved largely in terms of the magnetic field, which has only one component and is continuous at the interfaces.

4.2 Inhomogeneity in the Transverse Direction

4.2.1 Wave-Field Equations for Plane Geometry

The presentation of the propagation properties of SWs in nonuniform plasmas starts with effects due to density inhomogeneity in a direction perpendicular to that of the wave propagation (assumed again along the z axis). The simplest case of plane geometry (Fig. 3.1) is considered first. The plasma permittivity $\varepsilon(x)$ as given by (2.8) is space-dependent and a function of the x coordinate; $B_y(x)$ is the complex magnitude of the magnetic field of the wave $B_y(x, z) = B_y(x) \exp(\mathrm{i}\,kz)$.

The wave equation (2.16) which governs the variation of the magnetic-field amplitude $B_y(x) \equiv B$ is taken in the form

$$\varepsilon(x)\frac{\mathrm{d}}{\mathrm{d}x}\left(\frac{1}{\varepsilon(x)}\frac{\mathrm{d}B}{\mathrm{d}x}\right) - \varkappa_{\mathrm{p}}^2(x)B = 0\,, \tag{4.1}$$

where $\varkappa_{\mathrm{p}}(x)$ (3.9a) characterizes the depth of the wave-field penetration into the plasma and is now space-dependent.

The space distribution of the plasma permittivity $\varepsilon(x)$ is as follows: (i) in the plasma volume, $\varepsilon(x)$ is negative and varies weakly over the scale length of the wave penetration into the plasma; (ii) close to the interface there is a narrow (compared with the field penetration depth scale) region of strong inhomogeneity, in which the plasma permittivity changes its sign. Therefore, the plasma resonance, i.e. the point $x = x_{\mathrm{r}}$ where the real part of the plasma permittivity is equal to zero ($\varepsilon_{\mathrm{r}}(x = x_{\mathrm{r}}) = 0$), is located in this narrow region of strong inhomogeneity.

We start with a treatment of the regions of weak inhomogeneity. To determine the magnetic-field amplitude from (4.1), it is convenient to transform it to an equation for the function $\mathcal{B}(x)$ defined according to

$$B(x) = \sqrt{\varepsilon(x)}\mathcal{B}(x)\,. \tag{4.2}$$

After inserting (4.2) into (4.1), the latter becomes

$$\frac{\mathrm{d}^2\mathcal{B}}{\mathrm{d}x^2} - \left[\varkappa_{\mathrm{p}}^2(x) - \frac{1}{2\varepsilon(x)}\frac{\mathrm{d}^2\varepsilon(x)}{\mathrm{d}x^2} + \frac{3}{4\varepsilon^2(x)}\left(\frac{\mathrm{d}\varepsilon(x)}{\mathrm{d}x}\right)^2\right]\mathcal{B} = 0\,. \tag{4.3}$$

Since the variation of the plasma permittivity is slow on the scale of the characteristic depth of the field penetration into the plasma, the last two terms in (4.3) can be neglected, and therefore (4.3) reduces to

$$\frac{\mathrm{d}^2 \mathcal{B}}{\mathrm{d}x^2} - \varkappa_\mathrm{p}^2(x)\mathcal{B}(x) = 0. \tag{4.4}$$

With the assumption of weak inhomogeneity, the geometrical-optics approach [4.31] can be applied to (4.4). The solution is

$$\mathcal{B}(x) = \frac{1}{\sqrt{\varkappa_\mathrm{p}(x)}} \exp\left(\pm \int^x \varkappa_\mathrm{p}(x')\,\mathrm{d}x'\right) \tag{4.5}$$

and, thus, the magnetic-field amplitude obtained after using (4.2) is

$$B(x) = \sqrt{\frac{\varepsilon(x)}{\varkappa_\mathrm{p}(x)}} \exp\left(\pm \int^x \varkappa_\mathrm{p}(x')\,\mathrm{d}x'\right). \tag{4.6}$$

For a solution of (4.1) in a region of strong inhomogeneity, i.e. around the plasma resonance, a double integration is performed, yielding

$$B(x) = B(x = x_0) + \frac{1}{\varepsilon(x = x_0)} \left.\frac{\mathrm{d}B}{\mathrm{d}x}\right|_{x=x_0} \int_{x_0}^x \varepsilon(x')\,\mathrm{d}x'$$
$$+ \int_{x_0}^x \mathrm{d}x'\varepsilon(x') \int_{x_0}^{x'} \frac{\varkappa_\mathrm{p}^2(x'')}{\varepsilon(x'')} B(x'')\,\mathrm{d}x''. \tag{4.7}$$

Here x_0 is an arbitrary point inside the narrow region of strong inhomogeneity and, in particular, it could be located at the interface of this region. The distribution of the magnetic-field amplitude of the SW in the region of the plasma resonance can be obtained from the integral equation (4.7) by an iteration procedure based on the assumption that the last term is small. The zero-order approximation to (4.7) is

$$B(x) = B(x = x_0) + \frac{1}{\varepsilon(x = x_0)} \left.\frac{\mathrm{d}B}{\mathrm{d}x}\right|_{x=x_0} \int_{x_0}^x \varepsilon(x')\,\mathrm{d}x'. \tag{4.8}$$

The next approximation, which includes the first-order terms, is

$$B(x) = B(x = x_0) \left[1 + \int_{x_0}^x \mathrm{d}x'\,\varepsilon(x') \int_{x_0}^{x'} \frac{\varkappa_\mathrm{p}^2(x'')}{\varepsilon(x'')}\mathrm{d}x''\right]$$
$$+ \frac{1}{\varepsilon(x = x_0)} \left.\frac{\mathrm{d}B}{\mathrm{d}x}\right|_{x=x_0}$$
$$\times \int_{x_0}^x \mathrm{d}x'\varepsilon(x') \left[1 + \int_{x_0}^{x'} \mathrm{d}x'' \frac{\varkappa_\mathrm{p}^2(x'')}{\varepsilon(x'')} \int_{x_0}^{x''} \varepsilon(x''')\mathrm{d}x'''\right]. \tag{4.9}$$

The condition that the integral terms in the brackets are small compared with unity is the requirement for applicability of this solution. The assumption that this is true is justified if the characteristic length

$$L_n^{(\mathrm{r})} = \left.\left|\frac{\mathrm{d}\varepsilon_\mathrm{r}}{\mathrm{d}x}\right|^{-1}\right|_{x=x_\mathrm{r}} \approx n_\mathrm{c} \left.\left|\frac{\mathrm{d}n}{\mathrm{d}x}\right|^{-1}\right|_{x=x_\mathrm{r}} \tag{4.10}$$

of the plasma density inhomogeneity around the point $x = x_r$ of the plasma resonance is small compared with the effective depth $\varkappa_p^{-1}$ of the wave's magnetic-field penetration into the plasma, i.e. $L_n^{(r)} \ll \varkappa_p^{-1}$; here n_c ist the critical density (2.11). Cases where the assumptions made here are not valid are not of interest in the present context, since they represent situations where the width of the transition layer is comparable to the SW penetration length. In such situations there are no well-defined guiding surfaces for supporting SW propagation.

The longitudinal electric-field component E_z is

$$E_z(x) = \frac{\mathrm{i}\,c^2}{\omega\varepsilon(x)}\frac{\mathrm{d}B_y(x)}{\mathrm{d}x}\,. \tag{4.11}$$

With (4.9) inserted into (4.11) and small terms neglected, the E_z field amplitude is obtained as

$$E_z(x) = E_z(x = x_0) + \mathrm{i}\,\frac{c^2}{\omega}B(x = x_0)\int_{x_0}^{x}\frac{\varkappa_p^2(x')}{\varepsilon(x')}\,\mathrm{d}x'\,. \tag{4.12}$$

A comparison of (4.8) and (4.12) shows that – whereas the magnetic-field amplitude remains almost constant over the plasma resonance region – the amplitude of the tangential electric-field component $E_z(x)$ goes through a jump owing to the logarithmic singularity of the second term on the right-hand side of (4.12). This jump is associated with the transformation of the energy of the SW into energy of local plasma oscillations in the region of $\varepsilon(x) \approx 0$.

The electromagnetic energy dissipated in the vicinity of the plasma resonance can be obtained from the wave energy conservation law (2.27):

$$\frac{\mathrm{d}P_x(x)}{\mathrm{d}x} = -Q(x)\,, \tag{4.13}$$

where $P_x(x) = (2\mu_0)^{-1}\,\mathrm{Re}\{E_z^*(x)B(x)\}$ is the transverse component of the electromagnetic power flux, and $Q(x)$ is the power locally absorbed. Both quantities are defined over a unit length in the z direction. In order to obtain the total absorbed power, (4.13) must be integrated over a region $x_0 < x < x_0 + \Delta$ which includes the plasma resonance:

$$Q \equiv \int_{x_0}^{x_0+\Delta} Q(x)\,\mathrm{d}x = P_x(x = x_0) - P_x(x = x_0 + \Delta)\,. \tag{4.14}$$

Thus, the total power absorbed in a layer of thickness Δ is determined by the difference of the wave power fluxes $P_x(x = x_0)$ and $P_x(x = x_0 + \Delta)$. After inserting (4.12) and (4.8) into the definition of the flux component P_x, it turns out that the difference of the electromagnetic-power fluxes

$$P_x(x = x_0) - P_x(x = x_0 + \Delta)$$
$$= -\frac{c^2 k^2}{2\mu_0\omega}|B(x = x_0)|^2\,\mathrm{Im}\left\{\int_{x_0}^{x_0+\Delta}\frac{\mathrm{d}x}{\varepsilon(x)}\right\} \tag{4.15}$$

is determined by the jump of $E_z(x)$. The imaginary part of the integral in (4.15) can be calculated by using the well-known relation

$$\lim_{\eta' \to 0} \frac{1}{x + i\eta'} = \frac{\mathrm{Pr}}{x} - i\pi\,\mathrm{sign}(\eta')\,\delta(x)\,, \tag{4.16}$$

where Pr and $\mathrm{sign}(\eta')$ denote the principal value of the integral and the sign of the quantity η', respectively; $\delta(x)$ is the delta function. When (4.16) and the series expansion (for $\omega \gg \nu$)

$$\varepsilon(x) = (x - x_{\mathrm{r}}) \left(\frac{\mathrm{d}\varepsilon_{\mathrm{r}}}{\mathrm{d}x}\right)_{x=x_{\mathrm{r}}} + i\,\frac{\nu}{\omega} \tag{4.17}$$

of the plasma permittivity in the region of the plasma resonance $x = x_{\mathrm{r}}$ are used, one gets

$$\mathrm{Im}\left\{\int_{x_0}^{x_0+\Delta} \frac{\mathrm{d}x}{\varepsilon(x)}\right\} = -\frac{\pi}{|(\mathrm{d}\varepsilon_{\mathrm{r}}/\mathrm{d}x)_{x=x_{\mathrm{r}}}|}\,. \tag{4.18}$$

Thus, the absorbed power is

$$Q = \frac{\pi c^2 k^2}{2\mu_0\omega}|B(x = x_0)|^2 \frac{1}{|(\mathrm{d}\varepsilon_{\mathrm{r}}/\mathrm{d}x)_{x=x_{\mathrm{r}}}|} \equiv \frac{\pi c^2 k^2}{2\mu_0\omega} L_n^{(\mathrm{r})}|B(x = x_0)|^2. \tag{4.19}$$

4.2.2 Influence of Thermal Electron Motion on the Resonance Absorption of Electromagnetic Surface Waves

When the problem of resonance absorption of SWs was discussed in Sect. 4.2.1, the effects associated with the thermal motion of the electrons were neglected. However, such effects can play an important role if the electron temperature is high enough. This leads to the necessity of accounting for the plasma space dispersion [4.3] and its influence on the SW resonance absorption.

With thermal motion of the electrons, i.e. a space dispersion, included, the Maxwell equation which relates the transverse electric-field component E_x to the wave magnetic-field intensity becomes a differential equation

$$r_{\mathrm{D}}^2 \frac{\mathrm{d}^2 E_x}{\mathrm{d}x^2} + \varepsilon(x)E_x = \frac{c^2 k}{\omega}B\,, \tag{4.20}$$

where r_{D} is the Debye length (2.10) at $n(x) = n_{\mathrm{c}}$, and $\varepsilon(x)$ denotes the cold plasma permittivity (2.8). The form of (4.20) accounts for a Debye length that is small compared with the width of the resonance region and with the SW wavelength.

The first term in (4.20) is responsible for carrying the Langmuir waves out of the plasma resonance region. The importance of this term in comparison with the second one can be evaluated by using the following estimations:

$$\left|\frac{\mathrm{d}^2 E_x(x)}{\mathrm{d}x^2}\right| \approx \frac{c^2 k}{\omega}|B(x)| \left|\frac{\mathrm{d}^2}{\mathrm{d}x^2}\left(\frac{1}{\varepsilon(x)}\right)\right| \approx \frac{|E_x(x)|}{L_n^{(\mathrm{r})\,2}|\varepsilon(x)|^2}\,. \tag{4.21}$$

If $\varepsilon(x)$ is the cold plasma permittivity, this equation reduces to

$$\left| \frac{\mathrm{d}^2 E_x(x)}{\mathrm{d}x^2} \right| \approx \frac{|E_x(x)|}{L_{\mathrm{n}}^{(\mathrm{r})\,2}(\nu^2/\omega^2)} \,. \tag{4.22}$$

Thus the effects of the space dispersion can be neglected (i.e. the first term in (4.20) can be neglected in comparison with the second one) when the inequality

$$\frac{r_{\mathrm{D}}^2}{L_{\mathrm{n}}^{(\mathrm{r})\,2}} \ll \left(\frac{\nu}{\omega}\right)^3 \tag{4.23}$$

holds. At high enough temperatures, the plasma permittivity has – according to (4.20) – an operator form. When an inequality opposite to (4.23) is fulfilled, the finite value of the E_x field component at the resonance point is due to the space dispersion effects. The latter could be qualitatively described by the formulae for cold plasmas, with the electron–neutral elastic-collision frequency ν replaced by the quantity $\omega(r_{\mathrm{D}}/L_n^{(\mathrm{r})})^{2/3}$ [4.31]. For example, the E_x field component in the resonance region can be represented by

$$E_x(x) = \frac{c^2 k}{\omega} \frac{B(x)}{(\mathrm{d}\varepsilon_{\mathrm{r}}/\mathrm{d}x)_{x=x_{\mathrm{r}}}(x - x_{\mathrm{r}}) + \mathrm{i}\,(\nu_{\mathrm{eff}}/\omega)} \,, \tag{4.24}$$

where $\nu_{\mathrm{eff}} = \max[\nu, \omega(r_{\mathrm{D}}/L_n^{(\mathrm{r})})^{2/3}]$ gives a measure of the absorption in the resonance region of the plasmons (excited by the SW) by either collisional or Landau damping.

The total energy Q^{h} absorbed in the plasma resonance region is determined by the conductivity $\sigma_{\mathrm{p_{eff}}}(x) \approx \varepsilon_0\, \omega_{\mathrm{p}}^2(x)\nu_{\mathrm{eff}}/\omega^2$ ((2.13) with $\omega > \nu_{\mathrm{eff}}$):

$$Q^{\mathrm{h}} \approx \frac{1}{2} \int_{x_0}^{x_0+\Delta} \sigma_{\mathrm{p_{eff}}}(x)|E_x(x)|^2 \, \mathrm{d}x \,. \tag{4.25}$$

Using (4.24) with taking into account that the variation of the magnetic-field intensity in the resonance region (4.8) is slow, transforms (4.25) into

$$Q^{\mathrm{h}} = \frac{\nu_{\mathrm{eff}}}{2\mu_0} \frac{c^2 k^2}{\omega^2}|B(x = x_0)|^2 \int_{-\infty}^{\infty} \frac{\mathrm{d}x}{[(x - x_{\mathrm{r}})/L_n^{(\mathrm{r})}]^2 + (\nu_{\mathrm{eff}}/\omega)^2} \,. \tag{4.26}$$

After calculating the integral in (4.26) the latter reduces to (4.19), i.e.

$$Q^{\mathrm{h}} = Q \,, \tag{4.27}$$

where Q gives the energy losses in the resonance region for a cold plasma. Thus the total power absorbed in the plasma resonance region does not depend on the loss mechanism.

4.2.3 Surface Wave Propagation
Along an Inhomogeneous Plasma Slab

The wave-field solutions (4.6) and (4.9) will be used for deriving the dispersion law which governs SW propagation along a layered structure consisting of an inhomogeneous (in the transverse direction) plasma with a permittivity $\varepsilon(x)$ which is bounded by a homogeneous medium of permittivity ε_d (Fig. 4.1). It is assumed that in the central part ($|x| \le d - \Delta$, region I in Fig. 4.1) of the slab the plasma density is a slowly varying function of x, whereas close to the boundary ($d - \Delta < |x| < d$, region II in Fig. 4.1) it decays fast. "Slow" and "fast" variations of the density are defined with respect to the field penetration depth. The considerations here address the case of a symmetric density profile $\varepsilon(-x) = \varepsilon(x)$, with two points $x = \pm x_r$ (Fig. 4.1) of plasma resonances $\varepsilon(\pm x_r) = 0$ over the total slab width. Each of these two points is within a narrow transition layer $d - \Delta < |x| < d$ (with $\Delta \ll d$) close to the plasma–dielectric (vacuum) interface. Such assumptions model quite well the transverse density profile in many gas-discharge plasmas.

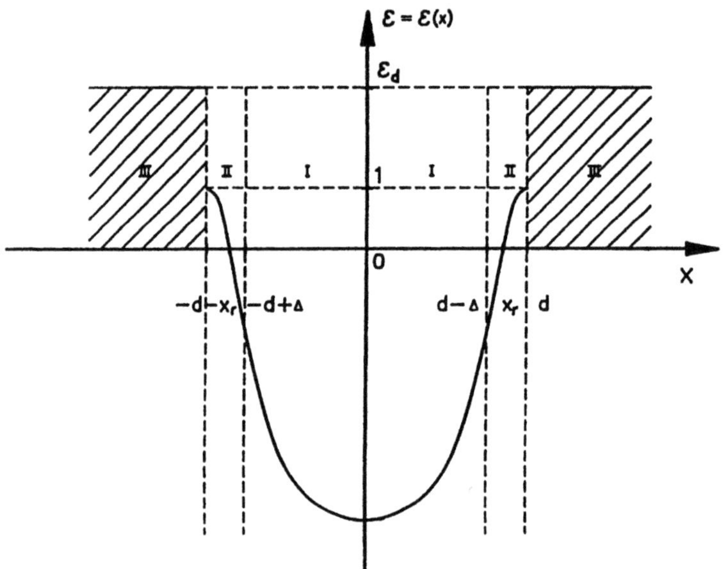

Figure 4.1. Geometrical configuration: inhomogeneous plasma slab of thickness $2d$ bounded by a dielectric with permittivity ε_d (region III); schematic representation in terms of the variation of the real part of the plasma permittivity $\varepsilon(x)$ over the slab width (region I: weakly inhomogeneous plasma; region II: strongly inhomogeneous plasma of thickness Δ); the plasma resonance points $\pm x_r$ are in regions II ($d - \Delta < |x| < d$), being the points where $\varepsilon(x)$ crosses the x axis ($\varepsilon_r(\pm x_r) = 0$) ([4.18], Fig. 3.1)

For the wide central part where the plasma is weakly inhomogeneous, the geometrical-optics method can be applied. The solution in this region, being a linear combination of (4.6), takes the following forms for the symmetric $(B(x=0)=0)$ and antisymmetric $(E_z(x=0)=0)$ modes, respectively:

$$B_{\{{}^s_a\}}(x) = \text{const.} \sqrt{\frac{\varepsilon(x)}{\varkappa_p(x)}} \left\{ \begin{array}{l} \sinh \int_0^x \varkappa_p(x')\,dx' \\ \cosh \int_0^x \varkappa_p(x')\,dx' \end{array} \right. . \tag{4.28}$$

The corresponding solutions for the E_z field component are

$$E_{z\{{}^s_a\}}(x)$$
$$= i\,\frac{c^2}{\omega}\,\frac{\varkappa_p(x)}{\varepsilon(x)}\,B_{\{{}^s_a\}}(x) \left\{ \begin{array}{l} \coth \int_0^x \varkappa_p(x')\,dx' \\ \tanh \int_0^x \varkappa_p(x')\,dx' \end{array} \right. + \frac{1}{\varkappa_p(x)}\,\frac{d}{dx}\ln\sqrt{\frac{\varepsilon(x)}{\varkappa_p(x)}} \right\} .$$
$$\tag{4.29}$$

For the transition layer where the plasma is strongly inhomogeneous (region II in Fig. 4.1), the solutions for the wave-field components are as described by (4.8) and (4.12) whereas the solution for the region of the homogeneous dielectric ($|x| > d$, region III in Fig. 4.1) is simply

$$\propto e^{-\varkappa_d|x|} \tag{4.30}$$

with $\varkappa_d$ given by (3.9b).

If the impedance $\mathcal{Z} = \mu_0 E_z/B_y$ (2.18a) is introduced, the SW dispersion relation can be obtained by matching its values at the boundaries $x = d - \Delta$ and $x = d$.

The surface impedance of the weakly inhomogeneous plasma at the $x = d - \Delta$ interface is

$$\mathcal{Z}^p_{\text{inh}}\bigg|^{\{{}^s_a\}}_{x=d-\Delta}$$
$$= \mathcal{Z}_p \left\{ \begin{array}{l} \coth \int_0^{d-\Delta} \varkappa_p(x)\,dx \\ \tanh \int_0^{d-\Delta} \varkappa_p(x)\,dx \end{array} \right. + \frac{1}{\varkappa_p(x)}\,\frac{d}{dx}\ln\sqrt{\frac{\varepsilon(x)}{\varkappa_p(x)}}\bigg|_{x=d-\Delta} \right\} , \tag{4.31}$$

where $\mathcal{Z}_p = [i\,\mu_0 c^2 \varkappa_p(x)/\omega\varepsilon(x)]|_{x=d-\Delta}$, for $\varepsilon(x) = \text{const.}$, is the surface impedance of a semi-bounded homogeneous plasma. The impedance of the layer of strongly inhomogeneous plasma, expressed according to (4.8) and (4.12) through the field values at an arbitrary point $x = x_0$ (inside or at the boundary of the layer), is

$$\mathcal{Z}(d - \Delta < |x| < d)$$
$$= \mu_0 \frac{E_z(x=x_0) + i\,(c^2/\omega)B(x=x_0)\int_{x_0}^x [\varkappa_p^2(x')/\varepsilon(x')]\,dx'}{B(x=x_0) - i\,(\omega/c^2)E_z(x=x_0)\int_{x_0}^x \varepsilon(x')\,dx'} . \tag{4.32}$$

The surface impedances at the interfaces $x = d - \Delta$ and $x = d$ can be obtained from (4.32) after giving x_0 the values d and $d - \Delta$, respectively:

$$\mathcal{Z}(x = d - \Delta)$$

$$= \mathcal{Z}(x = d)\,\frac{1 - \mathrm{i}\,(\mu_0 c^2/\omega)\mathcal{Z}^{-1}(x = d)\int_{d-\Delta}^{d}[\varkappa_{\mathrm{p}}^2(x)/\varepsilon(x)]\,\mathrm{d}x}{1 + (\omega/\mu_0 c^2)\mathcal{Z}(x = d)\int_{d-\Delta}^{d}\varepsilon(x)\,\mathrm{d}x} \tag{4.33a}$$

$$\mathcal{Z}(x = d)$$

$$= \mathcal{Z}(x = d - \Delta)\,\frac{1 + \mathrm{i}\,(\mu_0 c^2/\omega)\mathcal{Z}^{-1}(x = d - \Delta)\int_{d-\Delta}^{d}[\varkappa_{\mathrm{p}}^2(x)/\varepsilon(x)]\,\mathrm{d}x}{1 - (\omega/\mu_0 c^2)\mathcal{Z}(x = d - \Delta)\int_{d-\Delta}^{d}\varepsilon(x)\,\mathrm{d}x}\,.$$

$$\tag{4.33b}$$

The impedance of the third region (Fig. 4.1), occupied by the homogeneous dielectric, is

$$\mathcal{Z}_{\mathrm{d}}(|x| \geq d) = -\mathrm{i}\,\frac{\mu_0 c^2 \varkappa_{\mathrm{d}}}{\omega \varepsilon_{\mathrm{d}}}\,. \tag{4.34}$$

It appears as a surface impedance at $x = d$.

The right-hand sides of (4.33a,b) include $\mathcal{Z}(x = d)$ and $\mathcal{Z}(x = d - \Delta)$ as given by (4.34) and (4.31), respectively. The dispersion relation results from matching the surface impedances at one of the interfaces $x = d - \Delta$ or $x = d$, i.e. by equalizing (4.31) and (4.33a) or (4.33b) and (4.34):

$$\mathcal{Z}_{\mathrm{inh}}^{\mathrm{p}}\Big|_{x=d-\Delta}^{\{ \substack{\mathrm{s} \\ \mathrm{a}} \}} - \mathcal{Z}_{\mathrm{d}}(x = d)$$

$$= -\mathrm{i}\,\frac{\mu_0 c^2}{\omega}\int_{d-\Delta}^{d}\frac{\varkappa_{\mathrm{p}}^2(x)}{\varepsilon(x)}\,\mathrm{d}x - \mathrm{i}\,\frac{\omega}{\mu_0 c^2}\mathcal{Z}_{\mathrm{p}}\mathcal{Z}_{\mathrm{d}}\int_{d-\Delta}^{d}\varepsilon(x)\,\mathrm{d}x\,. \tag{4.35}$$

The terms on the right-hand side account for the jump of the impedance over the region of the plasma resonance. The first term is associated with the singularity of the electric-field component there and describes the damping of the SW because of excitation of local plasmons. The second term is a small correction to the dispersion.

If the effects of the weak inhomogeneity in the central part of the slab are neglected and the plasma density is replaced there by its averaged value (over the slab thickness), the dispersion relation (4.35) can be simplified to a form accounting only for the effect of the resonance absorption. In the case of a plasma slab separated from the vacuum by a diffuse boundary, the dispersion relation of the symmetric mode obtained from (4.35) by using (4.17) and (4.18) is

$$\frac{\varkappa_{\mathrm{p}}}{\varepsilon}\,\coth(\varkappa_{\mathrm{p}}d) + \varkappa_{\mathrm{v}} - \mathrm{i}\,\pi\beta^2 L_n^{(\mathrm{r})} = 0. \tag{4.36}$$

The effect of the linear mode transformation of the SW into local plasmons is described by a term (the third one in (4.36)) which is added to the dispersion relation (3.50a) of the symmetric SW in a homogeneous plasma slab with a sharp boundary.

For $\varkappa_{\rm p} d \gg 1$ (4.35) reduces to

$$\frac{\varkappa_{\rm p}(x = -\Delta)}{\varepsilon(x = -\Delta)} \left[1 + \frac{1}{2\varepsilon(x = -\Delta)} \frac{\rm d}{{\rm d}x} \left. \frac{\varepsilon(x)}{\varkappa_{\rm p}(x)} \right|_{x=-\Delta} \right] + \frac{\varkappa_{\rm d}}{\varepsilon_{\rm d}}$$

$$= -\frac{\varkappa_{\rm p}(x = -\Delta)}{\varepsilon(x = -\Delta)} \frac{\varkappa_{\rm d}}{\varepsilon_{\rm d}} \int_{-\Delta}^{0} \varepsilon(x)\,{\rm d}x - \int_{-\Delta}^{0} \frac{\varkappa_{\rm p}^2(x)}{\varepsilon(x)}\,{\rm d}x \ . \tag{4.37}$$

This describes SW propagation along a diffuse boundary between two semi-spaces occupied by a weakly inhomogeneous plasma ($x < -\Delta$) and a homogeneous dielectric ($x > 0$). With $\Delta \to 0$ and $\varepsilon(x) = $ const., (4.37) gives the dispersion relation (3.10a) of SW propagation in a homogeneous plasma semi-bounded by a dielectric.

A simplified form of (4.37) which accounts only for the effect of the resonance absorption is

$$\frac{\varkappa_{\rm p}}{\varepsilon} + \varkappa_{\rm v} - {\rm i}\,\pi\beta^2 L_n^{(\rm r)} = 0. \tag{4.38a}$$

In the region of fast SWs the dispersion is again given by (3.31a). However, now the space damping rate includes an additional term, by which the absorption due to mode conversion is added to the damping (3.33c) due to collisions:

$$\alpha = \frac{\omega^3}{c^3}\lambda_{\rm sk}^2 \left(\frac{\nu}{2\omega} + \frac{\pi L_n^{(\rm r)}}{\lambda_{\rm sk}} \right) . \tag{4.38b}$$

The region of slow SWs in the thick-slab approach ($\varkappa_{\rm p} d \gg 1$) is described by the quasi-static limit ($\omega/\beta \ll c$) of the wave dispersion behaviour in a semi-bounded plasma. In this case (4.37) gives, with $\varepsilon_{\rm d} = 1$, the following results for the frequency ω and the time damping rate γ:

$$\omega = \frac{\omega_{\rm p}(x = -\Delta)}{\sqrt{2}} \left[1 - \frac{\omega_{\rm p}^2}{8\beta^2 c^2} - \frac{1}{8\beta} \left. \frac{{\rm d}\ln\varepsilon(x)}{{\rm d}x} \right|_{x=-\Delta} \right.$$

$$\left. - \frac{\beta}{4} \int_{-\Delta}^{0} \varepsilon(x)\,{\rm d}x + \frac{\beta}{4} \fint_{-\Delta}^{0} \frac{{\rm d}x}{\varepsilon(x)} \right] , \tag{4.39a}$$

$$\gamma = -\frac{\pi}{4}\beta L_n^{(\rm r)}\omega - \frac{\nu}{2} \ . \tag{4.39b}$$

Excluding the second term between the brackets (see (3.31b)), the other dispersion corrections in (4.39a) are due to the effects of the plasma nonuniformity. The damping of SWs connected with the excitation of local plasmons is described by the first term on the right-hand side of (4.39b). The absorption due to this effect increases strongly when the SW wavelength becomes comparable with the scale length $L_n^{(\rm r)}$ of the density inhomogeneity in the resonance region. The second term in (4.39b) accounts for the damping due to collisions (3.33d). The result for the total space damping rate, which includes both the effects of collisions (3.33e) and resonance absorption, is

$$\alpha = \frac{\omega}{c} \left(\frac{\omega_{\rm p}^2}{\omega^2} - 2 \right)^{-3/2} \left[\frac{\nu}{\omega} + \frac{\pi L_n^{(\rm r)}}{2} \frac{\omega}{c} \left(\frac{\omega_{\rm p}^2}{\omega^2} - 2 \right)^{-1/2} \right]. \tag{4.39c}$$

SW propagation along a thin plasma slab should be treated separately, since in this case the thickness d of the entire slab is small compared with the field penetration depth and there is no region of weakly (in this sense) inhomogeneous plasma. Although the inhomogeneity of the density distribution could be arbitrary over the complete slab thickness, the position of the resonance point is assumed to be close to the plasma slab–dielectric interface. The dispersion relation can be obtained in a simple form by matching the surface impedance at the surface $x = d - \Delta$, chosen in much the same manner as in Fig. 4.1 (i.e. the region of the plasma resonance is separated from the other part of the slab). In the case of a thin plasma slab, the wave impedances in both region I and region II (i.e. the regions without ($0 < x \leq d - \Delta$) and with ($d - \Delta \leq x \leq d$) a plasma resonance point) can be described by (4.32), giving to x_0 the values 0 and d, respectively. In the case of a symmetric surface mode $B(x = 0) = 0$, the surface impedances at $x = d - \Delta$ calculated by accounting for the field distributions in the regions $0 < x \leq d - \Delta$ and $d - \Delta \leq x \leq d$ are

$$\left. \mathcal{Z}^{(\rm I)} \right|_{x=d-\Delta} = {\rm i} \frac{\mu_0 c^2}{\omega} \frac{1}{\int_0^{d-\Delta} \varepsilon(x)\, {\rm d}x}, \tag{4.40a}$$

$$\left. \mathcal{Z}^{(\rm II)} \right|_{x=d-\Delta} = \mathcal{Z}_{\rm d} - {\rm i} \frac{\mu_0 c^2}{\omega} \int_{d-\Delta}^{d} \frac{\varkappa_{\rm p}^2(x)}{\varepsilon(x)}\, {\rm d}x, \tag{4.40b}$$

respectively. Their matching yields the dispersion relation for SW propagation along a thin inhomogeneous plasma layer:

$$\frac{1}{d\bar{\varepsilon}} + \frac{\varkappa_{\rm d}}{\varepsilon_{\rm d}} + \int_{d-\Delta}^{d} \frac{\varkappa_{\rm p}^2(x)}{\varepsilon(x)}\, {\rm d}x = 0, \tag{4.41}$$

with $\bar{\varepsilon} = (1/d) \int_0^d \varepsilon(x)\, {\rm d}x$ being the plasma permittivity averaged over the slab thickness.

At the beginning of the region where the thin-plasma approach is valid on the phase diagram, where the SWs are still quite fast ($\beta \approx \omega/c$ for $\varepsilon_{\rm d} = 1$ according to (3.52b)), the solution of (4.41) gives the following result for the space damping rate:

$$\alpha = \frac{\omega^3}{c^3} \frac{\lambda_{\rm sk}^4}{d^2} \left(\frac{\nu}{\omega} + \frac{\pi L_n^{(\rm r)} d}{\lambda_{\rm sk}^2} \right). \tag{4.42}$$

The contribution of the resonance absorption (second term) is added to the damping (3.56c) due to collisions. In the region of slow SWs in a thin plasma slab, the quasi-static limit ($\beta c \gg \omega$) of (4.41) gives the following solutions for the frequency ω and the time damping rate γ of the wave (for $\varepsilon_{\rm d} = 1$):

$$\omega = \left(\overline{\omega_{\mathrm{p}}^2}\beta d\right)^{1/2} , \tag{4.43a}$$

$$\gamma = -\frac{\nu}{2} - \frac{\pi}{2}\frac{L_n^{(\mathrm{r})}}{d}\frac{\omega^3}{\overline{\omega_{\mathrm{p}}^2}} , \tag{4.43b}$$

where $\overline{\omega_{\mathrm{p}}^2}$ is the squared plasma frequency averaged over the slab thickness. In the case of an inhomogeneous plasma slab (4.43a) replaces (3.57c). The expression for the space damping rate which replaces (3.57e) is

$$\alpha = \frac{\omega^2}{\overline{\omega_{\mathrm{p}}^2}d}\left(\frac{\nu}{\omega} + \frac{\pi L_n^{(\mathrm{r})}}{d}\frac{\omega^2}{\overline{\omega_{\mathrm{p}}^2}}\right) . \tag{4.43c}$$

The group velocity is as given by (3.57d).

In the thin-slab approach the E_z field component can be taken as homogeneous. Owing to the existence of a plasma resonance close to the plasma boundary, the transverse field component E_x has the following resonant form:

$$E_x = \frac{E_{x0}}{1 - [n(x)/n_{\mathrm{c}}] + \mathrm{i}\,(\nu_{\mathrm{eff}}/\omega)} , \tag{4.44}$$

where $E_{x0} = (c^2 k/\omega)B(x = d)$. The Joule losses obtained after averaging over the wave period and integration over the slab thickness (see (2.29) and (2.32b)) are

$$\overline{Q} = \frac{e^2}{2m\omega^2}\frac{1}{2d}\int_{-d}^{d} n(\nu|E_z|^2 + \nu_{\mathrm{eff}}|E_x|^2)\mathrm{d}x$$

$$= \frac{e^2}{2m\omega^2}\left(\nu\bar{n}|E_z|^2 + n_{\mathrm{c}}\pi\frac{L_n^{(\mathrm{r})}}{d}\omega|E_{x0}|^2\right) . \tag{4.45}$$

The case of weak collisions is considered here, and ν_{eff} as introduced in (4.24) is used; $\bar{n} = (1/2d)\int_{-d}^{d} n(x)\mathrm{d}x$. The relation between $|E_z|^2$ and $|E_{x0}|^2$ can be obtained from the equation $\nabla \cdot (\varepsilon E) = 0$, for example, which in this case has the form

$$\frac{\partial}{\partial x}(\varepsilon E_x) + \mathrm{i}\,k\varepsilon E_z = 0 . \tag{4.46a}$$

The result is

$$|E_{x0}|^2 = (kd)^2\bar{\varepsilon}^2|E_z|^2 . \tag{4.46b}$$

Since the averaged electron density $\bar{n}$ of the plasma is somewhat larger than the critical density n_{c}, and thus $|\bar{\varepsilon}| \approx \bar{n}/n_{\mathrm{c}}$, (4.46b), after using (4.43a), reduces to

$$|E_{x0}|^2 = |E_z|^2. \tag{4.46c}$$

Therefore, (4.45) for Joule's losses finally becomes

$$\overline{Q} = \frac{e^2\nu\bar{n}}{2m\omega^2}|E_z|^2\left(1 + \pi\frac{L_n^{(\mathrm{r})}}{d}\frac{\omega}{\nu}\frac{n_{\mathrm{c}}}{\bar{n}}\right) . \tag{4.47}$$

In the situation of a more general density distribution such as that given by (2.65b), i.e.

$$n(x) = n_0 \cos \frac{\mu x}{d} \tag{4.48}$$

with $0 \leq \mu \leq \pi/2$, the following relation (instead of (4.41)) may be used as a useful approximation:

$$\frac{1}{d\bar{\varepsilon}} + \varkappa_{\mathrm{v}} + k^2 \int_0^d \frac{\mathrm{d}x}{\varepsilon(x)} = 0 \,. \tag{4.49a}$$

The origin of the damping in the collisionless case can also be visualized by means of (4.49a), since for $\nu \to 0$ and $\varepsilon(x)$ passing through zero at x_{r}, the integral in the last term can be calculated with the use of (4.16)–(4.18) so that (4.49a) can be written as

$$\frac{1}{d\bar{\varepsilon}} + \varkappa_{\mathrm{v}} + k^2 \left[\int_0^d \frac{\mathrm{d}x}{\varepsilon(x)} - \mathrm{i}\,\pi \left(\frac{\mathrm{d}\varepsilon}{\mathrm{d}x}\Big|_{x=x_{\mathrm{r}}} \right)^{-1} \right] = 0. \tag{4.49b}$$

The above analytical results for plasma–slab situations contain all the essential features caused by transverse inhomogeneity; even the simplification given by (4.49) is a good approximation . This will be demonstrated below by discussing step by step the dispersion relations obtained from exact numerical solutions of the Maxwell field equations.

The following two items may be emphasized, as effects of inhomoegeneity or of inhomogeneity acting together with collisions.

Firstly, formation of a maximum of $\omega/\overline{\omega_{\mathrm{p}}}$ in the dispersion/phase diagram and the appearance of a backward wave in the collisionless case ($\nu \to 0$), as known from previous studies [4.5,14,15] on SWs in cylindrical waveguides, is an essential feature introduced by transverse inhomogeneity. In fact, this maximum results from coupling in the collisionless case of two solutions (forward and backward waves) which extend above the maximum as complex conjugate solutions [4.17]. In the collisional case these two solutions separate from each other. The situation here, with formation of a maximum, is strongly suggestive of the discussion in Chap. 3 which addresses the antisymmetric mode in a homogeneous plasma slab (Fig. 3.13c) and the azimuthally symmetric mode in a plasma column with dielectric shielding (Fig. 3.25). A common feature of all three cases – of radial density nonuniformity, of an antisymmetric field configuration and of finite dielectric shielding – is that the maximum comes about because of the changing influence of spatial structure (in density, field or shielding) with growing β and decreasing penetration depth. Secondly, an essential feature introduced by appreciable density inhomogeneity is a second turn-around in the phase diagrams of SWs in nonuniform collisional plasmas. This appears when, in the transverse density profile, the resonance $\omega_{\mathrm{p}}(r) = \omega$ occurs at the plasma–dielectric interface with local peaks of $|E_{\mathrm{r}}|$ and is associated with increased damping via mode conversion. It should be mentioned that in regions of mode conversion and of complex solutions (in the collisionless case) the simple concept of group velocity $\partial\omega/\partial k$ should *not* be used [4.42].

The treatment proceeds by means of numerical solutions for the propagation of a symmetric surface mode along a plasma slab (of thickness $2d$) surrounded by vacuum. The transverse density profile is given by (4.48). Phase diagrams were obtained via numerical integration (in terms of complex variables) of the wave equations obtained from (2.15):

$$\frac{\mathrm{d}^2 E_z(x)}{\mathrm{d}x^2} + \frac{k^2}{\varkappa_\mathrm{p}^2(x)}\frac{\mathrm{d}\ln\varepsilon(x)}{\mathrm{d}x}\frac{\mathrm{d}E_z(x)}{\mathrm{d}x} - \varkappa_\mathrm{p}^2(x)E_z(x) = 0\,, \tag{4.50a}$$

$$E_x(x) = \mathrm{i}\,\frac{k}{\varkappa_\mathrm{p}(x)}\frac{\mathrm{d}E_z(x)}{\mathrm{d}x}\,. \tag{4.50b}$$

The equations for the vacuum correspond to (4.50), replacing ε by $\varepsilon_\mathrm{v} = 1$ and $\varkappa_\mathrm{p}$ with $\varkappa_\mathrm{v}$. The exponentially decreasing solution was taken. The complex wave number was calculated by fulfilling the boundary conditions at the plasma–vacuum interface, namely the continuity of E_z and D_x. A Runge–Kutta method was employed. The starting values for E_z on the axis could be chosen arbitrarily, and $\mathrm{d}E_z/\mathrm{d}x|_{x=0}$ was set to zero.

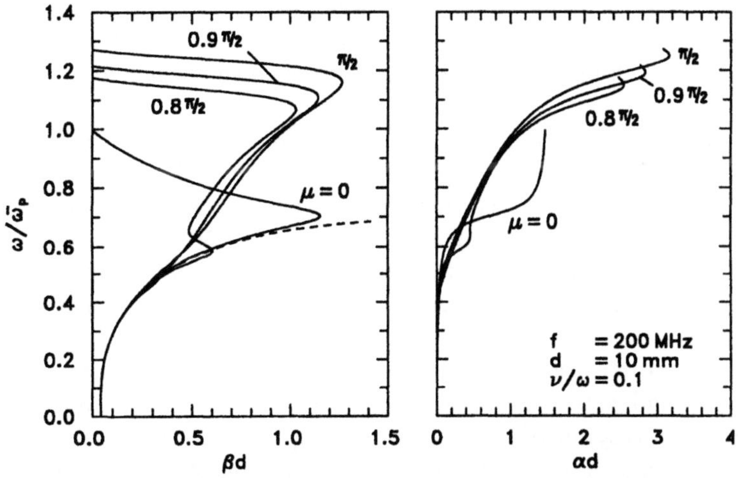

Figure 4.2. Phase diagrams (βd versus $\omega/\overline{\omega}_\mathrm{p}$) and space damping rate (αd versus $\omega/\overline{\omega}_\mathrm{p}$) of the symmetric mode in an inhomogeneous plasma slab for various μ values in the presence of collisions (*solid curves*); the homogeneous, collisionless case (*dashed curve*) is given for comparison. Wave frequency $f \equiv \omega/2\pi = 200$ MHz; slab thickness $d = 1$ cm ([4.17], Fig. 3)

Typical results are displayed in Fig. 4.2. In the homogeneous case ($\mu = 0$)) the turn-around of the β curve towards $\beta = 0$ and $\omega/\omega_\mathrm{p} = 1$ occurs at relatively large βd. This is connected solely to the presence of collisions ($\nu \neq 0$; Sect. 3.2). With increasing μ (e.g. the curve for $\mu = 0.8\,\pi/2$ on the left-hand figure) the turn-around shifts to higher densities, i.e. to smaller $\omega/\overline{\omega_\mathrm{p}}$. An additional phenomenon also shows up: another turn-around – in the $\omega/\overline{\omega_\mathrm{p}}$ range below the SW resonance in the homogeneous plasma – of the $\omega/\overline{\omega_\mathrm{p}}$ versus βd curve can be recognized, which depends on the degree of density inhomogeneity. This second turn-around of the curve always occurs just when the plasma frequency at the plasma border equals the wave frequency, i.e. $\omega_\mathrm{p}(d) = \omega$ (thus, when a resonance peak of the E_x component becomes possible within the plasma). With μ approaching $\pi/2$, the two points of turn-around approach each other, and the double turn-around becomes less and less evident. Finally, in the limit $\mu = \pi/2$, when the density at the plasma border tends to zero, a smooth curve results, and the E_x resonance is present everywhere in the phase diagram.

The changed behaviour of the phase diagram in the vicinity of $\omega/\overline{\omega_\mathrm{p}}$ values which correspond to $\omega_\mathrm{p}(d) = \omega$ is related to features discussed before in Sect. 3.2.3, when a homogeneous plasma is overlain by a thin plasma slab of lower density. There, a sharp loop appears in the β curves associated with the plasma resonance in the thin layer (see Fig. 3.10a), although the curves essentially return to the usual behaviour towards higher $\omega/\overline{\omega_\mathrm{p}}$. Now, in the case of a smooth inhomogeneity, the density profile can be considered as a sequence of thin layers from the wall to the plasma interior, each of which gives rise to a plasma resonance step by step with rising $\omega/\overline{\omega_\mathrm{p}}$. The sequence of effects from a multitude of layers results in total in the formation of a "plasmon" branch, which may significantly deviate from the usual curve with growing $\omega/\overline{\omega_\mathrm{p}}$.

Figure 4.2 also exhibits the contribution of the resonance absorption to the wave damping. In inhomogeneous plasmas the tendency of α to approach β is quite strong, and α dominates towards larger $\omega/\overline{\omega_\mathrm{p}}$. A comparison of the solutions of the exact numerical integrations via (4.50) with those of the approximated dispersion relation (4.49a) shows that the latter describes qualitatively quite well the features caused by the inhomogeneous density, as mentioned before.

In the presence of density inhomogeneities a second solution exists with different signs of α and β, in analogy with the considerations in Sect. 3.3.1 in connection with the antisymmetric slab mode. In Fig. 4.3 this second solution, a backward wave, is shown in addition to the first solution of a forward wave shown in Fig. 4.2. A case of rather low collisionality ($\nu/\omega = 0.01$) and a case of modest collisionality ($\nu/\omega = 0.1$) are considered. For $\nu/\omega = 0.01$ the forward (α_1, β_1) and backward (α_2, β_2) waves are rather close together near the region of the collisionless maximum. Mode conversion from the forward to the backward wave might indeed be expected. However, with increasing ν/ω the forward- and backward-wave solutions separate more and more from each other, and they may not cross before the α_1 of the forward wave is close to β_1. The $|\alpha_2|$ values of the backward wave tend to be larger than those of

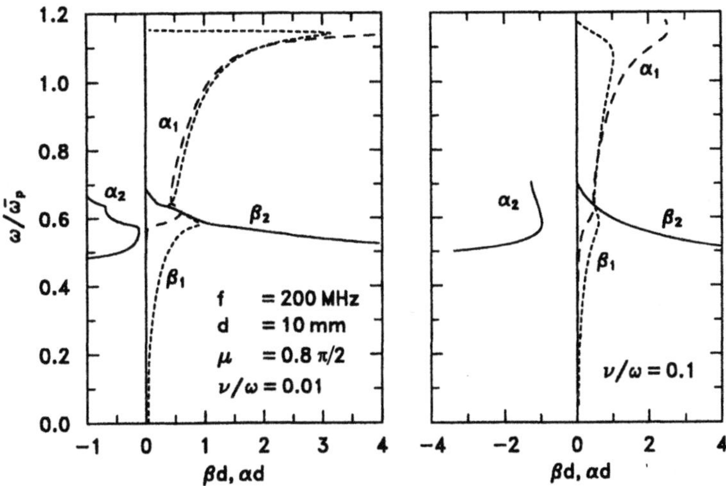

Figure 4.3. Effect of collisions ($\nu/\omega = 0.01$ on the *left* and $\nu/\omega = 0.1$ on the *right*) on the dispersion behaviour of the second solution – the solution for a backward wave – for an inhomogeneous plasma (*solid curves*). The first solution for the wave number β_1 (*dotted curve*) and the space damping rate α_1 (*dashed curve*) is given for comparison ([4.17], Fig. 4)

the forward wave. Thus, for sufficient collisionality the linear coupling to the backward wave may become insignificant [4.43].

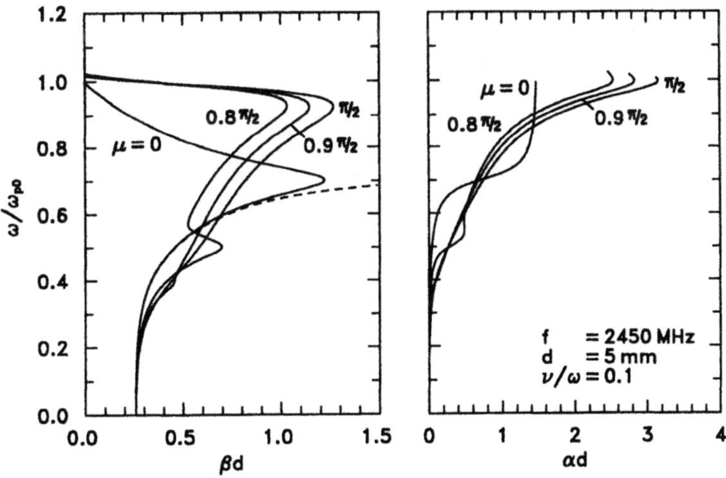

Figure 4.4. Dispersion behaviour at larger σ presented in terms of ω/ω_{p0} versus βd and αd. Phase diagrams for a wave frequency of $f \equiv \omega/2\pi = 2.45$ GHz and a slab thickness of $d = 0.5$ cm ([4.17], Fig. 5)

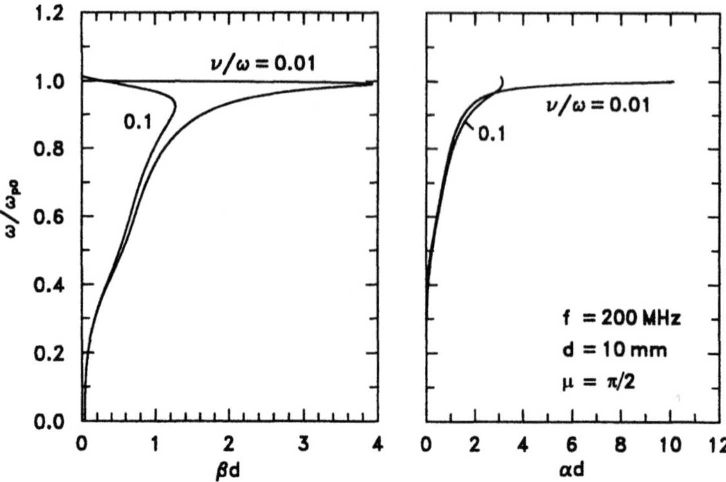

Figure 4.5. Effect of collisions on the phase diagram in the case of maximum density inhomogeneity ($\mu = \pi/2$), scalings as in Fig. 4.4. Phase diagrams for a wave frequency of $f \equiv \omega/2\pi = 200$ MHz and a slab thickness of $d = 1$ cm ([4.17], Fig. 6)

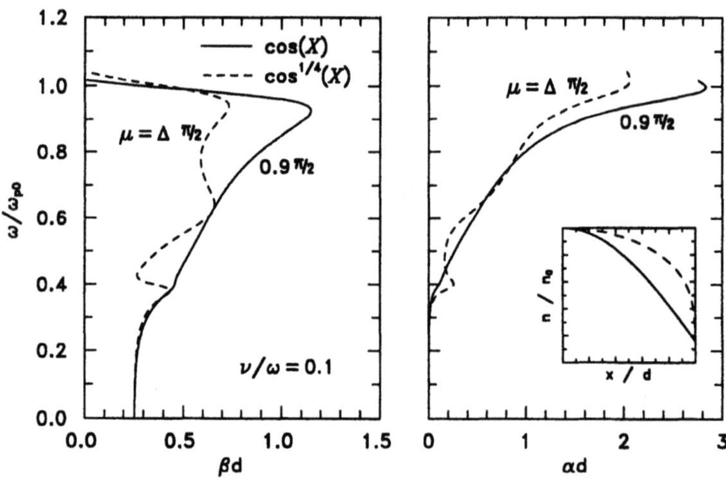

Figure 4.6. Influence of the shape of the density profile on the propagation characteristics (wave number β and damping rate α) of the symmetric mode in a plasma slab. Wave frequency $f \equiv \omega/2\pi = 2.45$ GHz, slab width $d = 5$ mm; $X = \mu x/d$ and $\Delta = 0.99962$; scalings as in Figs. 4.4 and 4.5 ([4.17], Fig. 7)

Figure 4.4 demonstrates that the behaviour described in Fig. 4.2 is not specific to the lower $\sigma = \omega d/c$ values, but is also exhibited for larger σ. The vertical scaling is chosen differently to illustrate the basic behaviour for larger α, even though this range is of little direct interest for SW-sustained disharges. Now ω/ω_{p_0} is plotted, where ω_{p_0} is the plasma frequency referring to the density at the slab axis, not as before to the density averaged over the cross-section ($\propto \overline{\omega_p^2}$). For $\omega/\omega_{p_0} \approx 1$, β tends to zero. It may be mentioned that the vertical asymptote of αd for $\mu = 0$ appears near the zero of $\cos(\alpha d)$, which can be confirmed by analysis of the dispersion relation in the case of large $|\alpha| \gg \beta$.

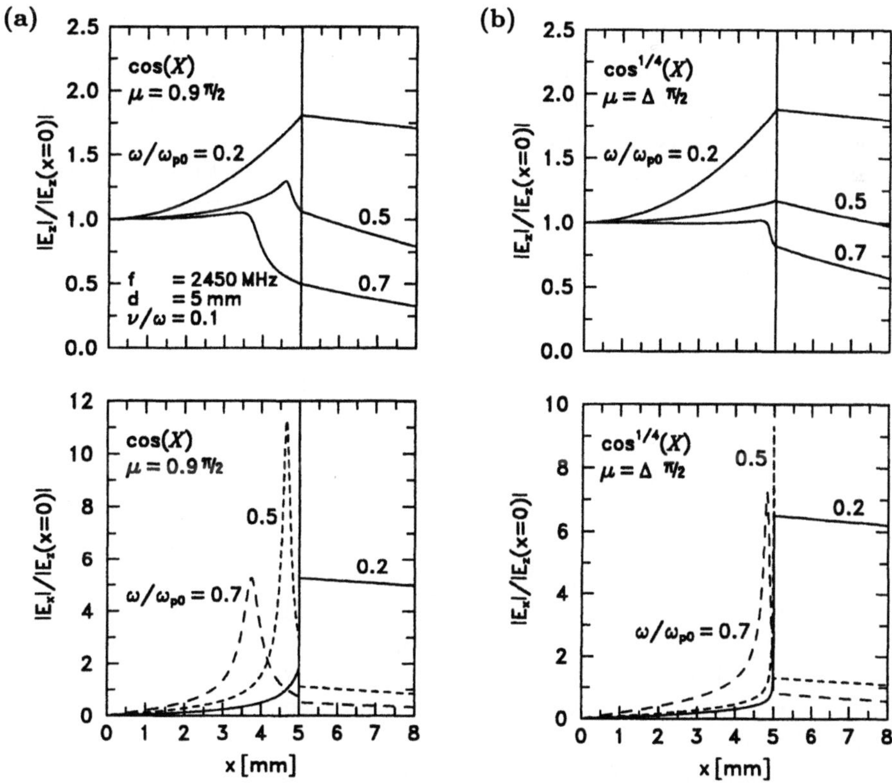

Figure 4.7. Transverse profiles for different ω/ω_{p_0} values: moduli of the E_z (*upper figures*) and E_x (*lower figures*) field components normalized to the E_z field modulus on the axis. **(a)** cosine profile corresponding to the *solid curve* in Fig. 4.6. **(b)** $\cos^{(1/4)}$ profile corresponding to the *dashed curve* in Fig. 4.6 ([4.17], Figs. 8a,b)

The effect of collision frequencies in the case of $\mu = \pi/2$ is depicted in Fig. 4.5. The value of the collision frequency influences the dispersion behaviour only in the range of very high ω/ω_{p_0}. For small ν/ω, the turn-around towards $\beta = 0$ occurs for rather larger β values. In the collisionless case $\nu \to 0$ this final turn-around moves to infinity (i.e. β achieves very large values). For lower values of $\omega/\omega_{\mathrm{p0}}$ and small ν/ω, α is not much changed, as is to be expected from (4.49b). For $\mu = \pi/2$, because of the resonance absorption, α is finite, even for the case $\nu = 0$, i.e. the solution is complex throughout.

The features discussed so far are in no way only typical of the cosine profiles chosen for the density. They reappear in a similar way, for example, for "steeper" profiles $\propto \cos^{(1/n)}(\mu x/d)$, with the modifications to be expected, as exhibited in Fig. 4.6 for the cases $n = 1, 4$. The parameter μ is taken here so that the density at $x = d$ has dropped by the same ratio from the central density at $x = 0$ in both cases. The corresponding transverse field profiles are depicted in Figs. 4.7a,b for three values of $\omega/\omega_{\mathrm{p0}}$, demonstrating the development of peaks in $|E_x|$ when $\omega_{\mathrm{p}}(x) = \omega$ becomes possible.

Obviously, negative ε_{r} and consequently a maximum of the electric-field intensity at the plasma/vacuum interface, do not characterize the whole phase diagram. As soon as a plasma resonance arises, the maximum electric field is shifted from the interface into the plasma volume. In the vicinity of the resonance the main contribution to the absolute field intensity is now dominated by the transverse component. The sign of the transverse field strength changes of course according to that of the permittivity.

4.2.4 Surface Wave Propagation
Along a Radially Inhomogeneous Plasma Column

The case of cylindrical geometry is now considered. The dispersion law for an azimuthally symmetric surface mode propagating along a plasma column of radially inhomogeneous density $\varepsilon = \varepsilon(r)$ is derived. Different scales for the density variation are assumed in order to ensure the existence of a region $(0 < r < R - \Delta)$ of a weakly inhomogeneous plasma in the central part of the column, and a region $(R - \Delta < r < R)$ of strongly inhomogeneous plasma which is limited to a narrow transition layer (of thickness $\Delta \ll R$) close to the interface. The plasma resonance point $r = r_{\mathrm{r}}$, where $\varepsilon(r_{\mathrm{r}}) = 0$, is in this layer. The plasma column is bounded by a homogeneous dielectric medium of permittivity ε_{d}.

The equation for the radial distribution of the amplitude $B(r)$ of the wave magnetic field $B_\varphi(r, z) = B(r)\exp(\mathrm{i}\,kz)$ in the plasma is

$$\frac{1}{r}\frac{\mathrm{d}}{\mathrm{d}r}\left(r\frac{\mathrm{d}B}{\mathrm{d}r}\right) - \frac{1}{\varepsilon(r)}\frac{\mathrm{d}\varepsilon(r)}{\mathrm{d}r}\frac{1}{r}\frac{\mathrm{d}}{\mathrm{d}r}(rB) - \left[\varkappa_{\mathrm{p}}^2(r) + \frac{1}{r^2}\right]B = 0 \qquad (4.51a)$$

or, equivalently,

$$\varepsilon(r)\frac{\mathrm{d}}{\mathrm{d}r}\left[\frac{1}{r\varepsilon(r)}\frac{\mathrm{d}}{\mathrm{d}r}(rB)\right] - \varkappa_{\mathrm{p}}^2(r)B = 0, \qquad (4.51b)$$

where $\varkappa_{\mathrm{p}}(r)$ is given by (3.9a) with $\varepsilon \equiv \varepsilon(r)$. The axial electric-field component $E_z(r)$ is related to the B field according to

$$E_z(r) = \mathrm{i}\,\frac{c^2}{\omega}\,\frac{1}{r\varepsilon(r)}\,\frac{\mathrm{d}}{\mathrm{d}r}(rB)\,. \tag{4.52}$$

In the region of a weakly inhomogeneous plasma $(0 < r < R - \Delta)$ the substitution

$$B(r) = \frac{\mathcal{U}(r)}{\sqrt{r}} \tag{4.53}$$

reduces (4.51a) to the following equation for the $\mathcal{U}(r)$ function:

$$\frac{\partial^2 \mathcal{U}}{\partial r^2} - \frac{1}{\varepsilon(r)}\,\frac{\mathrm{d}\varepsilon(r)}{\mathrm{d}r}\,\frac{\mathrm{d}\mathcal{U}}{\mathrm{d}r} - \left[\varkappa_{\mathrm{p}}^2(r) + \frac{3}{4r^2} + \frac{1}{2r}\,\frac{1}{\varepsilon(r)}\,\frac{\mathrm{d}\varepsilon(r)}{\mathrm{d}r}\right]\mathcal{U} = 0\,. \tag{4.54}$$

This has the solution

$$\mathcal{U}(r) = \mathrm{const.}\,\sqrt{\frac{\varepsilon(r)}{\varkappa_{\mathrm{p}}(r)}}\,\exp\left(\pm\int^r \varkappa_{\mathrm{p}}(r')\,\mathrm{d}r'\right)\,, \tag{4.55}$$

according to the geometrical-optics method. Since the last two terms in (4.54) are considered as second-order terms, the solution (4.55) is valid for comparatively large r values $(r > \varkappa_{\mathrm{p}}^{-1})$, i.e. far away from the column axis. Therefore, the results address the case of the thick-cylinder approximation. The solution for the magnetic field is

$$B(r < R - \Delta) = B(r = R - \Delta)P(r)\exp\left[-\int_r^R \varkappa_{\mathrm{p}}(r')\,\mathrm{d}r'\right]\,,$$
$$r > \varkappa_{\mathrm{p}}^{-1} \tag{4.56}$$

where $B(r = R - \Delta)$ is the field intensity at the interface between the weakly and strongly inhomogeneous plasmas and

$$P(r) = \sqrt{\frac{R}{r}\,\frac{\varkappa_{\mathrm{p}}(r = R - \Delta)}{\varepsilon(r = R - \Delta)}\,\frac{\varepsilon(r)}{\varkappa_{\mathrm{p}}(r)}}\,. \tag{4.57}$$

The $E_z(r)$ field component is

$$E_z(r < R - \Delta) = \mathrm{i}\,\frac{c^2}{\omega}\,\frac{1}{r\varepsilon(r)}\left\{1 + r\left[\varkappa_{\mathrm{p}}(r) + \frac{\mathrm{d}}{\mathrm{d}r}\ln P(r)\right]\right\}B(r)\,. \tag{4.58}$$

The distribution of the magnetic-field amplitude of the SW in the transition layer of a strongly inhomogeneous plasma can be obtained after double integration of (4.51b):

$$B(r) = \frac{r_0}{r}B(r = r_0)$$
$$+ \frac{1}{rr_0\varepsilon(r = r_0)}\,\frac{\mathrm{d}}{\mathrm{d}r}[rB(r)]\bigg|_{r=r_0}\int_{r_0}^r r'\varepsilon(r')\,\mathrm{d}r' \tag{4.59}$$
$$+ \frac{1}{r}\int_{r_0}^r r'\varepsilon(r')\,\mathrm{d}r'\int_{r_0}^{r'}\frac{\varkappa_{\mathrm{p}}^2(r'')}{\varepsilon(r'')}B(r'')\,\mathrm{d}r''\,,$$

where r_0 is an arbitrary point $r_0 \in [R - \Delta, R]$. Assuming that the width of the transition region is small in comparison with the depth of the field penetration into the plasma $\varkappa_p^{-1}$, an iteration procedure as used in Sect. 4.2.1 can be applied to obtain the magnetic-field amplitude. The solution of the zero-order approximation is

$$B(R - \Delta < r < R)$$
$$= \frac{r_0}{r} \left\{ B(r = r_0) + \frac{1}{r^2 \varepsilon(r_0)} \frac{\mathrm{d}}{\mathrm{d}r}[rB(r)]\Big|_{r=r_0} \int_{r_0}^{r} r' \varepsilon(r') \, \mathrm{d}r' \right\}, \qquad (4.60)$$

whereas the result of the next-order approximation is

$$B(R - \Delta < r < R)$$
$$= \frac{r_0}{r} \left\{ B(r = r_0) \left[1 + \frac{r_0}{r} \int_{r_0}^{r} r' \varepsilon(r') \, \mathrm{d}r' \int_{r_0}^{r'} \frac{\varkappa_p^2(r'')}{r'' \varepsilon(r'')} \, \mathrm{d}r'' \right] \right.$$
$$+ \frac{1}{r_0^2 \varepsilon(r_0)} \frac{\mathrm{d}}{\mathrm{d}r}[rB(r)]\Big|_{r=r_0}$$
$$\left. \times \int_{r_0}^{r} r' \varepsilon(r') \, \mathrm{d}r' \left[1 + \frac{r_0}{r} \int_{r_0}^{r'} \mathrm{d}r'' \frac{\varkappa_p^2(r'')}{r'' \varepsilon(r'')} \int_{r_0}^{r''} r''' \varepsilon(r''') \, \mathrm{d}r''' \right] \right\}. \quad (4.61)$$

The axial electric-field component in this transition layer, obtained after the first integration of (4.51b), is

$$E_z(R - \Delta < r < R) = E_z(r_0) + \mathrm{i} \frac{c^2}{\omega} B(r_0) \int_{r_0}^{r} \frac{r_0}{r'} \frac{\varkappa_p^2(r')}{\varepsilon(r')} \, \mathrm{d}r'. \qquad (4.62)$$

The solutions for the wave field in the region $r > R$ occupied by the homogeneous dielectric are

$$B(r > R) = B(r = R) \frac{K_1(\varkappa_d r)}{K_1(\varkappa_d R)}, \qquad (4.63a)$$

$$E_z(r > R) = -\mathrm{i} \frac{c^2}{\omega} \frac{\varkappa_d}{\varepsilon_d} B(r = R) \frac{K_0(\varkappa_d r)}{K_1(\varkappa_d R)}. \qquad (4.63b)$$

The surface impedances $\mathcal{Z} = \mu_0 E_z / B$ (2.18a) at the interfaces $r = R - \Delta$ and $r = R$, between the weakly and the strongly inhomogeneous plasma and between the transition plasma layer and the dielectric, respectively, are as follows:

(1) At $r = R - \Delta$, from (4.56) and (4.58):

$$\mathcal{Z} = \mathcal{Z}_{\mathrm{inh}}^{\mathrm{p}}(r = R - \Delta) \qquad (4.64a)$$

with

$$\mathcal{Z}_{\mathrm{inh}}^{\mathrm{p}}(r) = \mathcal{Z}_{\mathrm{p}}(r) \left(1 + \frac{1}{2R\varkappa_p(r)} + \frac{1}{\varkappa_p(r)} \frac{\mathrm{d}}{\mathrm{d}r} \ln \sqrt{\frac{\varepsilon(r)}{\varkappa_p(r)}} \right), \qquad (4.64b)$$

where $\mathcal{Z}_{\mathrm{p}}(r) = \mathrm{i}\,\mu_0 c^2 \varkappa_{\mathrm{p}}(r)/\omega\varepsilon(r)$ is the local impedance of a semi-bounded plasma. The second and third terms in (4.64b) are correction terms associated with the geometry under consideration and with the plasma inhomogeneity, respectively.

(2) At $r = R - \Delta$, from (4.60) and (4.62) with $r_0 = R$:

$$\mathcal{Z}(r = R - \Delta) = \mathcal{Z}(r = R)$$

$$\times \left[1 - \mathrm{i}\,\frac{\mu_0/c^2}{\omega}\,\mathcal{Z}^{-1}(r = R) \int_{R-\Delta}^{R} \frac{R}{r}\,\frac{\varkappa_{\mathrm{p}}^2(r)}{\varepsilon(r)}\,\mathrm{d}r \right] \tag{4.65a}$$

$$\times \left[1 + \mathrm{i}\,\frac{\omega}{\mu_0 c^2}\,\mathcal{Z}(r = R) \int_{R-\Delta}^{R} \frac{r}{R}\varepsilon(r)\,\mathrm{d}r \right]^{-1}.$$

(3) At $r = R$, from (4.60) and (4.62) with $r_0 = R - \Delta$:

$$\mathcal{Z}(r = R) = \mathcal{Z}(r = R - \Delta)$$

$$\times \left[1 + \mathrm{i}\,\frac{\mu_0 c^2}{\omega}\,\mathcal{Z}^{-1}(r = R - \Delta) \int_{R-\Delta}^{R} \frac{R-\Delta}{r}\,\frac{\varkappa_{\mathrm{p}}^2(r)}{\varepsilon(r)}\,\mathrm{d}r \right] \tag{4.65b}$$

$$\times \left[1 - \mathrm{i}\,\frac{\omega}{\mu_0 c^2}\,\mathcal{Z}(r = R - \Delta) \int_{R-\Delta}^{R} \frac{r}{R-\Delta}\varepsilon(r)\,\mathrm{d}r \right]^{-1}.$$

(4) At $r = R$, from (4.63):

$$\mathcal{Z}_{\mathrm{d}}(r = R) = -\mathrm{i}\,\frac{\mu_0 c^2}{\omega}\,\frac{\varkappa_{\mathrm{d}}}{\varepsilon_{\mathrm{d}}}\,\frac{\mathrm{K}_0(\varkappa_{\mathrm{d}}R)}{\mathrm{K}_1(\varkappa_{\mathrm{d}}R)}. \tag{4.66}$$

The dispersion relation of the azimuthally symmetric SWs obtained after equalizing (4.64) and (4.65a) (in which $\mathcal{Z}(r = R) = \mathcal{Z}_{\mathrm{d}}(r = R)$) or (4.66) and (4.65b) (in which $\mathcal{Z}(r = R - \Delta) = \mathcal{Z}_{\mathrm{inh}}^{\mathrm{p}}|_{r=R-\Delta}$) is

$$\mathcal{Z}_{\mathrm{inh}}^{\mathrm{p}}\big|_{r=R-\Delta} - \mathcal{Z}_{\mathrm{d}}(r = R)$$

$$= -\mathrm{i}\,\frac{\mu_0 c^2}{\omega} \int_{R-\Delta}^{R} \frac{\varkappa_{\mathrm{p}}^2(r)}{\varepsilon(r)}\,\mathrm{d}r - \mathrm{i}\,\frac{\omega}{\mu_0 c^2}\,\mathcal{Z}_{\mathrm{p}}\mathcal{Z}_{\mathrm{d}} \int_{R-\Delta}^{R} \varepsilon(r)\,\mathrm{d}r. \tag{4.67}$$

The ratios $(R - \Delta)/R$, $(R - \Delta)/r$ and r/R, with $r \in (R - \Delta, R)$, in (4.65) and (4.67) are approximated to unity. Relation (4.67) shows that the jump of the surface impedances between the interfaces $r = R - \Delta$ and $r = R$ is due to the resonance absorption of the EM energy in the vicinity of the plasma resonance point $r = r_{\mathrm{r}}$, where $\varepsilon(r_{\mathrm{r}}) = 0$. For $\varkappa_{\mathrm{p}}(R - \Delta)$, $\varkappa_{\mathrm{d}}R \to \infty$, (4.67) reduces to the dispersion law (4.37) for SW propagation along a diffuse boundary of a weakly inhomgeneous plasma in a semi-space.

When the local plasma resonance in the transverse direction close to the plasma column walls is the only effect under study, the smooth radial density profile can be replaced by a profile consisting of a homogeneous plasma in the central part of the column (with a density equal to the density averaged over the plasma-column cross-section) and an inhomogeneous plasma in a cylindrical layer close to the column boundary $(R - \Delta \leq r < R)$. This means that the geometrical-optics corrections due to the slow variation of the electron density in the central part of the column (i.e. the last term in (4.64b)) are not taken into account. The solutions for the wave field in $0 \leq r < R - \Delta$ are as given by (3.59) with ε_r replaced by $\bar{\varepsilon}_r = 1 - (\overline{\omega_p^2/\omega^2})$, the plasma permittivity averaged over the plasma-column cross-section. In the resonance region $R - \Delta \leq r < R$ the solution for the $E_r(r)$ field component with $\varepsilon(r) = 1 - [n(r)/n_c] + \mathrm{i}\,(\nu/\omega)$ is

$$E_r(r) = \frac{E_{r0}}{1 - [n(r)/n_c] + \mathrm{i}\,(\nu/\omega)}\,, \tag{4.68a}$$

where

$$E_{r0} = \frac{c^2 k}{\omega} B_\varphi(r = R) \tag{4.68b}$$

can be expressed in terms of the E_z field component at the boundary as

$$E_{r0} = -\mathrm{i}\,\frac{k\bar{\varepsilon}}{\varkappa_\mathrm{p}}\, E_z(r = R)\,\frac{\mathrm{I}_1(\varkappa_\mathrm{p} R)}{\mathrm{I}_0(\varkappa_\mathrm{p} R)}. \tag{4.68c}$$

In the vacuum region $(r > R)$, the field components are as given by (3.60).

Integration of (4.51b) over the cylindrical plasma layer $(R - \Delta \leq r < R)$ gives the boundary conditions

$$B_\varphi^\mathrm{v}(r = R) = B_\varphi^\mathrm{p}(r = R)\,,$$

$$\frac{1}{r}\,\frac{\mathrm{d}}{\mathrm{d}r}\,(rB_\varphi)\Big|_{r=R}^\mathrm{v} = \frac{1}{\varepsilon r}\,\frac{\mathrm{d}}{\mathrm{d}r}\,(rB_\varphi)\Big|_{r=R}^\mathrm{p} + B_\varphi(r = R)\int_{R-\Delta}^{R} \frac{\varkappa_\mathrm{p}^2(r)}{\varepsilon(r)}\,\mathrm{d}r\,. \tag{4.69}$$

These conditions lead to a dispersion relation that accounts for resonant absorption in the inhomogeneous plasma layer:

$$\frac{\varkappa_\mathrm{p}}{\bar{\varepsilon}}\,\frac{\mathrm{I}_0(\varkappa_\mathrm{p} R)}{\mathrm{I}_1(\varkappa_\mathrm{p} R)} + \varkappa_\mathrm{v}\,\frac{\mathrm{K}_0(\varkappa_\mathrm{v} R)}{\mathrm{K}_1(\varkappa_\mathrm{v} R)} + \int_{R-\Delta}^{R} \frac{\varkappa_\mathrm{p}^2(r)}{\varepsilon(r)}\,\mathrm{d}r = 0\,. \tag{4.70}$$

After introducing the notation $\tilde{x} = \varkappa_\mathrm{p} R$ and $\tilde{y} = \varkappa_\mathrm{v} R$ (Sect. 3.3.2) and performing the integration of the third term, (4.70) becomes

$$\frac{\tilde{x}}{\bar{\varepsilon}}\,\frac{\mathrm{I}_0(\tilde{x})}{\mathrm{I}_1(\tilde{x})} + \tilde{y}\,\frac{\mathrm{K}_0(\tilde{y})}{\mathrm{K}_1(\tilde{y})} - \mathrm{i}\,\pi L_n^{(\mathrm{r})} k^2 R = 0\,. \tag{4.71}$$

A term (the last one in (4.71)) appears additionally compared with the dispersion relation (3.61b) by accounting for wave damping due to resonance absorption. The real part of (4.71) yields (3.66c) for the wave's group velocity, whereas its imaginary part gives the time damping rate

$$\gamma = -\frac{\nu}{2}\left[1 - \frac{\sigma^2}{\tilde{x}^2}|\bar{\varepsilon}_{\mathrm{r}}|\tilde{M} + \pi\frac{\omega}{\nu}\frac{(\beta R)^2}{\tilde{x}}\frac{|\bar{\varepsilon}_{\mathrm{r}}|^2}{|\bar{\varepsilon}_{\mathrm{r}}|+1}\frac{L_n^{(\mathrm{r})}}{R}\frac{\mathrm{I}_1(\tilde{x})}{\mathrm{I}_0(\tilde{x})}\right]$$

$$\times\left[1 + \frac{|\bar{\varepsilon}_{\mathrm{r}}|}{|\bar{\varepsilon}_{\mathrm{r}}|+1}\sigma^2\tilde{T}\right]^{-1}. \tag{4.72}$$

This includes both collisional damping (3.66b) and SW damping due to plasmon generation (the term in (4.72) which involves $L_n^{(\mathrm{r})}$). The notation is the same as that used in (3.66).

With the electron–atom collision frequency for momentum transfer considered as a unified collision frequency (4.24), the Joule losses (2.32b) taken as a mean value over the cross-section

$$\overline{Q} \equiv \frac{Q}{\pi R^2} = \frac{e^2\nu}{m\omega^2}\frac{1}{R^2}\int_0^R n(|E_r|^2 + |E_z|^2)r\,\mathrm{d}r \tag{4.73a}$$

are described, using (3.52) and (4.68), by the following expression

$$\overline{Q} = \frac{e^2\nu}{2m\omega^2}\frac{2}{\tilde{x}}\frac{\mathrm{I}_1(\tilde{x})}{\mathrm{I}_0(\tilde{x})}\bar{n}\left(1 - \frac{\sigma^2}{\tilde{x}^2}|\bar{\varepsilon}_{\mathrm{r}}|\tilde{M} + P_r\frac{n_{\mathrm{c}}}{\bar{n}}\right)|E_z(r = R)|^2, \tag{4.73b}$$

where

$$P_{\mathrm{r}} = \pi\frac{(\beta R)^2}{\tilde{x}}|\bar{\varepsilon}_{\mathrm{r}}|^2\frac{\mathrm{I}_1(\tilde{x})}{\mathrm{I}_0(\tilde{x})}\frac{L_n^{(\mathrm{r})}}{R}\frac{\omega}{\nu}. \tag{4.73c}$$

The first two terms in (4.73b) represent the collisional losses associated with the axial and radial field components in the central part of the column. The third term, obtained by integrating $|E_r|^2$ over the cylindrical layer of inhomogeneous plasma close to the plasma-column walls, describes the losses due to the plasma resonance. Equation (4.73b) is more complicated than (4.47) since it covers the whole EM range of SW propagation along a plasma column.

The dispersion relation which governs the propagation of SWs along a thin radially inhomogeneous plasma column of radius R can be obtained by following the procedure applied for a thin slab in Sect. 4.2.3. Taking into account that the plasma-column radius is small compared with the field penetration into the plasma and assuming that the plasma resonance is close to the column radius R, (4.60) and (4.62) can be used to obtain the plasma impedances over the entire column radius:

$$\mathcal{Z}(r) = \mu_0\left[E_z(r = r_0) + \mathrm{i}\frac{c^2}{\omega}B(r = r_0)\int_{r_0}^r \frac{r_0}{r}\frac{\varkappa_{\mathrm{p}}^2(r')}{\varepsilon(r')}\,\mathrm{d}r'\right]$$

$$\times\left[\frac{r_0}{r}B(r = r_0) - \mathrm{i}\frac{\omega}{c^2}E_z(r = r_0)\int_{r_0}^r \frac{r'}{r}\varepsilon(r')\,\mathrm{d}r'\right]^{-1}. \tag{4.74}$$

The plasma impedance at the surface of r which separates the central part of the column from the resonance region can be obtained from (4.74) for r_0 either $r_0 = 0$ as

$$\mathcal{Z}(r) = \mathrm{i}\,\frac{\mu_0 c^2}{\omega}\,\frac{r}{\int_0^r r'\varepsilon(r')\,\mathrm{d}r'}\,, \tag{4.75}$$

or for $r_0 = R$ as

$$\mathcal{Z}(r) = \frac{r}{R}\mathcal{Z}(r = R) + \mathrm{i}\,\frac{\mu_0 c^2}{\omega}\,\frac{r}{R}\int_R^r \frac{R}{r'}\,\frac{\varkappa_\mathrm{p}^2(r')}{\varepsilon(r')}\,\mathrm{d}r'\,. \tag{4.76}$$

Here, we have used the fact that $B(r = 0) = 0$ for the azimuthally symmetric mode and that $\int_r^R r'\varepsilon(r')\,\mathrm{d}r' \to 0$. Equalizing (4.75) and (4.76) (where $\mathcal{Z}(r = R) \equiv \mathcal{Z}_\mathrm{d}$ as given by (4.66)) leads to the dispersion relation for SWs in a thin inhomogeneous plasma column:

$$\frac{2}{R\bar{\varepsilon}} + \frac{\varkappa_\mathrm{d}}{\varepsilon_\mathrm{d}}\,\frac{\mathrm{K}_0(\varkappa_\mathrm{d}R)}{\mathrm{K}_1(\varkappa_\mathrm{d}R)} + \int_{R-\Delta}^R \frac{\varkappa_\mathrm{p}^2(r)}{\varepsilon(r)}\,\mathrm{d}r = 0\,, \tag{4.77}$$

where $\bar{\varepsilon} = (2/R^2)\int_0^R r\varepsilon(r)\,\mathrm{d}r$ is the averaged (over the column cross-section) plasma permittivity. Equation (4.77) covers arbitrary plasma inhomogeneities of the waveguiding structure.

For quasi-static ($\omega \ll kc$) SW propagation along a thin inhomogeneous plasma column surrounded by a vacuum, the dispersion relation (4.77) yields the following results for the frequency ω and the time damping rate γ of SWs:

$$\omega = \left(\overline{\omega_\mathrm{p}^2}\,\frac{\beta R}{2}\,\frac{\mathrm{K}_0(\beta R)}{\mathrm{K}_1(\beta R)}\right)^{1/2}\,, \tag{4.78a}$$

$$\gamma = -\frac{\nu}{2} - \frac{\pi}{4}\beta^2 L_n^{(\mathrm{r})} R\,\frac{\overline{\omega_\mathrm{p}^2}}{\omega}\,, \tag{4.78b}$$

where the second term in (4.78b) accounts for the energy losses due to excitation of local plasmons.

The treatment of SW propagation along waveguides with cylindrical geometry and an inhomogeneous plasma is now complemented by considerations of smooth radial density profiles represented by Bessel functions (see (2.65c)):

$$n(r) = n_0 \mathrm{J}_0\left(\mu\frac{r}{R}\right)\,. \tag{4.79}$$

The phase diagrams for the azimuthally symmetric mode were calculated on the basis of numerical solutions of the wave equation (2.15), which in the plasma region reduces to

$$\frac{\mathrm{d}^2 E_z(r)}{\mathrm{d}r^2} + \left(\frac{1}{r} + \frac{k^2}{\varkappa_\mathrm{p}^2(r)}\,\frac{\mathrm{d}\ln[\varepsilon(r)]}{\mathrm{d}r}\right)\frac{\mathrm{d}E_z(r)}{\mathrm{d}r} - \varkappa_\mathrm{p}^2(r)E_z(r) = 0\,; \tag{4.80a}$$

$$E_r(r) = \mathrm{i}\,\frac{k}{\varkappa_\mathrm{p}^2(r)}\,\frac{\mathrm{d}E_z(r)}{\mathrm{d}r}\,. \tag{4.80b}$$

The behaviour of the propagation of the wave in the vacuum was described by explicit solutions of the corresponding equation via modified Bessel functions with the argument $\varkappa_\mathrm{v} r$.

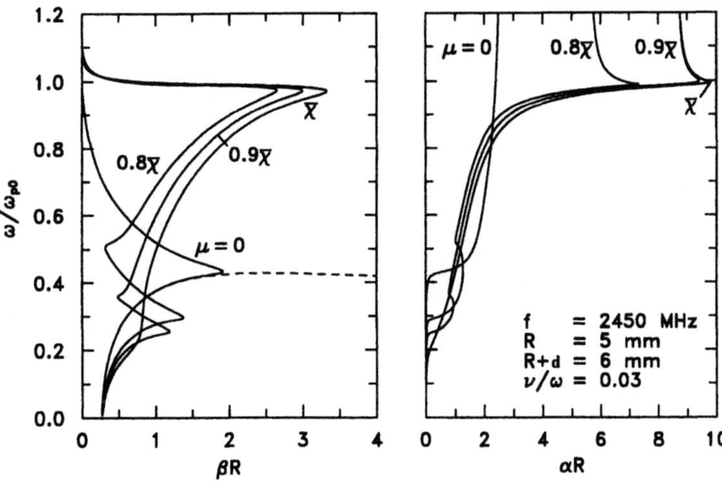

Figure 4.8. Characteristics (wave number β and space damping rate α) of SW propagation along an inhomogeneous plasma column of radius $R = 0.5$ cm surrounded by glass ($\varepsilon_d = 4.7$, thickness $d = 0.1$ cm) and vacuum for different values of $\mu = 0.8\bar{\chi}, 0.9\bar{\chi}, \bar{\chi}$ with $\bar{\chi} = 2.405$ (*solid curves*). A comparison with the dispersion behaviour in a cylindrical waveguide with a homogeneous collisional (*solid curve* for $\mu = 0$) and collisionless (*dashed curve*) plasma is shown. Normalization of the wave frequency ω to the plasma frequency ω_{p0} at the discharge axis; $f \equiv \omega/2\pi = 2.45$ GHz ([4.17], Fig. 10)

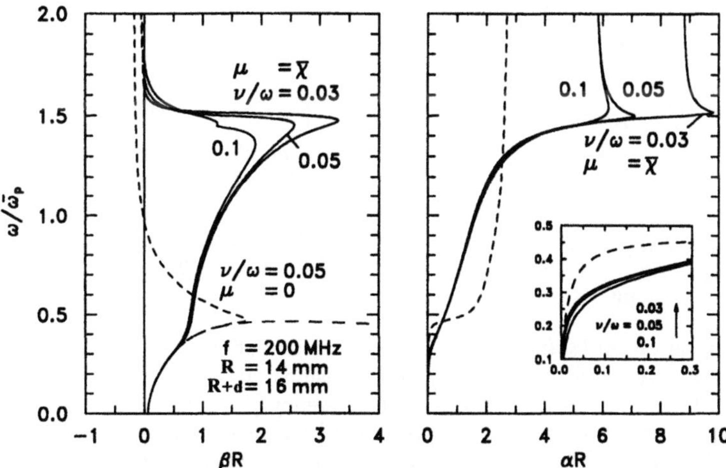

Figure 4.9. Influence of the collision frequency on the propagation behaviour of SWs in the case of maximum density inhomogeneity ($\mu = \bar{\chi} \equiv 2.405$) of the plasma column (*solid curves*) and comparison with propagation behaviour in homogeneous plasma ($\mu = 0$, *dotted curves*); the SW behaviour in a homogeneous collisionless plasma is also marked (*dashed curve*). Normalization of the wave frequency $\overline{\omega_p}$ defined by the density averaged over the cross-section; $f \equiv \omega/2\pi = 200$ MHz ([4.17], Fig. 11)

In all important aspects the effects of inhomogeneity, shown in Fig. 4.8, are found to be the same as for the slab configuration (Figs. 4.2–6, where the scaling ω/ω_{po} is the same). No basic changes of the propagation behaviour occur even when the presence of the dielectric glass tube enclosing the plasma ($\varepsilon_d = 4.7$, thickness $d = 1$ mm in Fig. 4.8) is taken into account. The inclusion of the dielectric tube condenses the curves, from the region of ω/ω_{po} $[0\ldots1/\sqrt{2}]$ to $[0\ldots1/(1+\varepsilon_d)^{1/2}]$, in the homogeneous collisionless case. Consequently, a lowering of the turn-around point of the β curve for $\mu = 0$ is noticeable. This is depicted for the case of a strong inhomogeneity $\mu = \bar{\chi} \equiv 2.405$ and a small σ with the scaling $\omega/\overline{\omega_p}$ in Fig. 4.9. The inset of Fig. 4.9 exhibits the contribution of the resonance absorption to the wave damping. For a strong inhomogeneity α is significantly larger than in the homogeneous case.

It should be mentioned that additional solutions always exist with larger $|\alpha|$, not considered here. Furthermore, the behaviour of the solutions with $\alpha \gg \beta$ for $\omega/\overline{\omega_p} \gtrsim 1$ included in Fig. 4.9 (in particular, for $\nu/\omega = 0.03$) is influenced by the fact that highly damped solutions are complex in the collisionless case.

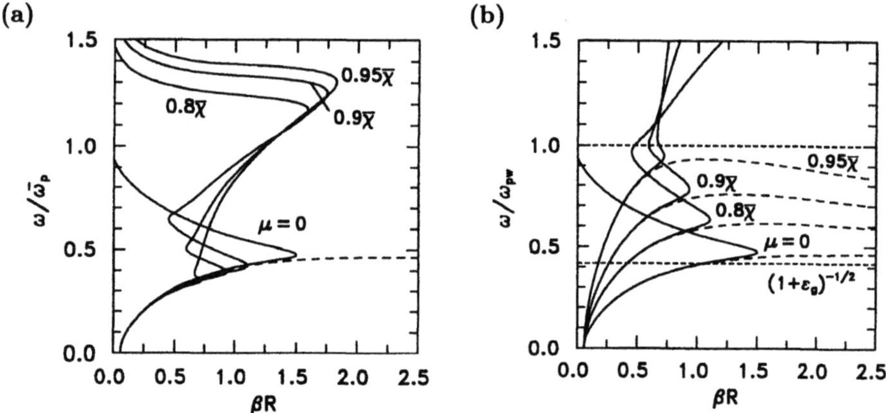

Figure 4.10. Presentation (*solid curves*) of the combined effects of radial inhomogeneity of a plasma (marked by μ values) and collisions ($\nu/\omega = 0.1$) on βR as a function of $\omega/\overline{\omega_p}$ (a) and of ω/ω_{p_w} (b); SW of frequency $f \equiv \omega/2\pi = 200$ MHz in a plasma column of radius $R = 1.4$ cm surrounded by a dielectric (with thickness $d = 0.2$ cm) and vacuum. The $\mu = 0$ curve shows a comparison with the case of homogeneous collisional plasmas. Comparison with the inhomogeneous collisonless plasma is also given (*dashed curves* in (b)), and the resonance in a homogeneous collisionless plasma is marked (*dotted lines* in (b) and *dashed curve* in (a) ([4.17], Fig. 12)

Finally, in Fig. 4.10a the basic behaviour of the phase diagram for different degrees of radial inhomogeneity is summarized, using the scaling βR versus $\omega/\overline{\omega_p}$. For different values of μ the second turn-around of βa with growing

μ shows up at lower $\omega/\overline{\omega_p}$ just when ω_p equals ω at the wall (ω_{p_w}). This is accentuated in the third manner of scaling with ω/ω_{p_w}, used in Fig. 4.10b (partially omitting the range of extreme attenuation). ν/ω is chosen to 0.1.

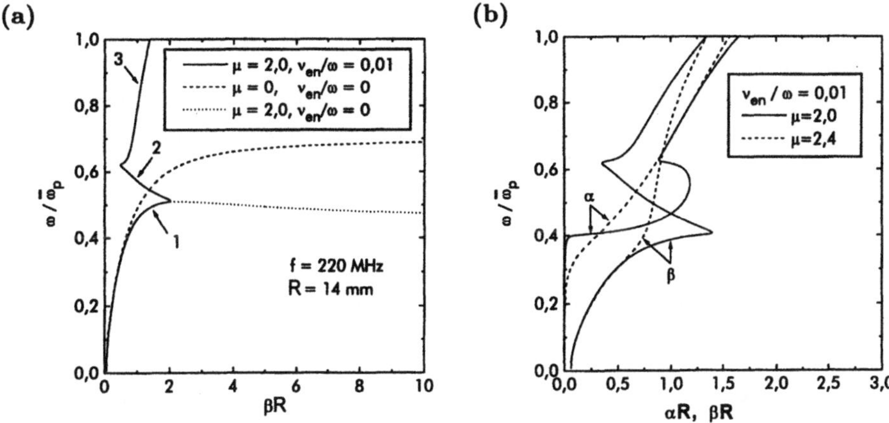

Figure 4.11. Phase diagrams of SWs of frequency $f \equiv \omega/2\pi = 220$ MHz in a plasma column of radius $R = 1.4$ cm surrounded by vacuum, demonstrating: **(a)** the branches (*solid curve*) of the surface mode (1), "evanescent" mode (2) and plasmon mode (3) in an inhomogeneous collisional plasma; a comparison with dispersion behaviour in collisionless inhomogeneous (*dotted curve*) and homogeneous (*dashed curve*) plasma is shown; **(b)** the extension of the plasmon branch over the entire phase diagram for $\mu = 2.4$ (*dashed curves* for β and α); a comparison with the case of $\mu = 2.0$ (*solid curves* for β and α) is shown ([4.39], Fig. 4.3)

A schematic summary of the features of the waves in a way related to the discussions in Sect. 3.2 but now accounting for the transverse inhomogeneity is presented in Fig. 4.11, with ν/ω smaller than in Fig. 4.10, no dielectric tube being present. The analogy in the behaviours shown in Figs. 3.2 and 4.11a (for the case of modest inhomogeneity $\mu = 2.0$) is evident. The section of the solid curve labelled (1) in Fig. 4.11a is the "usual" surface mode (region (a), Fig. 3.2, Sect. 3.2), which is shifted to a range of lower $\omega/\overline{\omega_p}$ values because of effects of weak inhomogeneity. Section (2) (Fig. 4.11a) is the "evanescent" mode (region (b) in Fig. 3.2, Sect. 3.2). Similarly to the case of a homogeneous plasma, the range of this mode is extended up to $\varepsilon(\omega) = 0$. However, in the case of inhomogeneous plasmas the $\omega/\overline{\omega_p}$ value at which the resonance $\omega_p(r_r = R) = \omega$ appears at the wall, for a given density profile, defines the end of section (2), determined by the zero value of ε there: $\varepsilon(r_r = R) = 0$. The absorption (Fig. 4.11b) in these two ranges – surface mode and "evanescent" mode – is determined by collisions (Fig. 3.2). The third mode (section (3) in Fig. 4.11a), which in the case of a homogeneous plasma (region (c) in Fig. 3.2, Sect. 3.2) was called the "radiative mode", existing for $\varepsilon(\omega) > 0$, is now a "plasmon mode" since it is characterized by $\varepsilon(r_r = R) > 0$ and the presence of a resonance, the position of which moves inside the plasma. The damping in this third range of the phase curve is quite different for homogeneous and inhomogeneous plasmas. Whereas in a homogeneous plasma the radiative mode is weakly damped, here – in an inhomogeneous plasma – the resonance absorption makes the plasmon mode a strongly damped wave. Owing to local mode interaction $\alpha \neq 0$ even as $\nu \to 0$. Section (3) in Fig. 4.11a starts with $(\omega/\overline{\omega_p})^2 = (\mu/2)\mathrm{J}_0(\mu)/\mathrm{J}_1(\mu)$. This value for the starting point is lowered by a factor of $\omega^2/(\omega^2 + \nu^2)$ if collisions are taken into account.

There is a subtle point to section (2) in Fig. 4.11a. Here the mode is not totally without propagation for $\nu \to 0$ after the turn-back at the maximum of $\omega/\overline{\omega_p}$, the appearance of which is caused here by density inhomogeneity. The mode is always complex, with both β and $\alpha \neq 0$ even for $\nu \to 0$. This is not quite so for the homogeneous plasmas treated in Chap. 3, where this region is strictly evanescent, i.e. with $\beta = 0$ and $\alpha \neq 0$ in the collisionless situation, unless there is another reason for the formation of a maximum, by field antisymmetry or a finite dielectric shield, as mentioned. The complex modes of section (2) with both β and $\alpha \neq 0$ for $\nu \to 0$ can be termed *reactive* modes as in Sect. 3.3.1, since again in any plane perpendicular to the discharge axis the power flux in the z direction in the *vacuum* is counterbalanced to zero by the backward flow in the *plasma*. This makes understandable that there is damping ($\alpha \neq 0$), though no absorption ($\nu = 0$). The actual turn-back point already before the maximum in Fig. 4.11 (section (1) and its continuation by the dotted curve) is determined by the strength of the collisions. Since the position of the maximum is largely determined by the degree of inhomogeneity, both collisions and inhomogeneity enter into the beginning of section (2). The horizontal position of the maximum is influenced usually to a lesser degree by a finite dielectric shield.

In both cases – homogeneous and weakly inhomogeneous plasmas – the two turn-around points on the phase diagrams, which determine the transitions between the different modes, have common features: the effect of collisions for the first turn-around and a zero value of the plasma permittivity for the second one. At a low degree of plasma inhomogeneity the resonance absorption appears too late on the phase diagram to influence the properties of the surface mode. A growing degree of inhomogeneity narrows the $\omega/\overline{\omega_p}$ range of this mode, which is accompanied by the formation of a maximum of $\omega/\overline{\omega_p}$ in the collisionless case. With increasing inhomogeneity (at fixed ν) the dispersion properties change smoothly over the entire diagram. The turn-around points shift to lower $\omega/\overline{\omega_p}$ values (Fig. 4.8) because the inhomogeneity narrows more and more the range of the surface mode and reduces the $\omega/\overline{\omega_p}$ value at which $\varepsilon(r_r = R) = 0$. As $\mu \to 2.4$, the plasmon mode extends down into the surface mode and the "evanescent" (reactive) mode completely disappears, as demonstrated in Figs. 4.8 and 11b. Now, when the plasma resonances are within the range of the surface mode, $\alpha < \beta$ and a weakly damped wave with resonant absorption present can be observed. At fixed μ, increased ν of course tends to smooth the structure of curves.

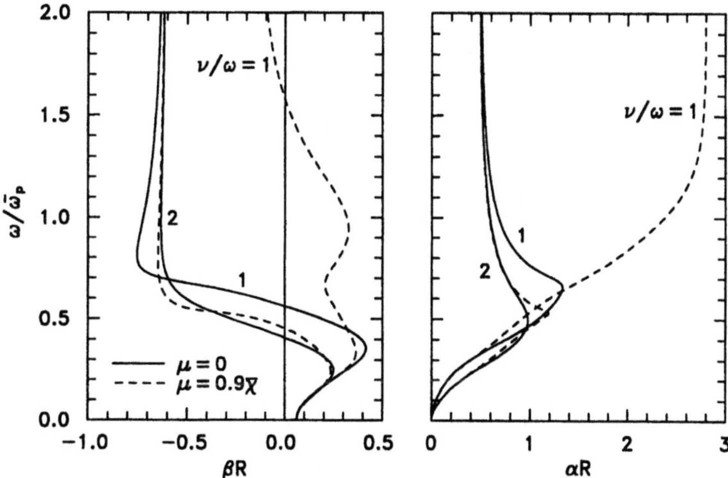

Figure 4.12. Effect of high values of ν/ω on the characteristics (wave number β on the *left* and space damping rate α on the *right*) of SWs of frequency $f \equiv \omega/2\pi = 200$ MHz in a cylindrical waveguide of an inhomogeneous plasma column of radius $R = 1.4$ cm surrounded by a dielectric of thickness $d = 0.2$ cm and vacuum (*dashed curves*). Comparison with the case of a homogneous plasma (*solid curves*) is shown ([4.17], Fig. 13)

The complementary situations with strong collisionality are displayed in Fig. 4.12. An increase of $\nu/\omega \geq 1$ has a relatively strong influence going beyond mere smoothening or shifting of bends in the structure of the curves of the phase diagrams. Nevertheless, the dependence on the degree of inhomogeneity (μ) remains recognizable. Since in these situations α approaches β relatively early, the interesting range of $\omega/\overline{\omega_\mathrm{p}}$ is restricted.

The inclusion of a metallic cylinder (of radius R_m) which terminates radially the vacuum region results in some weak corrections as a rule, unless R_m is close to $2R$. Figure 4.13 is an example of the influence of a metal cylinder on the phase diagram for the case of lower σ values, for which more significant changes might be expected than for higher σ. For smaller R_m the shifts indicated in Fig. 4.13 become more noticeable.

4.3 Inhomogeneity in the Longitudinal Direction

Plasma inhomogeneity in the longitudinal direction makes the problem of SW propagation essentially a two-dimensional one since the waveguide is in any case also inhomogeneous in the transverse direction. Besides, the SW field variation is also two-dimensional. The manner in which the geometrical-optics approach has been used in the previous section is applicable only to a one-dimensional problem. With the considerations below this method is extended to two-dimensional problems [4.18].

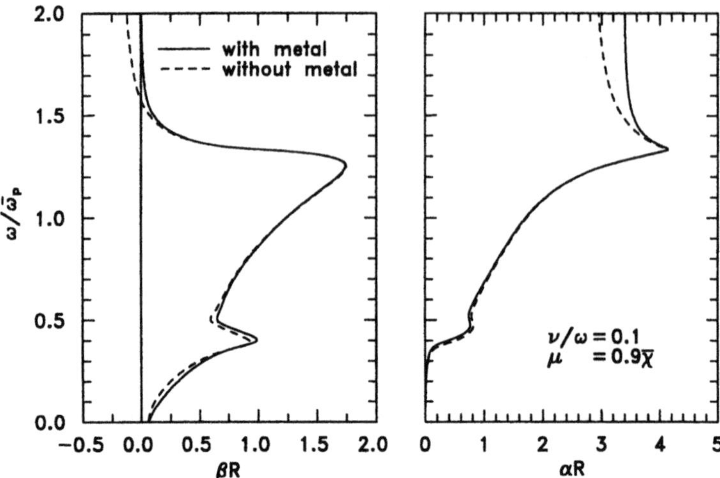

Figure 4.13. Comparison of the characteristics (wavenumber β on the *left* and space damping rate α on the *right*) of SWs of frequency $f \equiv \omega/2\pi = 200$ MHz in a cylindrical plasma–glass–vacuum waveguide system (*dashed curves*) and a plasma–glass–vacuum–metal system (*solid curves*); case of inhomogeneous collisional plasma. Plasma radius $R = 1.4$ cm, glass thickness $d = 0.2$ cm, radius of metal cylinder $R_{\mathrm{m}} = 4.5$ cm ([4.17], Fig. 14)

4.3.1 Single Interface

In order to analyse the basic effect of longitudinal plasma inhomogeneity on the behaviour of EM SWs, the simplest configuration of a single interface ($x = 0$) which separates two semi-spaces ($x > 0$ and $x < 0$) occupied by plasmas (Fig. 4.14) is considered first. The relative permittivities of the plasma semi-spaces, denoted by $\varepsilon_1 = \varepsilon_1(z)$ and $\varepsilon_2 = \varepsilon_2(z)$, respectively, are space-dependent functions of the z coordinate, which is in the direction of wave propagation.

The wave equations (see (2.16)) describing the distribution of the magnetic-field intensities $B_y^{(1,2)}(x,z) \equiv B^{(1,2)}$ in the two media ((1) and (2) in Fig. 4.14) are

$$\frac{\partial^2 B^{(1,2)}}{\partial z^2} - \frac{1}{\varepsilon_{1,2}(z)} \frac{\mathrm{d}\varepsilon_{1,2}(z)}{\mathrm{d}z} \frac{\partial B^{(1,2)}}{\partial z} + \frac{\partial^2 B^{(1,2)}}{\partial x^2} + \frac{\omega^2}{c^2} \varepsilon_{1,2}(z) B^{(1,2)}$$
$$= 0, \tag{4.81}$$

where $\varepsilon(z) = \varepsilon_{\mathrm{r}}(z) + \mathrm{i}\varepsilon_{\mathrm{i}}(z)$; the cold plasma permittivity (2.8) is now a function of z, through $\omega_{\mathrm{p}}(z)$.

With the assumption that the variations of the plasma permittivities are slow enough, the solutions of (4.81) are represented in the form

$$B^{(1)}(x,z) = B_0(z)\,\mathrm{e}^{-\varkappa_1(z)x}, \quad x > 0, \tag{4.82a}$$

$$B^{(2)}(x,z) = B_0(z)\,\mathrm{e}^{\varkappa_2(z)x}, \quad x < 0, \tag{4.82b}$$

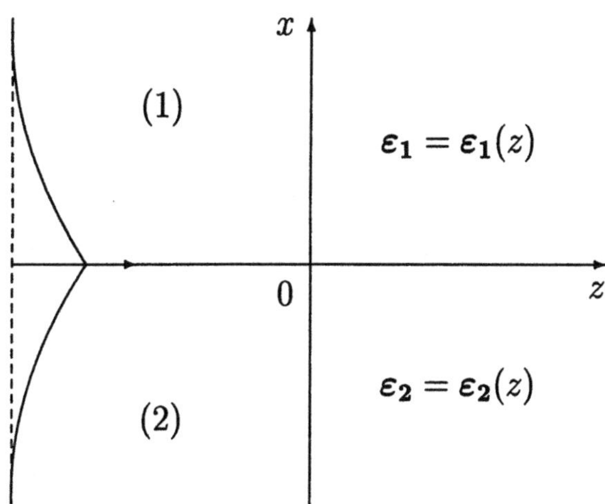

Figure 4.14. SW propagation along a single interface between two media of permittivities ε_1 and ε_2; the transverse field distribution of the SWs is schematically given on the *left* ([4.18], Fig. 2.1)

implying a local space dependence for the field distribution in the transverse (x) direction; $B_0(z)$ is the magnetic-field intensity at the interface ($x = 0$) and the functions $\varkappa_{1,2}(z)$ describing the variations of the magnetic field in the two media away from the interface are now unknown quantities. For surface modes, the real parts of $\varkappa_{1,2}(z)$ should be positive. The condition for continuity of the magnetic field across the boundary between the two media

$$\{B_0(z)\}|_{x=0} = 0 \tag{4.83}$$

has been taken into account in representing the solutions in the form (4.82). When (4.82) is inserted into (4.81), the following set of equations results:

$$\frac{\mathrm{d}^2 B_0}{\mathrm{d}z^2} - \frac{1}{\varepsilon_1(z)}\frac{\mathrm{d}\varepsilon_1(z)}{\mathrm{d}z}\frac{\mathrm{d}B_0}{\mathrm{d}z} + \left[\varkappa_1^2(z) + \frac{\omega^2}{c^2}\varepsilon_1(z)\right]B_0 = 0\,, \tag{4.84a}$$

$$\frac{\mathrm{d}^2 B_0}{\mathrm{d}z^2} - \frac{1}{\varepsilon_2(z)}\frac{\mathrm{d}^2\varepsilon_2(z)}{\mathrm{d}z^2}\frac{\mathrm{d}^2 B_0}{\mathrm{d}z^2} + \left[\varkappa_2^2(z) + \frac{\omega^2}{c^2}\varepsilon_2(z)\right]B_0 = 0\,. \tag{4.84b}$$

These equations have to be complemented by the boundary condition for continuity of the tangential electric-field components

$$E_z(x, z) = \frac{\mathrm{i}\,c^2}{\omega\varepsilon}\frac{\partial B_y(x, z)}{\partial x} \tag{4.85a}$$

at the interface $x = 0$, i.e.

$$\left\{\frac{1}{\varepsilon(z)}\frac{\partial B}{\partial x}\right\}\bigg|_{x=0} = 0. \tag{4.85b}$$

After introducing (4.82) into the boundary condition (4.85b), one gets the local dispersion relation

$$\mathcal{D} \equiv \frac{\varkappa_1(z)}{\varepsilon_1(z)} + \frac{\varkappa_2(z)}{\varepsilon_2(z)} = 0. \tag{4.86}$$

Equations (4.84) and (4.86) form a closed set of equations which determines the unknown functions $B_0(z)$, $\varkappa_1(z)$ and $\varkappa_2(z)$. This set can be reduced to an equation for the magnetic-field intensity $B_0(z)$ at the boundary by dividing (4.84a,b) by $\varepsilon_1^2(z)$ and $\varepsilon_2^2(z)$, respectively, and taking their difference. After using the relation

$$\frac{\varkappa_1^2(z)}{\varepsilon_1^2(z)} = \frac{\varkappa_2^2(z)}{\varepsilon_2^2(z)}, \tag{4.87}$$

which results directly from (4.86), the equation for $B_0(z)$ appears in the form

$$\frac{\mathrm{d}^2 B_0}{\mathrm{d}z^2} + \frac{1}{2}\left[\frac{\mathrm{d}}{\mathrm{d}z}\ln\left(\frac{1}{\varepsilon_1^2} - \frac{1}{\varepsilon_2^2}\right)\right]\frac{\mathrm{d}B_0}{\mathrm{d}z} + \frac{\omega^2}{c^2}\frac{\varepsilon_1\varepsilon_2}{\varepsilon_1+\varepsilon_2}B_0 = 0. \tag{4.88}$$

Its solution can be obtained by introducing the substitution

$$B_0(z) = \left(\frac{1}{\varepsilon_1^2} - \frac{1}{\varepsilon_2^2}\right)^{-1/4}\mathcal{V}(z). \tag{4.89}$$

In a sense, the second term in (4.88) is excluded by this procedure, and (4.88) is replaced by

$$\frac{\mathrm{d}^2\mathcal{V}}{\mathrm{d}z^2} + \left\{-\frac{1}{16}\left[\frac{\mathrm{d}}{\mathrm{d}z}\ln\left(\frac{1}{\varepsilon_1^2} - \frac{1}{\varepsilon_2^2}\right)\right]^2\right.$$
$$\left. -\frac{1}{4}\frac{\mathrm{d}^2}{\mathrm{d}z^2}\ln\left(\frac{1}{\varepsilon_1^2} - \frac{1}{\varepsilon_2^2}\right) + \frac{\omega^2}{c^2}\frac{\varepsilon_1\varepsilon_2}{\varepsilon_1+\varepsilon_2}\right\}\mathcal{V} = 0. \tag{4.90}$$

Because of the slow variation of the plasma permittivity assumed, the first two terms in the brackets can be neglected, and (4.90) simplifies to

$$\frac{\mathrm{d}^2\mathcal{V}}{\mathrm{d}z^2} + k_0^2(z)\mathcal{V} = 0, \tag{4.91}$$

where

$$k_0^2 = \frac{\omega^2}{c^2}\frac{\varepsilon_1\varepsilon_2}{\varepsilon_1+\varepsilon_2} \tag{4.92}$$

is the well-known dispersion law (3.10) governing SW propagation along a single interface between two media. In this case of space-dependent permittivities $\varepsilon_1(z)$ and $\varepsilon_2(z)$, (4.92) should be considered as a local law.

The geometrical-optics solution [4.31] of (4.91) is

$$\mathcal{V}(z) = \frac{1}{\sqrt{k_0(z)}}\exp\left(\pm\mathrm{i}\int^z k_0(z')\,\mathrm{d}z'\right). \tag{4.93}$$

Therefore, the magnetic-field intensity at the boundary, determined by (4.89) and (4.93), is

$$B_0(z) = \frac{1}{\sqrt{k_0}} \sqrt[4]{\frac{\varepsilon_1^2(z)\varepsilon_2^2(z)}{\varepsilon_2^2(z) - \varepsilon_1^2(z)}} \, \exp\left(\pm i \int^z k_0(z')\, dz' \right) . \qquad (4.94a)$$

With $B_0(z)$ known, $\varkappa_{1,2}(z)$ can be obtained straightforwardly from (4.84):

$$\varkappa_1^2(z) = k_0^2 - \frac{\omega^2}{c^2}\varepsilon_1 \pm i\,k_0\frac{d}{dz}\left(\ln\sqrt{1 - \frac{\varepsilon_1^2(z)}{\varepsilon_2^2(z)}} \right) , \qquad (4.94b)$$

$$\varkappa_2^2(z) = k_0^2 - \frac{\omega^2}{c^2}\varepsilon_2 \pm i\,k_0\frac{d}{dz}\left(\ln\sqrt{\frac{\varepsilon_2^2(z)}{\varepsilon_1^2(z)} - 1} \right) . \qquad (4.94c)$$

The last terms in (4.94b,c) are the geometrical-optics corrections.

In the case of propagation of an SW in a plasma semi-bounded by a vacuum ($\varepsilon_1 \equiv 1$, $\varepsilon_2(z) = \varepsilon(z)$), the geometrical-optics solution (4.94a) for the magnetic-field intensity at the interface ($x = 0$) is

$$B_0(z) = \frac{1}{\sqrt{k_0}} \left(1 - \frac{1}{\varepsilon^2(z)} \right)^{-1/4} \exp\left(\pm i \int^z k_0(z')\, dz' \right) . \qquad (4.95)$$

Correspondingly, the transverse component of the electric field in the plasma

$$E_x(x, z) = -i\,\frac{c^2}{\omega\varepsilon(z)}\,\frac{\partial B_y(x, z)}{\partial z} \qquad (4.96)$$

and the longitudinal component (4.85a) at $x = 0$ are

$$E_{x_0}(z) = \mp\frac{cB_0(z)}{\sqrt{\varepsilon(z)[\varepsilon(z) + 1]}} , \qquad (4.97a)$$

$$E_{z_0}(z) = -i\,c\sqrt{\frac{-1}{\varepsilon(z) + 1}}\,B_0(z) . \qquad (4.97b)$$

The geometrical-optics solutions are valid when the variations of the wave number $k_0(z)$ are slow enough, and their applicability requires fulfilment of the inequality

$$\left| \frac{dk_0^{-1}(z)}{dz} \right| \ll 1. \qquad (4.98a)$$

With (4.92), the requirement (4.98a) becomes

$$\frac{c}{2\omega}\left| \frac{d(\varepsilon_1 + \varepsilon_2)/dz}{\sqrt{\varepsilon_1\varepsilon_2(\varepsilon_1 + \varepsilon_2)}} \right| \ll 1. \qquad (4.98b)$$

Thus, the geometrical-optics approach could be invalid towards the point where $\varepsilon_1(z) + \varepsilon_2(z) = 0$, which corresponds to the quasi-static resonance (3.1) of the SWs. As mentioned before, for cases of SW-produced discharges, the end of the discharge approaches the region of the quasi-static resonance. The study of this region requires special attention, since, on the one hand, the wave still propagates and, on the other hand, the applicability of the geometrical-optics approach may be limited.

In the narrow region of the quasi-static resonance the distribution of the plasma density can be approximated by the linear function

$$n_{1,2}(z) = n_{1,2}(z = 0)\left(1 - \frac{z}{L_N^{(1,2)}}\right). \tag{4.99}$$

The position $z = 0$ is defined by the relation

$$\varepsilon_{1r}(z = 0) + \varepsilon_{2r}(z = 0) = 0, \tag{4.100}$$

and

$$L_N^{(1,2)} = \left(\frac{\mathrm{d}\ln n_{1,2}(z)}{\mathrm{d}z}\right)^{-1}\Bigg|_{z=0} \tag{}$$

denotes the scale lengths which characterize the longitudinal inhomogeneity of the plasma densities in this region. The corresponding expressions for the real parts of the permittivities are

$$\varepsilon_{1,2r}(z) = \varepsilon_{1,2r}(z = 0) + [1 - \varepsilon_{1,2r}(z = 0)]\frac{z}{L_N^{(1,2)}}. \tag{4.101}$$

Therefore, in the quasi-static-resonance region and with a low collison frequency ν, the quantity $\varepsilon_1 + \varepsilon_2$ can be represented as

$$\varepsilon_1(z) + \varepsilon_2(z) = \frac{z}{L_N} + 2\mathrm{i}\,\frac{\nu}{\omega}, \tag{4.102a}$$

where (4.100) has been used and L_N is defined by

$$L_N = \left(\frac{1 - \varepsilon_{1r}(z = 0)}{L_N^{(1)}} + \frac{1 - \varepsilon_{2r}(z = 0)}{L_N^{(2)}}\right)^{-1} \tag{4.102b}$$

(For simplicity, the same values of the electron–neutral collision frequencies are assumed for the two media.) The width Δ of the region of the quasi-static resonance is characterized by the length L_N of the plasma inhomogeneity and the electron collision frequency ν:

$$\Delta = 2\frac{\nu}{\omega}L_N. \tag{4.103}$$

Inside the resonance region, (4.92) results in equal values for the real (β) and imaginary (α) parts of the wave number k_0:

$$k_0 = \frac{\omega}{c}\,\frac{1}{\sqrt{-2\mathrm{i}\,(\nu/\omega)}} \equiv \beta + \mathrm{i}\,\alpha, \tag{4.104}$$

with

$$\beta \equiv \alpha = \frac{\omega}{2c}\sqrt{\frac{\omega}{\nu}}$$

when a plasma–vacuum interface is assumed.

Comparing the width Δ of the resonance region (4.103) with the local wavelength β^{-1} (4.104) leads to two inequalities which specify the conditions of comparatively high and low collisions, respectively, as follows.

(1) If the width Δ of the resonance exceeds the local wavelength β^{-1}, i.e. if $\Delta \gg \beta^{-1}$, which reduces to

$$\frac{\omega}{c}L_N \gg \sqrt{\frac{\omega}{\nu}} \, , \tag{4.105}$$

holds, the SWs are totally absorbed in passing into the region of resonance. Under such conditions (4.105) of high enough dissipation, the requirements for the application of the geometrical-optics approximation are also well fulfilled (it can be shown that the requirement (4.98) reduces to $\Delta \gg \beta^{-1}$). Moreover, the replacement of (4.90) by (4.91) is well justified. The range where (4.105) is reasonably fulfilled, as well as the transition region when the ratio $(\omega L_N/c)/\sqrt{\omega/\nu}$ decreases and even approaches unity, has been studied numerically [4.37,38] (Sect. 4.3.5). The results obtained show that even when (4.105) is only weakly fulfilled, i.e. $(\omega L_N/c) \gtrsim \sqrt{\omega/\nu}$, the trends and dependences predicted by geometrical optics may still be considered correct. However, near the resonance quantitative corrections should be expected.

(2) At comparatively low dissipation

$$\frac{\omega}{c}L_N \ll \sqrt{\frac{\omega}{\nu}} \, , \tag{4.106}$$

the width Δ of the quasi-static resonance is too narrow to lead to total absorption of an SW and the wave can penetrate beyond the turning point to positive z over a distance of the order of the vacuum wavelength c/ω. The local form (4.82) of the x dependence could still be used for a qualitative modelling – on the basis of (4.91) – of the wave behaviour in the resonance region. In this case, such a representation of the transverse field distribution is a good enough approximation in the vicinity of the boundary $x = 0$.

The geometrical-optics solution of (4.91) before the turning point corresponds to a propagating SW. Its form is

$$\mathcal{V}(z) = \frac{1}{\sqrt{k_0}} \exp\left[\mathrm{i}\int^z k_0(z')\,\mathrm{d}z'\right] , \quad z < 0 . \tag{4.107}$$

The solution behind the turning point, where $k_0^2 < 0$ ($k_0 = \mathrm{i}\,\varkappa_0$), describes a field decay:

$$\mathcal{V}(z) = \frac{1}{\sqrt{k_0}} \exp\left[-\int^z \varkappa_0(z')\,\mathrm{d}z'\right] , \quad z > 0 . \tag{4.108}$$

In the vicinity of the singular turning point $z = 0$, (4.102) reduces (4.91) to

$$\frac{d^2 \mathcal{V}}{dz^2} - \frac{\mu_n^2}{z + 2i\, L_N(\nu/\omega)}\mathcal{V} = 0, \tag{4.109}$$

where $\mu_n^2 = \omega^2 L_N/c^2$. The small $(\nu \to 0)$ imaginary contribution to the denominator of the second term in (4.109) determines how the solution passes the singular point. The solutions of (4.109) are

$$\mathcal{V}(z) = -i\,\mathcal{V}_0 \pi \mu_n \sqrt{-z}\, \mathrm{H}_1^{(2)}(2\mu_n \sqrt{-z}), \qquad z < 0, \tag{4.110a}$$

$$\mathcal{V}(z) = -\mathcal{V}_0 \pi \mu_n \sqrt{z}\, \mathrm{H}_1^{(1)}(2i\,\mu_n \sqrt{z}), \qquad z > 0, \tag{4.110b}$$

where $\mathrm{H}_1^{(1,2)}(\xi)$ are the first and second types of the Hankel functions of the first order, respectively, and $\mathcal{V}_0$ is introduced as the value of the $\mathcal{V}(z)$ function at the turning point $(z = 0)$. This is obtained by matching the limiting forms for small arguments of the $\mathrm{H}_1^{(1,2)}(\xi)$ functions. The asymptotic expansions at large argument of the functions $\mathrm{H}_1^{(2)}(\xi)$ and $\mathrm{H}_1^{(1)}(\xi)$ transform (4.110a,b) into the geometrical-optics solutions (4.107) and (4.108), respectively. Thus, the region of validity of the geometrical-optics solution is $|z| > (2\mu_n)^{-2} \equiv c^2/4\omega^2 L_N$. For the collisionless case $(\nu \to 0)$ the solutions (4.110) show that there is no reflection of the SW and that it totally transforms into a bulk EM wave which radiates towards the region $x > 0$ of the transparent plasma $(\varepsilon_1 > 0)$.

With the distribution (4.107), (4.108) and (4.110) of the magnetic-field intensity $B_0(z)$ at the interface $(x = 0)$ between the two media being known, the field distribution in the whole space can be deduced. For example, in the case of a longitudinally inhomogeneous plasma (occupying the semi-space $x < 0$) bounded by a vacuum, the field distribution in the vacuum semi-space $(x > 0)$ is

$$B(x, z) = \int_{-\infty}^{\infty} dz'\, B_0(z')\Omega(z - z', x), \tag{4.111}$$

where the kernel

$$\Omega(z, x) = \int_{-\infty}^{\infty} \frac{dk}{2\pi} \exp\left(i\,kz - \sqrt{k^2 - \frac{\omega^2}{c^2}}\,x\right)$$

satisfies (4.81) and the path of integration around the branch points $k = \pm\omega/c$ is chosen according to the condition of radiation $(\omega = \omega + i\gamma,\ \gamma \to +0)$. Such a method for determining the SW field corresponds to the Huygens principle in problems on diffraction of EM waves [4.44].

4.3.2 Layered Structures

Now attention is paid to SW propagation along a layered structure consisting of two plasma media with different densities, i.e. a plasma slab of permittivity $\varepsilon_2(z)$ surrounded by a plasma with permittivity $\varepsilon_1(z)$ (Fig. 4.15). The

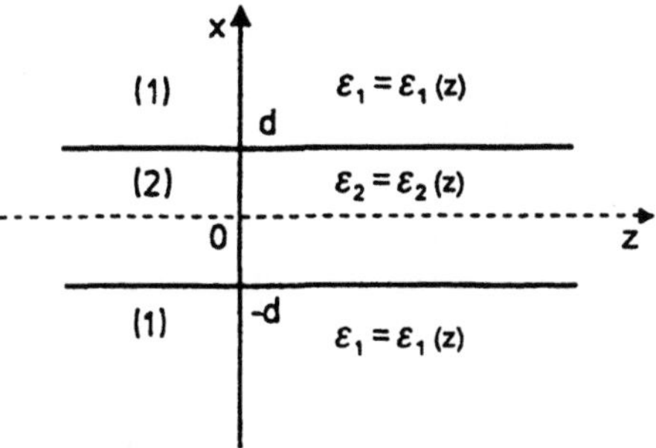

Figure 4.15. SW propagation along a layered structure ([4.18], Fig. 2.2)

equations (4.81) are those which should be solved to obtain the distribution of the magnetic-field intensities in the two media.

By analogy to the case of a semi-bounded plasma, the solutions for the magnetic field in the spaces denoted by (1) in Fig. 4.15 are of the form

$$B^{(1)}(x, z) = B_0(d, z) \exp[-\varkappa_1(z)(x - d)], \quad x > d, \tag{4.112a}$$

$$B^{(1)}(x, z) = B_0(-d, z) \exp[\varkappa_1(z)(x + d)], \quad x < -d, \tag{4.112b}$$

where $B_0(\pm d, z)$ are the magnetic-field intensities at the boundaries $x = \pm d$. The solutions for the field distributions of the symmetric $[B_0(d, z) = -B_0(-d, z)]$ and antisymmetric $[B_0(d, z) = B_0(-d, z)]$ modes in medium (2) $(-d < x < d)$ are sought in the following forms:

- symmetric mode:

$$B^{(2)}(x, z) = B_0(d, z) \frac{\sinh[\varkappa_2(z)x]}{\sinh[\varkappa_2(z)d]} \tag{4.113a}$$

- antisymmetric mode:

$$B^{(2)}(x, z) = B_0(d, z) \frac{\cosh[\varkappa_2(z)x]}{\cosh[\varkappa_2(z)d]}. \tag{4.113b}$$

The equations for the magnetic-field intensities $B_0(d, z)$ at the interface $x = d$ obtained by following the same procedure as applied to the case of SW propagation along a single interface (Sect. 4.3.1) exactly coincide with (4.84a,b). The condition (4.85b) for continuity of the tangential electric-field component at the interfaces $x = \pm d$ leads to the following relations:

- symmetric mode:

$$\mathcal{D} \equiv \frac{\varkappa_2(z)}{\varepsilon_2(z)} \coth[\varkappa_2(z)d] + \frac{\varkappa_1(z)}{\varepsilon_1(z)} = 0 \tag{4.114a}$$

- antisymmetric mode:

$$\mathcal{D} \equiv \frac{\varkappa_2(z)}{\varepsilon_2(z)} \tanh[\varkappa_2(z)d] + \frac{\varkappa_1(z)}{\varepsilon_1(z)} = 0. \tag{4.114b}$$

Thus, (4.84a,b) and (4.114a,b) constitute the closed set of equations, which determines the three unknown functions $\varkappa_{1,2}(z)$ and $B_0(z)$ and governs the propagation of the symmetric or antisymmetric EM SWs along the slab structure.

As an example, the propagation of the symmetric mode along a thin slab $|\varkappa_2(z)d| \ll 1$ is considered below. In this case the function $\varkappa_1(z)$ is straightforwardly defined by (4.114a):

$$\varkappa_1(z) = -\frac{\varepsilon_1(z)}{d\varepsilon_2(z)} \, .$$
(4.115)

After inserting (4.115) into (4.84a), one obtains the equation which governs the quantity $B_0(z)$:

$$\frac{\mathrm{d}^2 B_0}{\mathrm{d}z^2} - \frac{1}{\varepsilon_1(z)}\frac{\mathrm{d}\varepsilon_1(z)}{\mathrm{d}z}\frac{\mathrm{d}B_0}{\mathrm{d}z} + \varepsilon_1(z)\left(\frac{\omega^2}{c^2} + \frac{\varepsilon_1(z)}{d^2\varepsilon_2^2(z)}\right)B_0 = 0 \, .$$
(4.116)

Equation (4.116) does not exhibit a turning point in the region of validity of the thin-slab approximation, and its geometrical-optics solution in the case of $\varepsilon_{2\mathrm{r}}(z) < 0$, $\varepsilon_{1\mathrm{r}}(z) > 0$ is

$$B_0(z) = \sqrt{\frac{\varepsilon_1(z)}{k_0(z)}}\exp\left[\mathrm{i}\int^z k_0(z')\,\mathrm{d}z'\right] \, ,$$
(4.117a)

where

$$k_0^2(z) = \frac{\omega^2}{c^2}\varepsilon_1(z)\left(1 + \frac{c^2\varepsilon_1(z)}{\omega^2 d^2\varepsilon_2^2(z)}\right) \, .$$
(4.117b)

Replacing $B_0(z)$, as given by (4.117a), in (4.84b) yields the third unknown quantity, the function $\varkappa_2(z)$, which in the zero-order approximation is

$$\varkappa_{2_0}(z) = \sqrt{\left(\frac{\varepsilon_1(z)}{d\varepsilon_2(z)}\right)^2 + \frac{\omega^2}{c^2}[\varepsilon_1(z) - \varepsilon_2(z)]} \, .$$
(4.118)

Since the quasi-static resonance regions of both the symmetric and the antisymmetric modes of a slab structure are in the thick-slab limits, their behaviour at the resonance has been described in Sect. 4.3.1.

4.3.3 Cylindrical Geometry

The configuration to be considered now is that of a cylindrical waveguide (Fig. 4.16). A plasma column with permittivity $\varepsilon_2(z)$ is surrounded by a plasma medium of a different permittivity $\varepsilon_1(z)$. The density inhomogeneity of the two media is in the axial direction. The propagation of the azimuthally symmetric mode of the structure will be studied.

The equation which describes the variation of the magnetic-field intensities $B_\varphi^{(1,2)}(r, z) \equiv B^{(1,2)}$ in the two media (denoted by (1) and (2) in Fig. 4.16) is

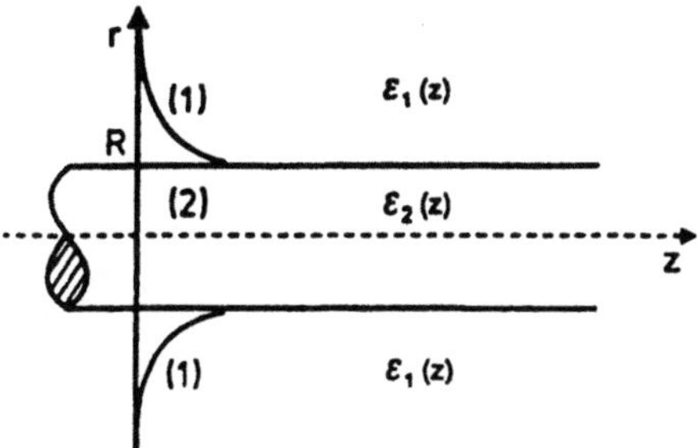

Figure 4.16. SW propagation along a cylindrical waveguide; the decay of the SW field in region (1) is schematically indicated on the *left* ([4.18], Fig. 2.3)

$$\frac{\partial^2 B^{(1,2)}}{\partial z^2} - \frac{1}{\varepsilon_{1,2}(z)} \frac{\mathrm{d}\varepsilon_{1,2}(z)}{\mathrm{d}z} \frac{\partial B^{(1,2)}}{\partial z} + \frac{1}{r} \frac{\partial}{\partial r}\left(r \frac{\partial B^{(1,2)}}{\partial r}\right)$$
$$+ \left[\frac{\omega^2}{c^2}\varepsilon_{1,2}(z) - \frac{1}{r^2}\right] B^{(1,2)} = 0 \,. \quad (4.119)$$

Its solution in the two media is sought in the form

$$B^{(2)}(r,z) = B_0(R,z)\frac{\mathrm{I}_1[\varkappa_2(z)r]}{\mathrm{I}_1[\varkappa_2(z)R]}, \qquad r < R, \tag{4.120a}$$

$$B^{(1)}(r,z) = B_0(R,z)\frac{\mathrm{K}_1[\varkappa_1(z)r]}{\mathrm{K}_1[\varkappa_1(z)R]}, \qquad r > R, \tag{4.120b}$$

where $B_0(z)$ is the magnetic-field intensity at the interface. Insertion of (4.120) into (4.119) shows that (4.84a,b) are again the equations which describe the behaviour of the magnetic-field intensity $B_0(z)$ at the boundary $r = R$. The condition for continuity of the tangential electric-field component

$$E_z(r,z) = \mathrm{i}\,\frac{c^2}{\omega\varepsilon}\frac{1}{r}\frac{\partial}{\partial r}[rB_\varphi(r,z)] \tag{4.121}$$

at the interface $r = R$,

$$\frac{1}{\varepsilon(z)}\frac{\partial}{\partial r}(rB)\bigg|_{r=R} = 0 \,, \tag{4.122}$$

results in

$$\mathcal{D} \equiv \frac{\varkappa_2(z)}{\varepsilon_2(z)}\frac{\mathrm{I}_0[\varkappa_2(z)R]}{\mathrm{I}_1[\varkappa_2(z)R]} + \frac{\varkappa_1(z)}{\varepsilon_1(z)}\frac{\mathrm{K}_0[\varkappa_1(z)R]}{\mathrm{K}_1[\varkappa_1(z)R]} = 0 \,. \tag{4.123}$$

Equation (4.123) together with (4.84) constitutes the closed set of equations for the functions $\varkappa_1(z)$, $\varkappa_2(z)$ and $B_0(z)$.

The limiting case of SW propagation in the thin-cylinder approximation ($\varkappa_{1,2}R \ll 1$) is taken as an example for applying the procedure. With $\varepsilon_{1\mathrm{r}} > 0$, $\varepsilon_{2\mathrm{r}} < 0$, (4.123) reduces to

$$\frac{2\varepsilon_1(z)}{\varepsilon_2(z)} + \varkappa_1^2(z)R^2|\ln \varkappa_1(z)R| = 0. \tag{4.124}$$

Within the limits of logarithmic accuracy the solution is

$$\varkappa_1(z) = \frac{2\sqrt{\varepsilon_1(z)}}{R\sqrt{-\varepsilon_2(z)\ln[-\varepsilon_2(z)/4\varepsilon_1(z)]}} \, . \tag{4.125}$$

The equation which governs the behaviour of $B_0(z)$ can be obtained by inserting (4.125) into (4.84a):

$$\frac{d^2 B_0}{dz^2} - \frac{1}{\varepsilon_1(z)} \frac{d\varepsilon_1(z)}{dz} \frac{dB_0}{dz}$$

$$+ \left\{ \frac{-4\varepsilon_1(z)}{R^2\varepsilon_2(z)\ln\left[-\varepsilon_2(z)/4\varepsilon_1(z)\right]} + \frac{\omega^2}{c^2}\varepsilon_1(z) \right\} B_0 = 0 \, . \tag{4.126}$$

Since this equation is in a thin-cylinder approximation, it has no turning point. Its geometrical-optics solution has the form

$$B_0(z) = \sqrt{\frac{\varepsilon_1(z)}{k_0(z)}} \exp\left(i \int^z k_0(z')\,dz' \right) , \tag{4.127}$$

where

$$k_0^2(z) = \frac{-4\varepsilon_1(z)}{R^2\varepsilon_2(z)\ln[-\varepsilon_2(z)/4\varepsilon_1(z)]} + \frac{\omega^2}{c^2}\varepsilon_1(z) \, .$$

The function $\varkappa_2(z)$ can be obtained directly from (4.84b) after inserting (4.127) there. Its zero-order approximation is

$$\varkappa_{2_0}(z) = \sqrt{k_0^2 - \frac{\omega^2}{c^2}\varepsilon_2(z)} \, . \tag{4.128}$$

For a thick cylinder $(\varkappa_{1,2}(z)R \gg 1)$, (4.123) transforms into (4.86), and thus the problem reduces to the case of SW propagation along a single interface. This is also the limit where the resonance behaviour of the wave field is pronounced.

4.3.4 Generalized Procedure for Obtaining the Geometrical-Optics Solutions

In addition to describing the problem in the framework of the geometrical-optics approach, the equations for $B_0(z)$ provide a possibility for modelling qualitatively the behaviour of SWs in the region around the turning point (Sect. 4.3.1). The results given there for the SW resonance region are also valid for the other configurations, since in all the cases the resonance point is the same.

Attention is now paid again to the geometrical-optics solutions. In fact, they can be obtained in a simpler manner which allows generalization to different geometries.

The equations (4.84) for the magnetic-field intensity $B_0(z)$ at the interface between the two media are independent of the specific details of the geometrical configuration. For all cases their geometrical-optics solution is of the form

$$B_0(z) = \sqrt{\frac{\varepsilon_{1,2}(z)}{k_{1,2}(z)}}\, \exp\left[\mathrm{i} \int^z k_{1,2}(z')\,\mathrm{d}z'\right]\,, \tag{4.129}$$

with

$$k_{1,2}^2(z) = \varkappa_{1,2}^2(z) + \frac{\omega^2}{c^2}\varepsilon_{1,2}(z)\,. \tag{4.130}$$

The condition for compatibility of (4.84) is

$$k_1(z) - \frac{\mathrm{i}}{2}\frac{\mathrm{d}}{\mathrm{d}z}\ln\left(\frac{\varepsilon_1(z)}{k_1(z)}\right) = k_2(z) - \frac{\mathrm{i}}{2}\frac{\mathrm{d}}{\mathrm{d}z}\ln\left(\frac{\varepsilon_2(z)}{k_2(z)}\right)\,. \tag{4.131}$$

This relation, together with one of (4.86), (4.114) or (4.123), chosen for the appropiate geometrical configuration, constitutes the closed set of two equations which determines the quantities $k_{1,2}(z)$.

Accounting for the slow variation (on the scale of the wavelength) of the permittivities of the media in the longitudinal direction, a perturbation scheme can be applied to determine the functions $k_{1,2}(z)$:

$$k_{1,2}(z) = k_0(z) + \delta k_{1,2}(z)\,. \tag{4.132}$$

In the zero-order approximation (4.131) leads to

$$k_1(z) = k_2(z) \equiv k_0(z)\,. \tag{4.133}$$

Thus, the local wave numbers determined by the SW dispersion relations are the zero-order approximation to the quantities $k_{1,2}(z)$.

The first-order approximation of (4.131) gives

$$\delta k_1(z) = \delta k_2(z) - \frac{\mathrm{i}}{2}\frac{\mathrm{d}}{\mathrm{d}z}\ln\left(\frac{\varepsilon_2(z)}{\varepsilon_1(z)}\right)\,. \tag{4.134}$$

The second equation which relates $\delta k_1(z)$ and $\delta k_2(z)$, i.e.

$$\delta k_1 \frac{\partial \mathcal{D}}{\partial \varkappa_1}\frac{\partial \varkappa_1}{\partial k_1} = -\delta k_2 \frac{\partial \mathcal{D}}{\partial \varkappa_2}\frac{\partial \varkappa_2}{\partial k_2}\,, \tag{4.135}$$

is obtained from the SW dispersion law $\mathcal{D} = 0$. With

$$\frac{\partial \varkappa_{1,2}(z)}{\partial k_{1,2}} = \frac{k_0}{\sqrt{k_0^2 - (\omega^2/c^2)\varepsilon_{1,2}(z)}} \tag{4.136}$$

obtained from (4.130), the solutions of (4.134) and (4.135) are

$$\delta k_1(z) = -\frac{\mathrm{i}}{2}\frac{1}{1 + G(z)}\frac{\mathrm{d}}{\mathrm{d}z}\left[\ln\left(\frac{\varepsilon_2(z)}{\varepsilon_1(z)}\right)\right]\,, \tag{4.137a}$$

$$\delta k_2(z) = \frac{\mathrm{i}}{2}\frac{G(z)}{1 + G(z)}\frac{\mathrm{d}}{\mathrm{d}z}\left[\ln\left(\frac{\varepsilon_2(z)}{\varepsilon_1(z)}\right)\right]\,, \tag{4.137b}$$

where

$$G(z) = \frac{\varkappa_{2_0}}{\varkappa_{1_0}} \frac{\left(\dfrac{\partial D}{\partial \varkappa_1}\right)\bigg|_{\varkappa_{1_0}}}{\left(\dfrac{\partial D}{\partial \varkappa_2}\right)\bigg|_{\varkappa_{2_0}}} \tag{4.138}$$

and

$$\varkappa_{(1,2)_0}(z) = \sqrt{k_0^2 - \frac{\omega^2}{c^2}\varepsilon_{1,2}(z)}\,. \tag{4.139}$$

Therefore, the solution for the magnetic field (4.129) satisfying the condition (4.83) is

$$B_0(z) = \sqrt{\frac{\varepsilon_1(z)}{k_0(z)}}\, \exp\left[i \int^z \delta k_1(z')\,\mathrm{d}z'\right] \exp\left[i \int^z k_0(z')\,\mathrm{d}z'\right]$$

$$\equiv \sqrt{\frac{\varepsilon_2(z)}{k_0(z)}}\, \exp\left[i \int^z \delta k_2(z')\,\mathrm{d}z'\right] \exp\left[i \int^z k_0(z')\,\mathrm{d}z'\right]\,. \tag{4.140}$$

The functions $\varkappa_{1,2}(z)$ which describe the field distribution in the transverse direction can be obtained from (4.130) by using (4.132) and (4.137).

Now the procedure will be applied to different geometrical configurations.

SW Propagation Along a Single Interface Between Two Media. The solution of the dispersion relation (4.86) in the zero-order approximation gives the local wave number

$$k_0 = \frac{\omega}{c}\sqrt{\frac{\varepsilon_1(z)\varepsilon_2(z)}{\varepsilon_1(z) + \varepsilon_2(z)}}\,. \tag{4.141}$$

With the $G(z)$ function (4.138)

$$G(z) = -\frac{\varkappa_{2_0}^2(z)}{\varkappa_{1_0}^2(z)} \equiv -\frac{\varepsilon_2^2(z)}{\varepsilon_1^2(z)}\,, \tag{4.142}$$

obtained from (4.86), (4.137) for the quantities $\delta k_{1,2}$ reduces to

$$\delta k_1(z) = \frac{i}{2}\frac{\varkappa_{1_0}^2}{\varkappa_{2_0}^2 - \varkappa_{1_0}^2}\frac{\mathrm{d}}{\mathrm{d}z}\ln\left(\frac{\varepsilon_2}{\varepsilon_1}\right) \equiv \frac{i}{2}\frac{\varepsilon_1^2}{\varepsilon_2^2 - \varepsilon_1^2}\frac{\mathrm{d}}{\mathrm{d}z}\ln\left(\frac{\varepsilon_2}{\varepsilon_1}\right)\,, \tag{4.143a}$$

$$\delta k_2(z) = \frac{i}{2}\frac{\varkappa_{2_0}^2}{\varkappa_{2_0}^2 - \varkappa_{1_0}^2}\frac{\mathrm{d}}{\mathrm{d}z}\ln\left(\frac{\varepsilon_2}{\varepsilon_1}\right) \equiv \frac{i}{2}\frac{\varepsilon_2^2}{\varepsilon_2^2 - \varepsilon_1^2}\frac{\mathrm{d}}{\mathrm{d}z}\ln\left(\frac{\varepsilon_2}{\varepsilon_1}\right)\,. \tag{4.143b}$$

The result for the magnetic-field intensity $B_0(z)$ at the interface obtained from (4.140) and (4.143) is as given by (4.94a).

SW Propagation Along Layered Structures. The dispersion relations of the symmetric and antisymmetric modes, which correspond to the structure considered in Sect. 4.3.2 (Fig. 4.15), are represented by (4.114a,b), respectively. Thus, the $G(z)$ functions which determine the quantities $\delta k_{1,2}(z)$ are

$$G_{\left\{ {s \atop a} \right\}}(z) = -\frac{\varkappa_{2_0}^2(z)}{\varkappa_{1_0}^2(z)} K_{\left\{ {s \atop a} \right\}}(z) , \tag{4.144a}$$

where

$$K_{\left\{ {s \atop a} \right\}}(z) = \left[1 \mp \frac{2\varkappa_{2_0}(z)d}{\sinh\left[2\varkappa_{2_0}(z)d \right]} \right]^{-1} . \tag{4.144b}$$

The subscripts "s" and "a" refer to the symmetric and antisymmetric modes, respectively. Using (4.144a), (4.137), the equation for $\delta k_{1,2}(z)$ (which appear in the slowly varying amplitude of the magnetic field at the boundary (4.140)) reduces to

$$\delta k_{1\left\{ {s \atop a} \right\}}(z) = \frac{i}{2} \frac{\varkappa_{1_0}^2(z)}{K_{\left\{ {s \atop a} \right\}}\varkappa_{2_0}^2(z) - \varkappa_{1_0}^2(z)} \frac{d}{dz} \ln\left(\frac{\varepsilon_2(z)}{\varepsilon_1(z)} \right) , \tag{4.145a}$$

$$\delta k_{2\left\{ {s \atop a} \right\}}(z) = \frac{i}{2} \frac{K_{\left\{ {s \atop a} \right\}}\varkappa_{2_0}^2(z)}{K_{\left\{ {s \atop a} \right\}}\varkappa_{2_0}^2(z) - \varkappa_{1_0}^2(z)} \frac{d}{dz} \ln\left(\frac{\varepsilon_2(z)}{\varepsilon_1(z)} \right) . \tag{4.145b}$$

The results for the magnetic-field intensity at the boundary can be obtained analytically in explicit form in the limiting cases of the thick- and thin-slab approximations, as follows.

(1) For a thick slab $(\varkappa_{2_0}(z)d \gg 1)$, the dispersion relation (4.114) tends to that for SW propagation along a single interface (4.86), $K_{\left\{ {s \atop a} \right\}} \to 1$, (4.145) for $\delta k_{1,2\left\{ {s \atop a} \right\}}$ reduces to (4.143) and the result for the magnetic-field intensity is as given by (4.94a).

(2) The thin-slab limit $(\varkappa_{2_0}d \ll 1)$ requires a separate treatment of the symmetric and antisymmetric modes.

   (i) For the case of symmetric-mode propagation the dispersion relation (4.114a) reduces to

$$\varkappa_1^2(z) = \frac{\varepsilon_1^2(z)}{\varepsilon_2^2(z)d^2} , \tag{4.146}$$

with a zero-order solution for the local wave number given by (4.117b). Since the $K_s(z)$ and $G_s(z)$ functions obtained from (4.144) have the form

$$K_s(z) = \frac{3}{2}[\varkappa_{2_0}(z)d]^{-2} , \tag{4.147a}$$

$$G_s(z) = -\frac{3}{2}[\varkappa_{1_0}(z)d]^{-2} , \tag{4.147b}$$

the results for $\delta k_{1,2}(z)$ are

$$\delta k_1(z) = \frac{\mathrm{i}}{3} \frac{\varepsilon_1^2(z)}{\varepsilon_2^2(z)} \frac{\mathrm{d}}{\mathrm{d}z} \ln \left(\frac{\varepsilon_2(z)}{\varepsilon_1(z)} \right) , \tag{4.148a}$$

$$\delta k_2(z) = \frac{\mathrm{i}}{2} \frac{\mathrm{d}}{\mathrm{d}z} \ln \left(\frac{\varepsilon_2(z)}{\varepsilon_1(z)} \right) . \tag{4.148b}$$

Equations (4.148b) and (4.140) determine straightforwardly the magnetic-field intensity at the boundary:

$$B_0(z) = \sqrt{\frac{\varepsilon_1(z)}{k_0(z)}} \exp \left[\mathrm{i} \int^z k_0(z')\, \mathrm{d}z' \right] . \tag{4.149}$$

This result coincides with (4.117). Inserting (4.148a) for $\delta k_1(z)$ (for $\varepsilon_1^2(z) \ll \varepsilon_2^2(z)$) into (4.140) yields the same result for $B_0(z)$.

(ii) In the case of an antisymmetric mode, considered in the electromagnetic region of its existence, the dispersion relation (4.114b) is

$$\frac{\varkappa_2^2(z)d}{\varepsilon_2(z)} = -\frac{\varkappa_1(z)}{\varepsilon_1(z)} \tag{4.150}$$

and the local wave number is determined by

$$k_0^2 = \frac{\omega^2}{c^2} \varepsilon_1(z) \left[1 + \left(\frac{\omega d}{c} \right)^2 \varepsilon_1(z) \right] . \tag{4.151}$$

The expressions (4.144) reduce to

$$K_{\mathrm{a}}(z) = \frac{1}{2} , \tag{4.152a}$$

$$G_{\mathrm{a}}(z) = -\frac{1}{2} \left(\frac{\varkappa_{2_0}(z)}{\varkappa_{1_0}(z)} \right)^2 , \tag{4.152b}$$

determining (with $\varkappa_{1_0} d \ll |\varepsilon_2(z)/\varepsilon_1(z)|$) the following results for the $\delta k_{1,2}(z)$ functions:

$$\delta k_1(z) = -\mathrm{i} \frac{\varepsilon_1^2(z)}{\varepsilon_2(z)} \left(\frac{\omega d}{c} \right)^2 \frac{\mathrm{d}}{\mathrm{d}z} \ln \left(\frac{\varepsilon_2(z)}{\varepsilon_1(z)} \right) , \tag{4.153a}$$

$$\delta k_2(z) = \frac{\mathrm{i}}{2} \frac{\mathrm{d}}{\mathrm{d}z} \ln \left(\frac{\varepsilon_2(z)}{\varepsilon_1(z)} \right) . \tag{4.153b}$$

With the latter relation the integration in (4.140) can be performed straightforwardly, giving

$$B_0(z) = \sqrt{\frac{\varepsilon_1(z)}{k_0(z)}} \exp \left[\mathrm{i} \int^z k_0(z')\, \mathrm{d}z' \right] . \tag{4.154}$$

The result obtained for the magnetic-field intensity at the boundary (4.154) has the same form as for the symmetric mode, i.e. (4.149).

SW Propagation in Structures with Cylindrical Geometry. The last configuration to be considered is that shown in Fig. 4.16. In the case of an azimuthally symmetric surface mode the dispersion relation is given by (4.123). The $K(z)$ and $G(z)$ functions have the following form:

$$G_{\mathrm{c}}(z) = -\frac{\varkappa_{2_0}^2(z)}{\varkappa_{1_0}^2(z)} K_{\mathrm{c}}(z) , \tag{4.155a}$$

$$K_{\mathrm{c}}(z) = \frac{1 - \varkappa_{1_0}(z) R\,\Psi_1}{1 + \varkappa_{2_0}(z) R\,\Psi_2}, \tag{4.155b}$$

with

$$\Psi_1 = \frac{\mathrm{K}_1[\varkappa_{1_0}(z)R]}{\mathrm{K}_0[\varkappa_{1_0}(z)R]} - \frac{\mathrm{K}_0[\varkappa_{1_0}(z)R]}{\mathrm{K}_1[\varkappa_{1_0}(z)R]} - \frac{1}{\varkappa_{1_0}(z)R}, \tag{4.155c}$$

$$\Psi_2 = \frac{\mathrm{I}_1[\varkappa_{2_0}(z)R]}{\mathrm{I}_0[\varkappa_{2_0}(z)R]} - \frac{\mathrm{I}_0[\varkappa_{2_0}(z)R]}{\mathrm{I}_1[\varkappa_{2_0}(z)R]} + \frac{1}{\varkappa_{2_0}(z)R}. \tag{4.155d}$$

The results for $\delta k_{1,2}$ are given by (4.145) with $K_{\{{}^{\mathrm{s}}_{\mathrm{a}}\}}(z)$ replaced by $K_{\mathrm{c}}(z)$ as given by (4.155b).

In order to show some analytical results for the magnetic-field intensity at the boundary in an explicit form, the three limiting cases of SW dispersion behaviour are considered.

(1) In the limit of the thick-cylinder approximation ($\varkappa_{1,2}(z)R \gg 1$, (3.47c)), $K_{\mathrm{c}} \to 1$ and $\delta k_{1,2}(z)$ reduce to (4.143). The magnetic-field intensity at the boundary is as described by (4.94a): SW propagation along a single interface.

(2) In the limiting case of a thin cylinder ($\varkappa_{1,2}(z)R \ll 1$, (3.47b)), which describes the azimuthally symmetric mode in the region of comparatively small σ values ($\sigma < 0.3$) [4.45–47], i.e. a region close to the quasi-static one, the zero-order approximation of the dispersion relation (4.123)

$$\varkappa_{1_0}^2(z) = -\frac{4\varepsilon_1(z)}{R^2\varepsilon_2(z)} \left[\ln\left(-\frac{\varepsilon_2(z)}{4\varepsilon_1(z)} \right) \right]^{-1} \tag{4.156a}$$

reduces to

$$k_0^2(z) = \frac{\omega^2}{c^2}\varepsilon_1(z) \left\{ 1 - \frac{c^2}{\omega^2}\frac{4}{R^2\varepsilon_2(z)} \left[\ln\left(-\frac{\varepsilon_2(z)}{4\varepsilon_1(z)} \right) \right]^{-1} \right\} \tag{4.156b}$$

for the local wave number. The expressions for the $K_{\mathrm{c}}(z)$ and $G_{\mathrm{c}}(z)$ functions (4.155) are

$$K_{\mathrm{c}}(z) = \frac{8}{[\varkappa_{2_0}(z)R]^2}, \tag{4.157a}$$

$$G_{\mathrm{c}}(z) = -\frac{8}{[\varkappa_{1_0}(z)R]^2}, \tag{4.157b}$$

leading to

$$\delta k_1(z) = -\frac{\mathrm{i}}{4}\frac{\varepsilon_1(z)}{\varepsilon_2(z)} \left[\ln\left(-\frac{\varepsilon_2(z)}{4\varepsilon_1(z)} \right) \right]^{-1} \frac{\mathrm{d}}{\mathrm{d}z}\ln\left(\frac{\varepsilon_2(z)}{\varepsilon_1(z)} \right), \tag{4.158a}$$

$$\delta k_2(z) = \frac{\mathrm{i}}{2}\frac{\mathrm{d}}{\mathrm{d}z}\ln\left(\frac{\varepsilon_2(z)}{\varepsilon_1(z)} \right). \tag{4.158b}$$

Accounting for the slow variation of $\ln[-\varepsilon_2(z)/4\varepsilon_1(z)]$, (4.158) yields the following result for the magnetic-field intensity at the boundary

$$B_0(z) = \sqrt{\frac{\varepsilon_1(z)}{k_0(z)}} \exp\left[i \int^z k_0(z')\,dz'\right], \tag{4.159}$$

which coincides with (4.127).

(3) In the third limiting case ($\varkappa_1(z)R \ll 1$, $\varkappa_2(z)R \gg 1$, i.e. region (3.47a)), which describes SW behaviour at comparatively large σ values ($\sigma > 0.3$), and the region of high electron densities at $\sigma < 0.3$, i.e. the bottom of the phase diagrams [4.47], the zero-order approximation to the dispersion relation (4.123)

$$\varkappa_{2_0}(z)R = -(\varkappa_{1_0}R)^2 \frac{\varepsilon_2(z)}{\varepsilon_1(z)} |\ln \varkappa_{1_0}(z)R| \tag{4.160a}$$

reduces to

$$k_0^2(z) = \frac{\omega^2}{c^2}\varepsilon_1(z) \left\{ 1 + \frac{2c}{R\omega\sqrt{-\varepsilon_2(z)}} \left[\ln\left(\frac{c}{2\omega R}\frac{\sqrt{-\varepsilon_2(z)}}{\varepsilon_1(z)}\right)\right]^{-1} \right\} \tag{4.160b}$$

for the local wave number. The $K_c(z)$ and $G_c(z)$ functions (4.155) assume the form

$$K_c(z) = 1, \tag{4.161a}$$

$$G_c(z) = -\frac{\varkappa_{2_0}^2(z)}{\varkappa_{1_0}^2(z)}, \tag{4.161b}$$

leading to

$$\delta k_1(z)$$

$$= i\,\frac{c}{\omega R}\frac{\varepsilon_1(z)}{[-\varepsilon_2(z)]^{3/2}} \left[\ln\left(\frac{c}{2\omega R}\frac{\sqrt{-\varepsilon_2(z)}}{\varepsilon_1(z)}\right)\right]^{-1} \frac{d}{dz}\ln\left(\frac{\varepsilon_2(z)}{\varepsilon_1(z)}\right), \tag{4.162a}$$

$$\delta k_2(z) = \frac{i}{2}\frac{d}{dz}\ln\left(\frac{\varepsilon_2(z)}{\varepsilon_1(z)}\right). \tag{4.162b}$$

The latter expression determines straightforwardly the following result for the magnetic-field intensity at the boundary:

$$B_0(z) = \sqrt{\frac{\varepsilon_1(z)}{k_0(z)}} \exp\left[i \int^z k_0(z')\,dz'\right]. \tag{4.163}$$

This is of the same form as that for the thin-cylinder approximation (4.159).

Thus, in all the geometrical configurations under consideration here, the form of the results obtained for the magnetic-field intensity at the boundary in the frequency range of fast SWs is the same ((4.94a) for $|\varepsilon_2(z)| \gg |\varepsilon_1(z)|$

and (4.149), (4.154), (4.159) and (4.163)), with $k_0(z)$ being the proper local wave number (for the given structure). Since these regions are far from the resonance point, the variations of the amplitude of the magnetic-field intensity associated with the effects of inhomogeneity are slow.

4.3.5 Numerical Results for a Single Interface and for Cylindrical Waveguides

In this subsection evaluations for some examples are outlined on the basis of an assumed linear axial density profile (according to (4.99)). The accuracy of the geometrical-optics approximation is demonstrated by some comparisons with exact solutions which have been obtained by using an eigenfunction method. The qualitative behaviour is always correctly described by the geometrical-optics approach, quantitative differences from the eigenfunction results become noteworthy towards rather low ν/ω.

First the case of SW propagation along a single interface is considered. The eigenfunction method is based on the separation ansatz

$$B(x,z) = \mathcal{X}(x)\mathcal{Y}(z) \tag{4.164}$$

for (4.81), with $\varepsilon(z)$ in the plasma according to (2.8) and $\varepsilon = 1$ outside. The linear axial profile (4.59) of the plasma density assumed,

$$n(z) = n(z = 0)\left(1 - \frac{z}{L_N}\right), \tag{4.165a}$$

results in

$$\omega_{\mathrm{p}}^2(z) = 2\omega^2\left(1 - \frac{z}{L_N}\right). \tag{4.165b}$$

Equation (4.81) reduces to

$$\frac{\mathrm{d}^2\mathcal{X}}{\mathrm{d}x^2} - l\mathcal{X} = 0, \tag{4.166a}$$

$$\frac{\mathrm{d}^2\mathcal{Y}}{\mathrm{d}z^2} - \frac{1}{\varepsilon}\frac{\mathrm{d}\varepsilon}{\mathrm{d}z}\frac{\mathrm{d}\mathcal{Y}}{\mathrm{d}z} + \frac{\omega^2}{c^2}\varepsilon\mathcal{Y} + l\mathcal{Y} = 0. \tag{4.166b}$$

The aim is to find the final solutions of the problem, satisfying the continuity conditions at the plasma/vacuum interface, by a superposition of the solutions $B_n = \mathcal{X}_n\mathcal{Y}_n$. With $\mathcal{X} \to 0$ for $x \to \pm\infty$, (4.166a) has solutions of the type

$$\mathcal{X} \propto \mathrm{e}^{\pm\sqrt{l}\,x}. \tag{4.167}$$

The eigenvalue equation (4.166b) is solved with the following boundary conditions.

(1) At a position z_1 which is on the high-density side of the plasma, i.e. far away from the (resonance) region (3.1b) near $\omega_{\mathrm{p}}^2(z)/\omega^2 \approx 2$, the transverse dependence of E_z is taken to be described by

$$E_z(z_1, x) \propto \mathrm{e}^{\sqrt{l_0}\, x},\tag{4.168}$$

where l_0 is determined from the local dispersion relation of geometrical optics (4.92). For $n \neq 0$, all $\mathcal{Y}_n(z_1) = 0$.

(2) At a position z_{+0} which is behind the region of the quasi-static resonance $(\omega_{\mathrm{p}}^2(z)/\omega^2 \approx 2)$, reflection conditions are taken: $\mathrm{d}\mathcal{Y}_n/\mathrm{d}z = 0$. The exact choice of z_{+0} turns out not to be essential owing to the very low field intensities finally obtained there.

The complex eigenvalues l_n are determined numerically, as well as the eigenfunctions $\mathcal{Y}_n$. In the vacuum they can be determined explicitly.

The magnetic-field strengths in plasma and vacuum are described by

$$B^{\mathrm{P}}(x, z) = \sum_0^N a_n^{\mathrm{P}} \mathcal{Y}_n^{\mathrm{P}}(z) \mathcal{X}_n^{\mathrm{P}}(x),\tag{4.169a}$$

$$B^{\mathrm{v}}(x, z) = \sum_0^N a_n^{\mathrm{v}} \mathcal{Y}_n^{\mathrm{v}}(z) \mathcal{X}_n^{\mathrm{v}}(x).\tag{4.169b}$$

The $2N$ unknown coefficients a_n^{P}, a_n^{v} are determined from the continuity conditions for B and E_z at the plasma/vacuum interface at N equidistant points chosen between z_1 and z_{+0}. This leads to a $2N$-dimensional system of linear equations, the inhomogeneous part of the system given by the zero term in the superpositions (4.169). N is chosen high enough (in the cases described below; $40 \leq N \leq 100$) so that the continuity conditions are also well fulfilled at intermittent positions and that the resulting curves are smooth. More details were given in [4.37,38].

In Fig. 4.17 the real part and modulus of B_y, E_x and E_z at the interface $(x = 0)$ are shown for a set of conditions with medium collisional damping $(\nu/\omega = 0.1)$. The position $z = 0$ was chosen so that $\overline{\omega_{\mathrm{p}}^2}(z = 0)/\omega^2 = 2$. The dashed curves represent the eigenfunction results, the solid curves the ones for the geometrical-optics approximation according to (4.95) and (4.97).

As ν/ω becomes smaller the differences between the eigenvalue solution and the geometrical-optics result become more noteworthy, as Fig. 4.18 demonstrates for the modulus of E_z for $\nu/\omega = 1/30$, although the same general trend is given by both methods. More results for various parameters were presented in [4.37,38]; these show, for example, that reducing L_N or ω tends to make the resonances more pronounced.

For some cylindrical cases some numerical evaluations of the geometrical-optics approximations and comparisons with exact solutions from an eigenfunction method will again be given. Moreover – with a view to practical cases – in some cases considered below the presence of dielectric walls (and metal shielding) is taken into account.

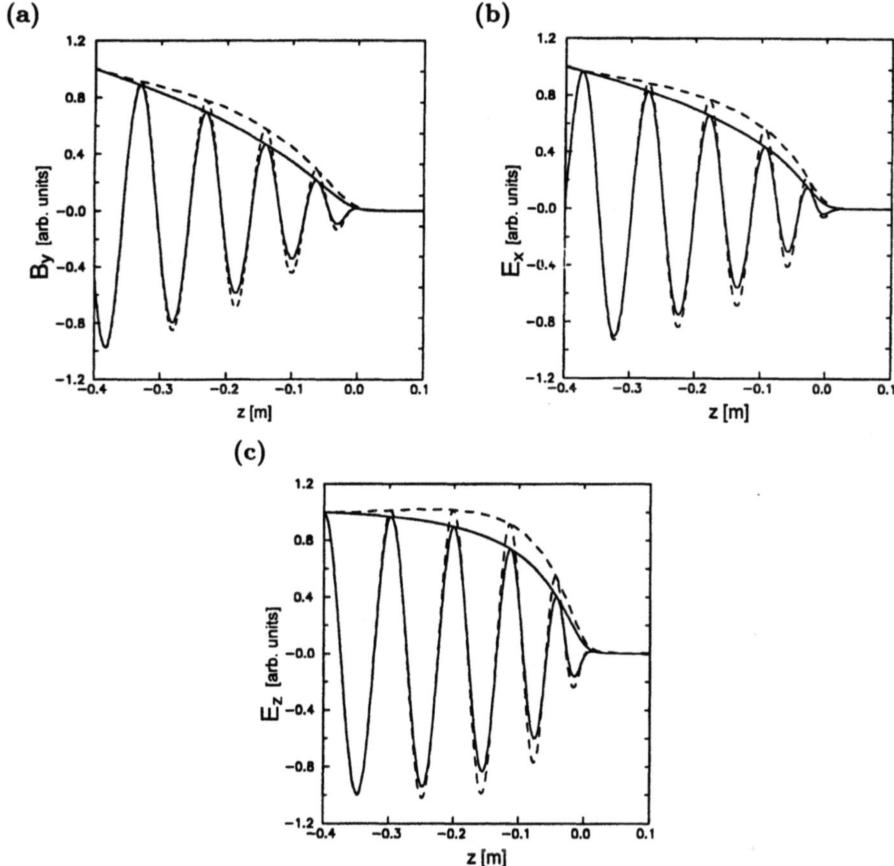

Figure 4.17. Variation of the real parts of the field components (*oscillating curves*) and their moduli (*smooth curves*) in the longitudinal direction as obtained from the eigenfunction method (*dashed curves*) and the geometrical-optics method (*solid curves*): **(a)** B_y field, **(b)** E_x field and **(c)** E_z field. Values of the parameters: $\nu/\omega = 0.1$, $f \equiv \omega/2\pi = 2.45$ GHz, $L_N = 0.3$ m ([4.38], Figs. 4.3.1a,b,c)

The extension of the eigenfunction method to the cylindrical case [4.38,48] is quite analogous to the procedure for the semi-space case given above and starts with an ansatz and equations similar to (4.164) and (4.166):

$$B_\varphi = \mathcal{X}(r)\mathcal{Y}(z)\,, \tag{4.170}$$

$$\frac{1}{r}\frac{\partial}{\partial r}\left(r\frac{\partial \mathcal{X}}{\partial r}\right) - \left(l + \frac{1}{r^2}\right)\mathcal{X} = 0\,, \tag{4.171a}$$

$$\frac{\mathrm{d}^2\mathcal{Y}}{\mathrm{d}z^2} - \frac{1}{\varepsilon}\frac{\mathrm{d}\varepsilon}{\mathrm{d}z}\frac{\mathrm{d}\mathcal{Y}}{\mathrm{d}z} + \frac{\omega^2}{c^2}\varepsilon\mathcal{Y} + l\mathcal{Y} = 0\,. \tag{4.171b}$$

With R and $R + d$ being the plasma radius and the outer radius of the dielectric tube, respectively, the solutions for $\mathcal{X}(r)$ are of the type

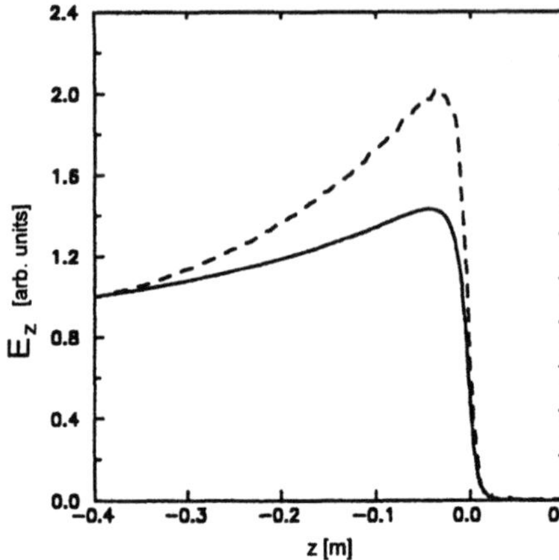

Figure 4.18. Variation of the E_z field (modulus) in the longitudinal direction as obtained by the eigenfunction method (*dashed curve*) and by geometrical optics (*solid curve*) with the conditions of Fig. 4.17, but $\nu/\omega = 1/30$ ([4.38], Fig. 4.3.2)

$$\mathfrak{X}(r) = A_1 \mathrm{I}_1(\sqrt{l_\mathrm{p}}\, r) \qquad\qquad (r < R)\,, \tag{4.172a}$$

$$\mathfrak{X}(r) = A_2 \mathrm{I}_1(\sqrt{l_\mathrm{d}}\, r) + A_3 \mathrm{K}_1(\sqrt{l_\mathrm{d}}\, r) \quad (R < r < R+d)\,, \tag{4.172b}$$

$$\mathfrak{X}(r) = A_4 \mathrm{K}_1(\sqrt{l_v}\, r) \qquad\qquad (r > R+d)\,. \tag{4.172c}$$

Consequently, the final solutions for B_φ are of the form

$$B^\mathrm{p}(r,z) = \sum_0^N a_n^\mathrm{p} \mathcal{Y}_n^\mathrm{p}(z) \mathfrak{X}_n^\mathrm{p}(r)\,, \tag{4.173a}$$

$$B^g(r,z) = \sum_0^N \mathcal{Y}_n^d(z) \left[a_n^{1\mathrm{d}} \mathfrak{X}_n^{1\mathrm{d}}(r) + a_n^{2\mathrm{d}} \mathfrak{X}_n^{2\mathrm{d}}(r) \right]\,, \tag{4.173b}$$

$$B^\mathrm{v}(r,z) = \sum_0^N \mathcal{Y}_n^\mathrm{v}(z) \left[a_n^{1\mathrm{v}} \mathfrak{X}_n^{1\mathrm{v}}(r) + a_n^{2\mathrm{v}} \mathfrak{X}_n^{2\mathrm{v}}(r) \right]\,. \tag{4.173c}$$

The procedure for determining the eigenfunctions $\mathcal{Y}_n^\mathrm{p}$, $\mathcal{Y}_n^\mathrm{d}$, $\mathcal{Y}_n^\mathrm{v}$ and the $5N$ unknown coefficients a_n^p, $a_n^{1\mathrm{d}}$, $a_n^{2\mathrm{d}}$, $a_n^{1\mathrm{v}}$, $a_n^{2\mathrm{v}}$ is quite analogous to that described for the semi-space case, as are the assumptions for the zero terms and the points z_1 and z_{+0} of the interval over which this solution is developed.

With a view to experiments performed in lower-frequency ranges [4.48], examples are given below for frequencies close to 200 MHz and a correspondingly low axial density scale length near 0.03 m. But for higher frequencies, in the GHz range, the trends are not basically altered.

Figure 4.19 depicts results of the eigenfunction method for B_φ, E_r, E_z (modulus) in the vacuum just outside the glass tube. E_z has a stronger trend towards showing a resonance peak than E_r, which is particularly apparent for the relatively strong damping $\nu/\omega = 0.2$ chosen. In Fig. 4.20a the result for E_r (modulus) is displayed for $\nu/\omega = 0.2$ and compared with the result from the corresponding geometrical-optics approximation. The eigenfunction result obviously exhibit a slightly sharper bend near the "resonance" point

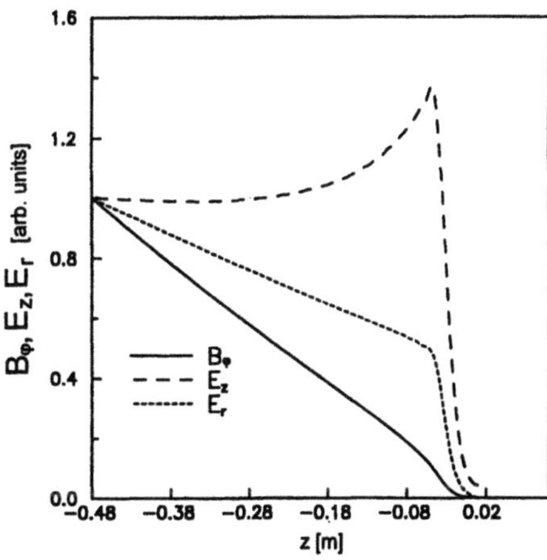

Figure 4.19. Variation of the modulus of the B_φ (*solid curve*), E_z (*dashed curve*) and E_r field component (*dotted curve*) in the longitudinal direction obtained from the eigenfunction method for $\nu/\omega = 1/5$, $f \equiv \omega/2\pi = 220$ MHz and $L_N = 0.0274$ m. Configuration of the waveguide: plasma column of radius $R = 1.4$ cm surrounded by a dielectric ($\varepsilon_d = 4.7$, thickness $d = 0.2$ cm), vacuum and metal cylinder of radius $R_m = 4.5$ cm. Full resolution for E_z is achieved in Fig. 4.21 by using a higher number of eigenfunctions ([4.38], Fig. 4.1.12)

(a)

(b)

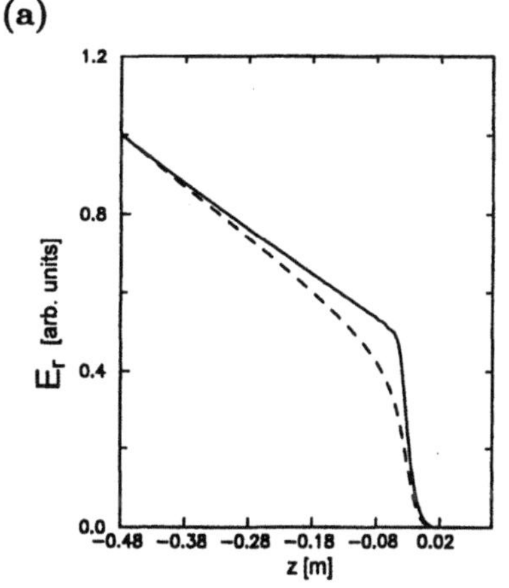

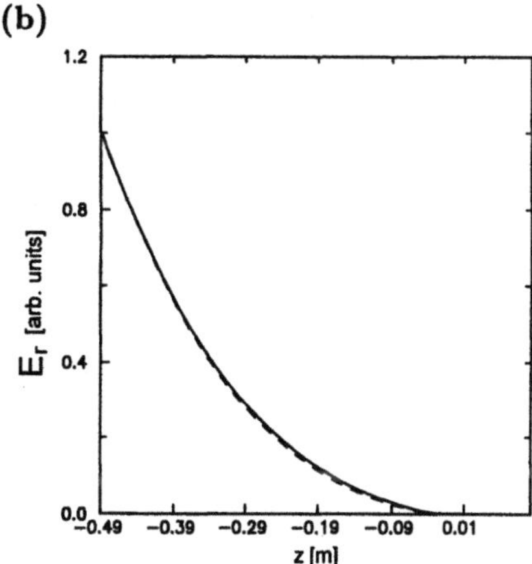

Figure 4.20. Variation of the E_r field (modulus) in longitudinal direction obtained from the eigenfunction method (*solid curves*) and from geometrical optics (*dashed curves*) for $f \equiv \omega/2\pi = 220$ MHz, $L_N = 0.0274$ m and $\nu/\omega = 1/5$ (a) and 2/3 (b) ([4.38], Fig. 4.3.4)

$z = 0$. For stronger damping, $\nu/\omega = 0.66$ (Fig. 4.20b), E_r has a behaviour of a stronger decay towards $z = 0$ with virtually no difference between the results from the two methods. For strong damping, $\nu/\omega = 0.66$, even E_z exhibits such a decaying behaviour, as Fig. 4.21 demonstrates; this figure presents plots obtained from the eigenfunction method for different values of ν/ω, the conditions otherwise being as before. For comparison, results from the geometrical-optics approximation for various ν/ω are depicted in Fig. 4.22. The tendency shown by the eigenfunction solutions to develop resonant-like behaviour with decreasing ν/ω also appears in the geometrical-optics solutions. However, in the latter the development is weaker. For example, for $\nu/\omega = 0.2$ the geometrical-optics result shows no peaking, whereas the eigenfunction result reveals it quite visibly. It takes values as small as $\nu/\omega \approx 0.105$

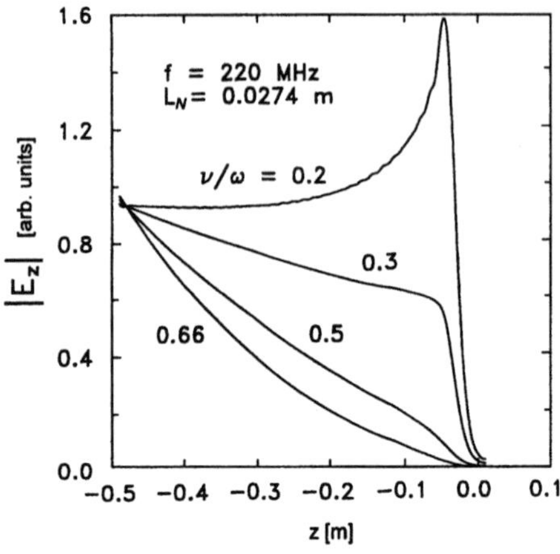

Figure 4.21. Variation of the E_z field (modulus) in the longitudinal direction obtained from the eigenfunction method for several values of ν/ω ([4.38], Fig. 4.3.5a)

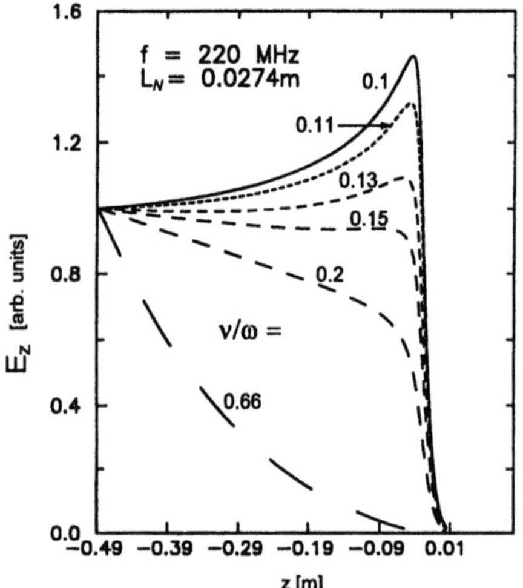

Figure 4.22. Variation of the E_z field (modulus) in the longitudinal direction obtained from the geometrical-optics approach for several values of ν/ω ([4.38], Fig. 4.3.5b)

to exhibit a resonance behaviour in the geometrical-optics solution; this is similar to that given by the eigenfunction result for $\nu/\omega = 0.2$.

It should be noted that in discussing the influence of parameters such as L_N and ν/ω, these parameters have largely been considered as independent, as would be the case for low-amplitude SW test signals in target plasmas, for example. In SW-sustained discharges they are, however, interconnected by self-consistency, as treated in detail in the next chapter.

The axial variation of the field strength (modulus), with a resonant behaviour, was shown above for a radial position at the plasma edge (the interface with the glass). However, inspection of the axial variation of E_z and E_r (modulus) at different radial positions in the vacuum and the plasma, respectively, also reveals peaking further inside the plasma, though it is strongest at the interface. The resonant-like behaviour is not solely associated with a pushing of the field intensity towards the interface at the expense of field intensity disappearing from the areas radially further away, i.e. it is not caused merely by a kind of skin effect when the resonance position is approached. It is a truly resonant-like behaviour with, for example, $\int |E_z|^2 r\, dr$ clearly peaking.

As demonstrated above in this section, outside the region around the SW resonance in a radially homogeneous plasma waveguide the influence of axial density inhomogeneity as described by the eigenfunction method is in good agreement with the first-order geometrical-optics (WKB) approximation given in Sect. 4.3.3. There are some interesting points in the comparison with the zero-order geometrical-optics approximation.

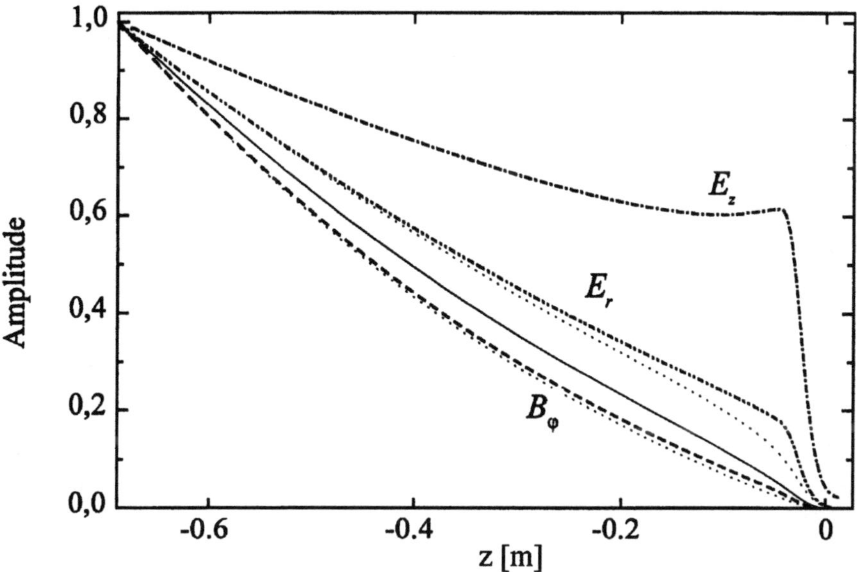

Figure 4.23. Normalized moduli of E_z, E_r and B_φ (*dashed* and *dashed–dotted* curves) in the vacuum just outside the glass tube, obtained by the eigenfunction approach for a radially homogeneous plasma cylinder, versus the axial coordinate z. E_r and B_φ (*dotted curves*) obtained from the first-order geometrical-optics approximation and the zero-order geometrical-optics solution (*solid curve*) for E_z, E_r, B_φ (coincident) are also presented. Conditions: $f \equiv \omega/2\pi = 166$ MHz, $\nu/\omega = 0.32$, $L_N = 0.0228$ m; waveguide configuration of plasma column of radius $R = 1.4$ cm surrounded by a dielectric ($\varepsilon_{\rm d} = 4.7$, thickness $d = 0.2$ cm), vacuum and metal shielding ($R_{\rm m} = 4.5$ cm)

Figure 4.23 depicts, for a radially homogeneous plasma cylinder with a glass tube, the moduli of E_z, E_r and B_φ at a radial position in the vacuum just outside the glass tube versus the axial coordinate z. A linear axial density profile was used, and 100 eigenfunctions were employed. The results for the modulus of E_r and the modulus of B_φ obtained from the first-order geometrical-optics approximation of Sect. 4.3.3 are also presented. The figure obviously shows good agreement between the eigenfunction and the first-order geometrical-optics solutions except in the region of the SW resonance in a radially homogeneous plasma. The normalized plots for the moduli of E_z, E_r and B_φ obtained from the zero-order geometrical-optics solution are also shown. Figure 4.24 gives the corresponding phases. Excluding

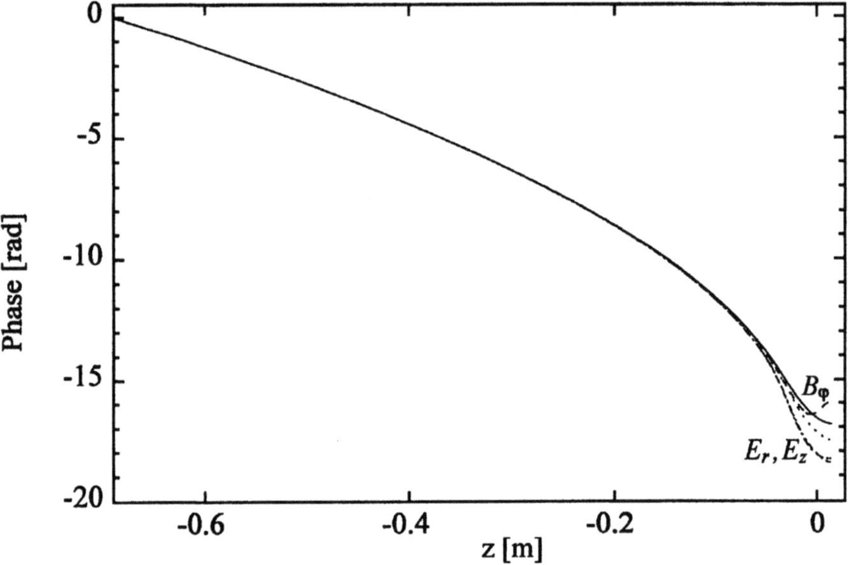

Figure 4.24. Phases of E_z, E_r and B_φ for a radially homogeneous plasma cylinder: conditions as in Fig. 4.23.

again the region of the SW resonance in a radially homogenous plasma, the phases of all three components agree no matter whether the eigenfunction or the geometrical-optics method in zero or first order is taken. Obviously, the corrections to the ks beyond zero order (as described in Sect. 4.3.3) do not essentially concern the wave numbers (i.e. β), but yield different corrections for the three field components via α. As a consequence the radial profiles of the field components are also virtually unaffected by corrections, as long as α remains below β, and are not noticeably changed even when radial inhomogeneity is also accounted for, as the subsequent section will show.

The relatively weak influence of the axial inhomogeneity on the phases and the βs is of course helpful in diagnostics, when for instance E_r in the vacuum region and its axial change are recorded an analysed, see Sect. 7.2.1 on diagnostics. However, as Fig. 4.23 demonstrates, this influence is generally more important for an analysis of the αs, since E_r exhibits an amplitude decay with values of α different from the zero-order values.

An interesting point relevant to the modelling of SW-sustained discharges is the fact that the axial behaviours of the moduli of E_r and B_φ are corrected in opposite directions as compared with the zero-order result. In fact the product $|E_r||B_\varphi|$ in the first-order treatment is virtually the same as in zero order, except again for the region of the SW resonance in a radially homogeneous plasma (which anyhow can hardly be reached, because of the combined effects of collisions and inhomogeneity in the transverse direction described in Sect. 4.2). This also applies with good accuracy for situations different from the one of Fig. 4.23 and holds well when integration of the product is performed across the whole plane perpendicular to the z axis. Therefore the Poynting flux is virtually the same in zero and first order everywhere, yielding virtually no differences when differentiation with respect to z is performed. In a way the two first-order α corrections to E_r and B_φ cancel. Because of this cancellation effect, there is no stringent need to go to first order in the wave-power equation (2.27), as has been mentioned already [4.49]. This simplifies the use of the equation later in this book for determining the self-consistent axial structure of SW-sustained discharges.

The aspects just discussed are not essentially altered by the additional presence of radial density inhomogeneity, although the role of the SW resonance is diminished and slightly altered. The field profiles inside the plasma may of course be drastically altered, but again good agreement between one- and two-dimensional results is found to a large extent, as the subsequent section will demonstrate from code calculations.

4.4 Calculations with Both Transverse and Longitudinal Inhomogeneities

In this section the simultaneous presence of transverse and longitudinal density nonuniformities is addressed. Some trends are pointed out on the basis of numerical solutions of Maxwell's field equations for a given electron density profile. This was assumed as before to be represented by

$$n(r, z) = n(r = 0, z = 0) \left(1 - \frac{z}{L_N}\right) J_0\left(\mu \frac{r}{R}\right) \tag{4.174}$$

where $n(r = 0, z = 0)$ is the density at the axis at the position of the SW resonance as defined by (4.165ab). Again the cold-plasma permittivity (2.8) was employed, with $\omega_p(r, z)$ governed by the inhomogeneity parameters μ and L_N. Solutions $\propto \exp(i\omega t)$ were obtained by a finite-integration technique from the integral formulation of Maxwell's equations. As described in some detail in [4.39–41], a numerically modelled SW launcher (surfatron device, Chap. 7) was used to excite the SWs. Since only azimuthally symmetric solutions were sought, the size of the discretization grid was reduced by employing ideal magnetic azimuthal boundary conditions and variable grids.

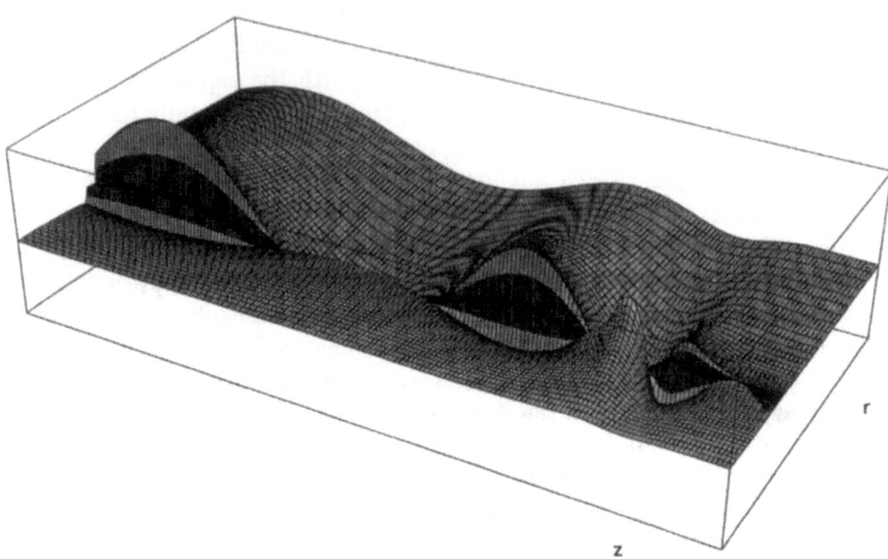

Figure 4.25. Re$\{E_r\}$ in relative units versus z and r for an axial density profile with $L_N = 0.0274$ m, $\mu = 2$; $f \equiv \omega/2\pi = 220$ MHz, $\nu/\omega = 0.2$, $R = 5$ mm, $R + d = 6$ mm, $R_m = 45$ mm, shown from $z = -44$ cm (*left*) to $z = 1$ cm (*right*) with $z = 0$ at $\overline{\omega_p^2/\omega^2} = 2$. r runs from 2.4 mm in the plasma (*front*) to $R_m = 45$ mm at the metallic shield (*back*) ([4.39], Fig. 6.7)

Figure 4.25 illustrates an example of a two-dimensional representation (in the (r, z) plane of Re$\{E_r\}$) for an SW calculated in this way. In Fig. 4.26 the modulus of E_r and in Fig. 4.27 the modulus of E_z are depicted.

Calculations of this type confirm that the assumption made in the above modelling – that the influence of axial density inhomogeneity is usually weak – is warranted, even in the case of strong changes in the radial field profile brought about by pronounced radial density inhomogeneity. This is demonstrated by radial sections in Fig. 4.28 for a radially inhomogeneous case and in Fig. 4.29 for a radially homogeneous situation. Comparisons are made with simplified calculations neglecting z dependences. Differences only become noticeable when z positions close to the system ("quasi-static") resonance are considered. The differences are more pronounced in the radially inhomogeneous case. In a situation with a rather strong radial density variation (profile parameter $\mu = 2.4$), but with increased damping such that $\nu/\omega = 0.6$ (suppressing the formation of a system resonance connected to the axial nonuniformity), field peaks due to radial plasma resonances, where $\omega_p(r) = \omega$, appear clearly resolved. They are satisfactorily described in both cases, as can be inferred from Fig. 4.30.

Figure 4.31 exhibits axial sections of the $|E_r|$ field strength when a (modest) radial inhomogeneity ($\mu = 2.0$ as in Figs. 4.25–28) is present. As com-

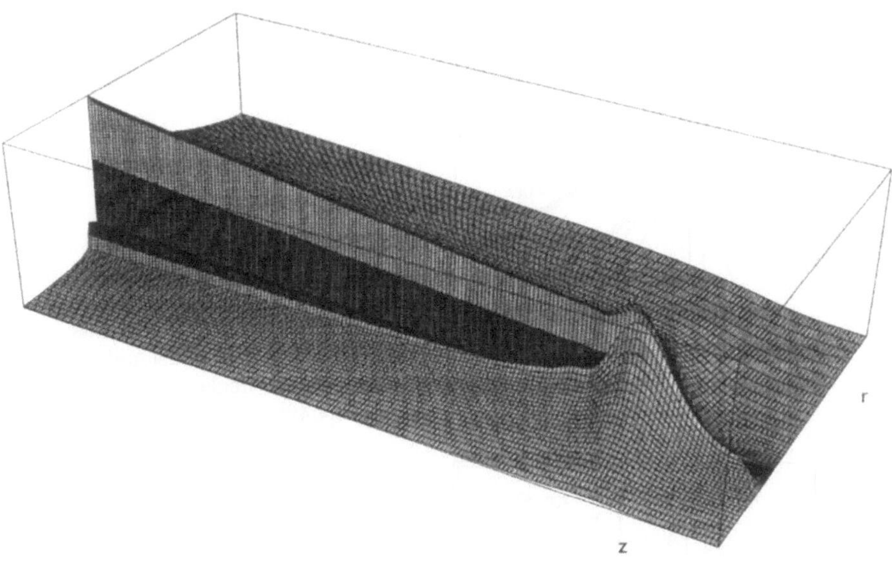

Figure 4.26. Conditions as for Fig. 4.25: the modulus of E_r is depicted ([4.39], Fig. 6.9)

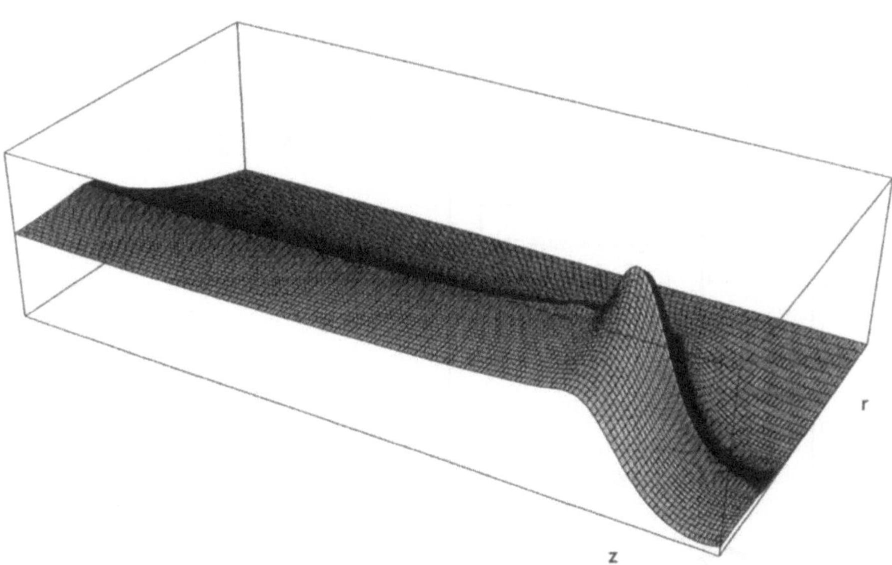

Figure 4.27. Conditions as for Fig. 4.25: the modulus of E_z is depicted ([4.39], Fig. 4.10)

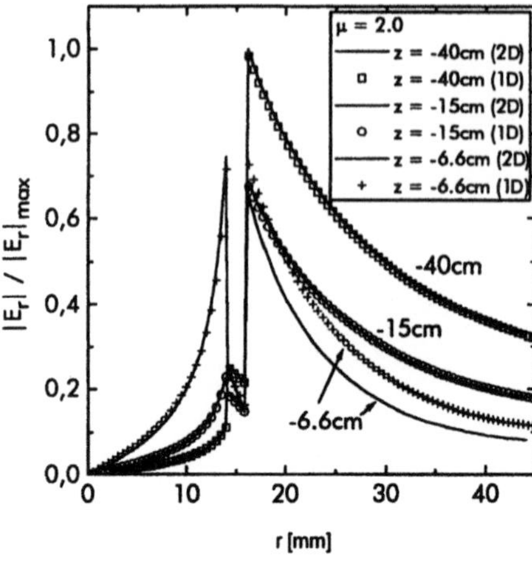

Figure 4.28. Radial sections at different z positions for the conditions of Fig. 4.25 comparing full two-dimensional (2D) solutions (*solid lines*) with one-dimensional (1D) solutions which neglect z dependences; $\mu = 2$ ([4.40], Fig. 4)

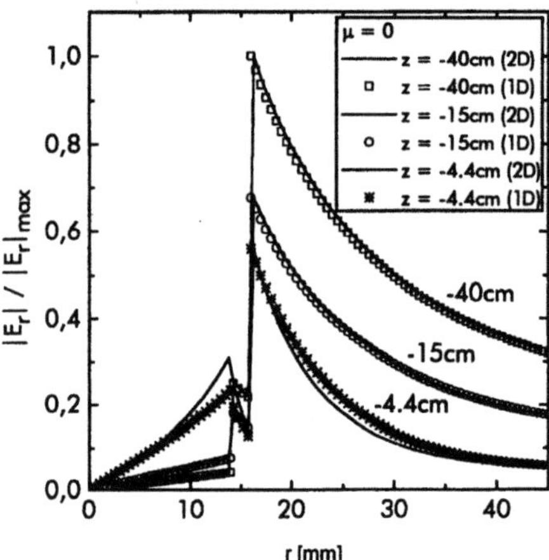

Figure 4.29. As in Fig. 4.28, but the radially homogeneous situation ($\mu = 0$) is considered ([4.40], Fig. 3)

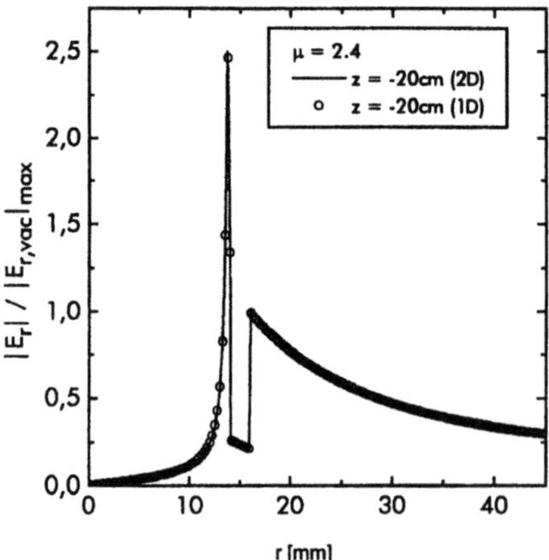

Figure 4.30. Radial section at $z = -20$ cm for conditions as in Figs. 4.28 and 4.29, but $\mu = 2.4$ and $\nu/\omega = 0.6$ ([4.40], Fig. 5)

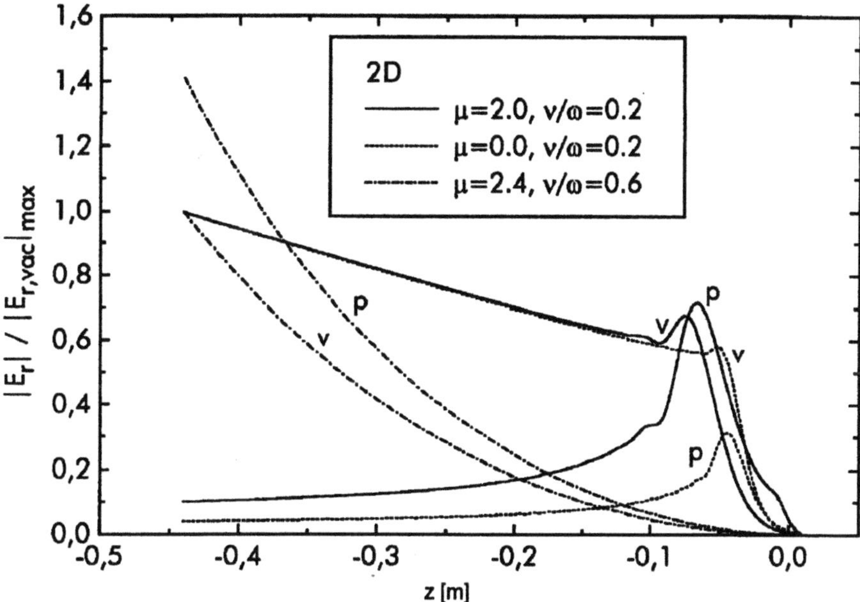

Figure 4.31. Axial sections for $|E_r|$ near the wall in the plasma (p, $r = R$) and in the vacuum (v, $r = R + d$), for $\mu = 0$ and $\mu = 2$ with $\nu/\omega = 0.2$, with a system resonance recognizable, and for $\mu = 2.4$ and $\nu/\omega = 0.6$ without a system resonance recognizable ([4.40], Fig. 4.2)

pared with calculations for a radially homogeneous density of an averaged value, the plasma–vacuum field distribution is, of course, changed. Moreover the position of the system resonance starts to be affected, since towards the SW resonance the density values near the wall become more relevant than the value simply averaged over the cross-section. Included in Fig. 4.31 is a situation of stronger radial inhomogeneity ($\mu = 2.4$) and larger damping ($\nu/\omega = 0.6$), where no quasi-static (system) resonance shows up.

In summary, the procedures that neglect a longitudinal density nonuniformity when calculating the influence of strong transverse inhomogeneity by radial integration techniques seem largely to be warranted. When strong quasi-static (system) resonances appear towards the end of the discharge does this procedure become inaccurate, and consideration of the longitudinal nonuniformity becomes essential. In such situations, however, inclusion of a radial nonuniformity becomes necessary, since it modifies the radial field profiles and suppresses the appearance of the quasi-static resonance. These latter features are quite in line with qualitative expectations that towards the end of the discharge the densities near the discharge wall become more influential than are implied in a cross-section-averaged value, since the field intensities become more concentrated in the outer plasma regions towards the end of the discharge.

5. Fluid Theory
of Surface-Wave-Produced Plasmas

An early concept of gas ionization by EM wave fields was introduced by Bailey [5.1] in studies on the propagation of strong radio waves in the ionosphere. In later years, with the start of extended research on problems involving high-power microwave transmission (radar waveguides and antenna systems), space–earth communication and HF heating of thermonuclear plasmas, the phenomenon of microwave breakdown of neutral gases achieved growing importance. Studies on the influence of additional ionization on the absorption of EM waves propagating through nonuniform plasmas, on gas breakdown in wave-beam fields and on the formation of waveguide channels by ionizing TM modes made ionization nonlinearity the basis of self-sustained wave–plasma interactions [5.2–6]. The electron heating in HF fields and the temperature dependence of the ionization frequency revealed the dependence of the plasma density on the field intensity. Local mechanisms of particle losses have usually been considered: there is a local increase of the plasma density with the field intensity, implying a nonlinear plasma permittivity. The dynamics of the breakdown and the effects of nonlocality are the recent directions in the development of the research on production of plasmas in wave-beam fields [5.6–8]. An ionization nonlinearity not connected with electron heating was also shown to be possible in superstrong EM wave fields, when the energy of the oscillating electron motion is higher than the ionization energy. In this case the decrease of the ionization frequency with increasing field intensity offers possibilities to create waveguide channels by means of transverse electric modes [5.9]. In some applications the breakdown phenomenon has been used for protection of systems (for instance, radar and communication receivers against high-power microwave pulses with extremely short rise times [5.10]).

The first studies of the effects of ionization nonlinearity in the field of SWs dealt with the derivation of the nonlinear dispersive law of waves with constant amplitude under an assumption of a local increase of the density with the field amplitude [5.11–13].

The possibilities for the production and maintenance of gas discharges by propagating SWs [5.14,15] directed the theory of ionization nonlinearity in SW fields to modelling of this type of discharge [5.16–20]. The fluid model set out in this chapter and the numerical modelling of the discharge kinetics

presented in the next chapter outline the present status of the theory of
SW-sustained discharges.

5.1 Surface-Wave-Sustained Discharges: Nonlinear Systems Unifying Plasma and Wave Field

When a powerful wave field is applied to a neutral gas and the electric-field
intensity exceeds a threshold value for breakdown, which is determined by
the losses of charged particles by free-fall diffusion to the discharge vessel (or
attachment in the case of electronegative gases), the electron density starts
growing, initially exponentially, in time. However, on reaching the critical
value (2.11) of the electron density, the plasma permittivity (2.9a) becomes
negative, and further propagation of EM bulk waves becomes impossible (see
Sect. 2.1). Therefore, gas discharges with high-density plasmas cannot be
created by bulk EM waves. This can only be done by a wave which can
propagate in a dense plasma. The surface mode (Sect. 3.2, region (a) in
Fig. 3.2) is such a wave. This feature of the SW mode is its advantage,
because there is no upper limit to the plasma density produced. This has
been exploited for SW-sustained discharges.

Although both regimes – diffusion-controlled and recombination-controlled – are covered, more attention is paid in this chapter to discharge
maintenance under conditions of diffusion losses. This is because experiments
on SW-sustained discharges are usually performed in this regime.

As has been mentioned, the production of a discharge is an effect of a
strong ionization nonlinearity. Its final result is a plasma density related to
the amplitude of the HF field which produces the discharge. A thermal nonlinearity, relating the electron temperature to the heating-field intensity, is
the first stage in the development of the ionization nonlinearity. The strong
dependence of the ionization frequency on the temperature and, through the
latter, on the field intensity makes the process of ionization a mechanism of
nonlinearity. Since the dependences of the diffusion and recombination coefficients on the temperature are relatively weak, they can be considered as
constants. However, the role of the processes of charged-particle losses – diffusion and recombination – is important since they determine the type of
nonlinear response: local or nonlocal. Since the thermal nonlinearity can also
appear as a local or a nonlocal process, the locality/nonlocality of the total
phenomenon of ionization nonlinearity, i.e. of the dependence of the plasma
density on the electric-field intensity, is governed by the characteristic lengths
of diffusion and of thermal conduction (Sect. 2.2.2).

As a process of a dissipative nonlinearity, the discharge maintenance has
a threshold. In general, the determination of the threshold field intensity is
a key to the understanding of the nonlinear processes. In the cases of stationary discharges produced in an ambipolar diffusion regime, the Schottky

solution (2.67) has commonly been used as a condition which determines the threshold field intensity E_{th} for discharge maintenance [5.21]. As mentioned in Sect. 2.2.2, this condition results from balancing the charged-particle gain through direct ionization, which is strictly linear in the electron density, with ambipolar diffusion losses, which are also linear in the electron density. This condition indeed links the electron temperature to the electric-field intensity. However, in that solution there is no link to the electron density. In spite of this the Schottky condition gives a good estimate of the field intensity in discharges with a comparatively low plasma density, for example DC discharge plasmas. The discussions of these discharges normally involve a maintenance field equal to the Schottky threshold value, which is not connected to the density; the density is determined by the discharge current [5.22]. Increasing the input power, which in DC discharges means a higher discharge current, is considered to lead to an increase of the density under the condition of a constant value of the maintenance field and, therefore, of the temperature. In a way, the discharge current appears as an external parameter which determines the plasma density, and the maintenance field amplitude is kept at a constant value.

The direct transfer of the concept of a constant temperature and, consequently, of a constant maintenance-field intensity from the theory of DC discharges to that of SW-sustained discharges causes a lack of self-consistency in the description of the discharge. A link between field intensity and plasma density should exist, since the discharge production is clearly a nonlinear interaction of the EM fields and the plasma. In the case of SW-sustained plasmas the waves exist because of the presence of the plasma, and the plasma exists only because of the presence of the waves.

When the Schottky condition is used in the modelling of SW-sustained discharges, the fact that it is only an approximation shows up to the extent of missing the physical basis of the discharge production. To remedy this, one should go back to the point that the discharge production is an effect of a self-consistent ionization nonlinearity. Self-consistency and the link of the electron density to the field intensity throughout the whole plasma is provided by the contributions of the nonlinear processes, in particular those involved in the balance equation for the charged-particle density. This can be realized in models by including a density dependence in the ionization frequency that accounts for step ionization and also recombination losses. The resultant changes in, for instance, the value of the maintenance-field intensity may be quantitatively small, but are sufficient to restore self-consistency throughout.

The fact that SW-sustained discharges produce high-density plasmas can be considered as a reason for taking the nonlinear processes in the particle balance equation as sources of the "field intensity – plasma density" link. As discussed in Chaps. 3 and 4, the surface mode (Sect. 3.2, region (a) in Fig. 3.2) is a proper mode of overdense plasmas. In order to sustain a discharge by this mode, the breakdown of the neutral gas in the region of the wave launcher should be at a plasma density which is not less than that given by the resonance condition (3.1). Therefore, if a discharge is produced by a bulk EM mode (2.24b), n_c (2.11) is the maximum possible value of the density, whereas in SW-sustained discharges the value (3.28)

$$n_c^* = n_c(1 + \varepsilon_d) \tag{5.1a}$$

is its minimum possible value. Moreover, for producing the discharge, propagation of the wave should be ensured, i.e. the condition for weak damping

$$\alpha \lesssim \beta \tag{5.1b}$$

should hold. Inequality (5.1b) for a minimum possible value of the density is a stronger requirement than (5.1a). This minimum density is the threshold density in SW-sustained discharges. The field strength E_{Th} which actually ensures balance between the processes of losses and gain of charged particles at that value of the density should be considered a threshold intensity for these discharges. Since the threshold density (5.1) itself depends on the wave frequency, the frequency not only enters into the effective field for HF discharge production, which is involved in the mechanism of thermal nonlinearity, but also appears separately as a parameter which determines the threshold for breakdown of the neutral gas inside the launcher. An increase of the applied field far above E_{Th} leads to an increase of the density there. This ensures conditions for the propagation of the waves and "moving" the minimum possible threshold value of the density to the end of the column. Therefore, the threshold field E_{Th} and the threshold density n_c^* related to it appear both in the dynamics of the discharge production, as mutually connected parameters of the start of the discharge, and in the stationary state of the discharge, as parameters characterizing the end part of the discharge. Movement from the end of the discharge towards the launcher corresponds to a movement on the corresponding phase diagram (Fig. 3.16), fixed by the σ value, from the region of the SW resonance (3.1) to the bottom of the phase diagram.

In the Schottky approximation (2.67) the threshold field E_{th} is fixed in a simple manner by the discharge conditions: the nature and pressure of the gas and the size of the discharge vessel. This is not the case with the threshold field E_{Th}. The threshold density (5.1) to which E_{Th} is related can vary over a wide range because SW discharges can be produced over a wide span of wave frequencies. Since in the various density ranges different mechanisms influence the discharge production, the dependence of E_{Th} on the discharge conditions is complicated. E_{Th} is not, like E_{th}, fixed at a constant value in the complete electron-density range of existence of discharges in a diffusion-controlled regime. In fact, when the wave frequency changes, E_{Th} follows the same variation which the maintenance field exhibits with density changes along the discharge length. Therefore, using E_{Th} as a normalizing parameter is not convenient, and in the following presentation the maintenance field is normalized to E_{th} given by the Schottky approximation. Moreover, the situation that determines E_{Th} is even more complicated than was outlined in connection with (5.1a). The latter gives the threshold density of the discharge if the plasma is homogeneous in the transverse direction. Since transverse plasma inhomogeneity is a feature of diffusion-controlled discharges, (5.1b) should be used for estimating the threshold density and the threshold field intensity. The determination of the threshold density and field intensity from this equation involves the propagation characteristics of SWs in an inhomogeneous plasma for the given configuration of the waveguide.

Since variation of the plasma density along the discharge length is a basic behaviour of SW-sustained discharges, longitudinal diffusion can be another candidate – besides the nonlinear processes in the particle balance – for ensuring the self-consistent variation of the field intensity and the plasma density. In discharges without an external magnetic field such a "nonlocal" nonlinearity can be important in the launcher region, owing to the higher density in that area and consequent axial diffusion there. The description of a self-consistency ensured by longitudinal diffusion would involve the charged-particle flux from the launcher as a boundary condition. It would result in a nonlocal dependence of the electron density (and of the plasma permittivity) on the electron temperature and, through it, on the field intensity. Nonlocal nonlinearity associated with the charged-particle balance is not included in the following presentation.

In this chapter stationary discharges will be considered. The wave producing the discharge is that branch of the SWs which is classified in Chap. 3 as a surface mode. Having a weak damping in a large range over the phase diagrams, the surface mode, called from here onwards simply an SW, can indeed ensure discharge maintenance and create long plasma columns which extend far from the wave launcher.

The presentation in this chapter does not aim primarily at a precise quantitative determination of the maintenance field intensity or the related power lost on average per electron (Θ) in the diffusion regime. This would actually call for atomic data and for kinetic methods such as those discussed in the next chapter. This chapter, rather, aims at stressing the very aspect of self-consistency and to set it into relief conceptually. To achieve self-consistency in a diffusion-controlled regime, only modest changes of Θ with the electron density, i.e. along the discharge length, are required. The largely analytical approach performed here, using a fluid model, employs some simplifying assumptions, but it has the advantage of preserving more transparency in concentrating on the main aspects of self-consistency, as compared with a more detailed and complete kinetic analysis. In order to connect the modelling to a concrete discharge one needs one base parameter characterizing the specific case under consideration, for instance the plasma density at the beginning of the discharge section of interest or near the launcher; in the latter case the density could conceivably be influenced by intricate processes in the launcher region. Instead of the plasma density, of course, the maintenance field intensity at the start of the discharge section may be employed, or the total power transferred to the whole discharge or discharge section. With a view to measures taken to ensure matching of the power source to the discharge, in some situations the choice of the total power transferred as a parameter may have some practical value. In any case only one parameter is needed and in a totally self-consistent theory the parameters mentioned are conceptually interchangeable.

The organization of this chapter is as follows.

First the set of equations is presented on which the fluid model of the discharge is based (Sect. 5.2). Then the solutions describing the two stages of the ionization nonlinearity – the temperature and density dependence of the intensity of the SW field which maintains the discharge – are considered in Sect. 5.3. With respect to the thermal nonlinearity, cases of both local and nonlocal heating are covered for Joule heating in the volume and in regions of resonance absorption of the field energy. The final results for the ionization nonlinearity are expressed by the transverse density profiles, obtained in terms of the electron temperature and thus of the maintenance field intensity, in both recombination- and diffusion-controlled regimes. Nonlinearity associated with the recombination losses and with step ionization, and a generalization of the treatment which includes the transition between them governed by saturation [5.23] in the step ionization are introduced next. The resulting nonlinear plasma permittivity is given for all of these cases. Up to this point the model deals with the gas-discharge part of the problem of discharge maintenance. The presentation (Sect. 5.4) of the electrodynamic part of the problem is largely based on the results already given in Chaps. 3 and 4. The coupling of the gas-discharge part to the electrodynamic part of the model (Sect. 5.5) is first done for discharge production by slow EM SWs, i.e. the case of discharge maintenance at comparatively small σ values ($\sigma \leq 0.3$) and at values of the density of the plasma produced which keep the wave characteristics within the limits of the thin-cylinder approach ((3.47b), region I in Fig. 3.16). The axial structure of the discharge resulting from self-consistent wave–plasma behaviour appearing in terms of mutually related variations of the field intensity and electron density along the discharge length is the final outcome from the model. The results presented cover all the cases of ionization nonlinearity under consideration: diffusion and recombination regimes with Joule heating in the plasma volume and in regions of resonance absorption. In this way the first part of the chapter is completed.

The second part (Sect. 5.6) starts with general relations describing the axial structure of travelling-wave-sustained discharges. Then axial density profiles in discharges produced by SWs in their complete electromagnetic region of existence are discussed. Cases of discharge maintenance under conditions of both weak and strong collisions are treated. The formation of axial density profiles by consecutively located regions (along the discharge length) of Joule heating in the volume and Joule heating by resonance absorption are described.

5.2 The Set of Equations

The discharge is produced by a travelling wave. The wave propagates owing to the plasma produced. These two statements form the picture of the maintenance of SW-produced discharges. Gas-discharge physics is combined with the physics of EM wave propagation. Neutral-gas breakdown is realized at a given pressure p in a given discharge tube by applying HF power at a given frequency ω to a launcher which covers only a small part of the discharge (Fig. 1.1). The plasma parameters (density n, electron temperature T_e) and wave characteristics (field intensity $|E|^2$, wave number k) are interrelated all over the discharge. In terms of equations which model the phenomenon of discharge maintenance it is necessary to have an adequate set of equations combining the description of both the plasma production and the wave propagation. The structure of the discharge, especially the axial structure expressed by mutually related variations of plasma and wave characteristics, is the result aimed at in the modelling.

The stationary forms of the particle balance equation (2.56)

$$\Delta(D_A n) + \nu_i n + \varrho_{si} n^2 - \varrho_r n^2 = 0 \tag{5.2}$$

and the electron energy balance equation (2.70)

$$-\frac{5}{2}\nabla.(D_e n_e \nabla T_e) + \frac{3}{2} n_e \nu_* U_* = Q \tag{5.3}$$

complete the presentation of the gas-discharge part of the problem. In (5.3) Q is the electron heating as given by the last term in (2.70). The first and second terms on the left-hand side account for energy losses through thermal conduction and inelastic collisions with excitation (2.73), taken as the main contributor to the local energy losses. Neglecting losses by elastic collisions, ionization, etc. [5.24] because these are usually smaller terms of course decreases the accuracy of the analysis, but this does not basically impair the aspect of self-consistency pursued here. Therefore n, T_e and the intensity $|E|^2$ of the heating electric field are the unknown quantities involved in (5.2) and (5.3). Equation (5.3), relating T_e to $|E|^2$, accounts for the thermal nonlinearity. After replacing T_e, determined in terms of $|E|^2$ in (5.2), the final result – the relation of the density n to the field intensity $|E|^2$, i.e. the nonlinear plasma permittivity – can be obtained.

The electrodynamic part of the model for the discharge involves the stationary form of the wave power balance equation (2.32)

$$\frac{dP(|E|^2, n, k)}{dz} = -Q(|E|^2, n) \tag{5.4}$$

and the local dispersion law of SWs

$$\mathcal{D}[\omega, k(z), \varepsilon(|E|^2)] = 0. \tag{5.5}$$

In (5.4) P is the energy flux (2.28b) in the axial (z) direction (i.e., the direction of the wave propagation) and Q is the Joule losses (2.29) of the wave. Both P and Q are defined per unit axial length after integration (2.32b) over the total cross-section $S_\perp$ of the waveguiding structure. Equation (5.4) brings into the model the wave-power influx to the discharge and its distribution over the discharge length. It involves n and $|E|^2$ and adds k as a fourth unknown quantity requiring the use of the dispersion relation.

Equations (5.4) and (5.5) are the set of equations for the nonlinear geometrical-optics to describe wave propagation under stationary conditions [5.25,26], see also the discussion in Sect. 4.3.5.

The plasma permittivity (2.8) involved in (5.5) is a nonlinear plasma permittivity obtained from the gas-discharge part of the problem, because $n = n(|E|^2)$.

Maintenance of a cylindrical plasma column of radius R sustained by an azimuthally symmetric SW of frequency ω, and the field components E_r, H_φ, $E_z \neq 0$ (Sect. 3.3.2) with a phase variation of the form $\propto \exp(\mathrm{i} \int^z k(z')\,\mathrm{d}z' - \mathrm{i}\omega t)$ (Sect. 4.3) are considered. The expressions for P and Q are as given by (3.70) and (3.72).

Equations (5.2)–(5.5) constitute a closed set for the unknown quantities n, $|E|^2$, T_e and k required for a self-consistent description of SW-sustained discharges. Two of the unknown quantities, n and $|E|^2$, appear in both the gas-discharge and the electrodynamic parts of the problem. The connection between these two parts is obvious: the electron energy balance equation (5.3) and the wave power balance equation (5.4) include the same quantity Q ($Q = \int_{S_\perp} Q\,\mathrm{d}S_\perp$), once as Joule heating of the electrons in the wave field and then as Joule losses of the wave power.

5.3 Ionization Nonlinearity

Equations (5.2) and (5.3), used as the gas-discharge part of the model for discharge maintenance in SW fields, are common to all HF discharges (and to discharges in general). Their solution represents a relation between plasma density and field intensity, i.e. a result for the plasma permittivity with a strong ionization nonlinearity – a case in which the plasma density as a whole is determined by the heating-field intensity.

5.3.1 Electron Temperature in Terms
of the Maintenance Field Intensity

According to the discussion at the end of Sect. 2.2.2, the effect of the thermal nonlinearity appears in either a local or a nonlocal approach.

The presentation of the results on the thermal nonlinearity starts here with the Joule heating through collisions in the plasma volume as being a mechanism of heating common to all kinds of discharges. Moreover, this type of electron heating is the most important one in SW-sustained discharges; it covers the main, longer part of the discharge length. Then the results on the thermal nonlinearity occuring through the Joule heating in the regions of the resonance absorption of SWs in plasmas with inhomogeneity in the transverse direction – specific to SWs – (Sect. 4.2) are given.

Joule Heating in the Plasma Volume. Conditions of *local heating* (Sect. 2.2.2, [5.17,20,27,28]) cause the thermal-conductivity term to drop from (5.3) and reduce (5.3) to a simple form

$$\frac{3}{2}n\nu_*U_* = Q, \tag{5.6a}$$

from which, after using (2.75) for the excitation frequency, the electron temperature is easily obtained as

$$T_e = -\frac{U_*}{\ln(|E|^2/E_i^2)}. \tag{5.6b}$$

Here

$$E_i = \left[\frac{3m}{e^2}\frac{\mathring{\nu}_*}{\nu}(\omega^2 + \nu^2)U_*\right]^{1/2} \tag{5.7}$$

is a normalizing field which determines the efficiency of the thermal nonlinearity and $|E|^2$ is the heating-field intensity. The locality of the linkage (5.6b) between T_e and $|E|^2$ means – in the case of electron heating in SW fields – radial inhomogeneity of the temperature, with higher values close to the tube walls.

When (2.72a) – instead of (2.73) – is employed for describing the electron energy losses through collisions, the result for the temperature as given by (5.6b) is the same. However, in (5.7) U_* is replaced by some value of $T_e \rightarrow \bar{T}_e$ estimated as a guess in advance, which enters the slow variation of ν_* with T_e (not the strong, exponential variation).

When energy transfer through elastic collisions can be considered as being the predominant mechanism (see (2.72a)), the electron temperature is [5.29]

$$T_e = T_g + \frac{e^2|E|^2}{3m\delta(\omega^2 + \nu^2)}. \tag{5.8}$$

Such a case will not be treated further.

Now the regime of *nonlocal heating* will be considered.

In a diffusion-controlled regime the characteristic length of the thermal conduction L_χ (2.76) is usually larger than the discharge radius ($L_\chi > R$). This means (Sect. 2.2.2) that the discharge is under conditions of *nonlocal heating* [5.30–35].

The effect of the thermal conduction, given by the first term in (5.3), is to ensure radial uniformity of the plasma heating and leads to an almost homogeneous radial distribution of T_e. This means that the solution of the electron energy balance equation (5.3) should be represented in the form

$$T_e(r, z) = T_{e_0}(z) + \Delta T_e(r, z),\tag{5.9}$$

where $T_{e_0}(z)$ is constant over the discharge cross-section and $\Delta T_e(r, z)$ is a small correction which depends on the radial profiles of both the electric-field intensity and the electron density.

The integration of (5.3) over the cross-section of the discharge

$$U_* \int_{S_\perp} n \nu_* \, \mathrm{d}S_\perp = \frac{e^2 \nu}{3m(\omega^2 + \nu^2)} \int_{S_\perp} n|\boldsymbol{E}|^2 \, \mathrm{d}S_\perp \tag{5.10}$$

is performed with the assumption that heat flux to the wall

$$w = -\frac{5}{2} R D_e(r = R) \left[n \left(\frac{\mathrm{d}T_e}{\mathrm{d}r} \right) \right]_{r=R} \tag{5.11}$$

is negligibly small.

Substitution of (5.9) into (5.10) for a small enough value of $\Delta T/T_0$ ($\Delta T/T_0 \ll T_0/U_*$) leads to an expression for the averaged energy gain per electron

$$3U_* \nu_*(T_{e_0}) = \frac{e^2 \nu \left\langle |\boldsymbol{E}|^2 \right\rangle}{m(\omega^2 + \nu^2)} \tag{5.12}$$

and gives the zero-order approximation to the electron temperature as

$$T_{e_0} = -\frac{U_*}{\ln\left(\langle E^2 \rangle / E_i^2\right)}. \tag{5.13a}$$

Here

$$\langle E^2 \rangle = \frac{\int_{S_\perp} n|\boldsymbol{E}|^2 \, \mathrm{d}S_\perp}{\int_{S_\perp} n \, \mathrm{d}S_\perp} \tag{5.13b}$$

is the heating-field intensity. Now, in the nonlocal regime, the averaged field intensity (over the transverse density profile) – and not directly the field intensity, as in the case of local heating – appears as the maintenance field intensity. The normalizing field E_i is the same (5.7) as in the local regime. Equation (5.13a) describes the effect of the thermal nonlinearity, linking plasma and wave properties.

The correction ΔT_e, obtained in [5.31] for the case of a cylindrical discharge, is

$$\Delta T_e(r) = \Delta T_e(r = 0) + \frac{1}{5D_e(T_{e_0})} \frac{e^2 \nu}{m(\omega^2 + \nu^2)} \int_0^r \frac{\mathrm{d}r'}{r' n(r')}$$

$$\times \int_0^{r'} \left[\langle |\boldsymbol{E}|^2 \rangle - |\boldsymbol{E}(r'')|^2 \right] r'' n(r'') \, \mathrm{d}r'', \tag{5.14a}$$

where the value of the integration constant $\Delta T_e(r = 0)$,

$$\Delta T_e(r = 0) = - \left[5 D_e(T_{e0}) \int_0^R r n(r)\, dr \right]^{-1} \frac{e^2 \nu}{m(\omega^2 + \nu^2)}$$

$$\times \int_0^R r n(r)\, dr \int_0^r \frac{dr'}{r' n(r')} \int_0^{r'} \left[\langle |\boldsymbol{E}|^2 \rangle - |\boldsymbol{E}(r'')|^2 \right] r'' n(r'')\, dr'' \,,$$

$$(5.14b)$$

results from the normalization condition

$$\int_0^R \Delta T(r) n(r) r\, dr = 0 \,. \tag{5.14c}$$

Therefore, for a radially uniform plasma density and an electric-field intensity monotonically decaying towards the discharge axis, the contribution of the second term on the right-hand side of (5.14a) decreases towards the discharge centre, and the temperature has a minimum at the discharge axis. However, this minimum is much less pronounced than that which would result from the local approach.

For spatially uniform heating-field intensities, $\langle |\boldsymbol{E}|^2 \rangle = |E|^2$ and $\Delta T_e(r) = 0$.

The zero-order solution (5.13a) of (5.3) is the one taken into account in the considerations below. It is obtained from (5.10), which represents the electron energy balance in a form unified with that of the wave energy balance (5.4).

Joule Heating in Regions of Resonance Absorption. This is a mechanism of heating related to the mechanism of SW damping in a plasma with an inhomogeneity in the transverse direction: damping which occurs through the linear mode transformation of the SW into volume plasmons at a point on the transverse density profile where the local density $n(r = r_r)$ is equal to the critical density (2.11) and the local permittivity (2.9) has zero real part (Sect. 4.2). This heating occurs in the resonance region (i.e. close to the discharge walls) and acts mainly through the resonantly enhanced E_r field component. Because of the strong variation of the field, the regime of the heating is *nonlocal.*

The electron temperature is obtained from (5.10) for the electron energy balance after integration of (5.3) over the cross-section of the discharge. The introduction of (4.73) for the Joule losses Q due to resonance absorption of SWs in a radially inhomogeneous plasma column in (5.3) (and in the corresponding term in (5.10)) leads to the relation

$$T_e = - \frac{U_*}{\ln(E_{(r)}^2 / E_i^2)} \,. \tag{5.15a}$$

between the electron temperature and the heating field. Here E_i is the same as (5.7) and

$$E_{(r)}^2 = P_r \frac{n_c}{\bar{n}} \frac{2}{\tilde{x}} \frac{I_1(\tilde{x})}{I_0(\tilde{x})} |E_z(r = R)|^2 \tag{5.15b}$$

contains P_r as given by (4.73c). $E_{(r)}$ appears here as an effective heating field [5.35]. In fact, it is the field for discharge maintenance through the resonance absorption mechanism of heating. Therefore, the field intensity $\langle E^2 \rangle$, which in the case of Joule heating in the plasma volume appears in (5.13a) for the electron temperature, is now replaced by $E_{(r)}^2$ (see (5.15a)). The characteristic length of the density inhomogeneity $L_n^{(r)} = |(d\varepsilon_r/dr)^{-1}|_{r=r_r}$ at the resonance point, which determines the efficiency of the resonance absorption, determines the maintenance field intensity as well.

5.3.2 Power Absorbed on Average per Electron

The quantity $\Theta = Q/\bar{n}$, expressing the power absorbed on average per electron, due to Glaude et al. [5.16], Ferreira [5.18] and Zakrzewski [5.19], is widely used in the literature on SW-sustained discharges. It is proportional to the maintenance field intensity, i.e. to the wave field $|E|^2$ in the case of local Joule heating in the plasma volume and to the effective fields $\langle |E|^2 \rangle$ and $E_{(r)}^2$, which are the maintenance fields in the nonlocal regimes (nonlocal Joule heating in the plasma volume and Joule heating in regions of resonance absorption, respectively).

5.3.3 Ionization Frequency in Terms of the Maintenance Field Intensity

Power laws govern the dependence of the ionization frequency ν_i on the maintenance electric field. The dependences obtained after inserting the results for T_e (Sect. 5.3.1) in the expressions for ν_i are as listed below.

Introduction of (5.6b), which gives T_e when the discharge is sustained by local Joule heating in the plasma volume, into (2.59b) and (2.60b) leads to

$$\nu_i^{(1)} = \mathring{\nu}_i^{(1)} \frac{|E|^2}{E_i^2}, \tag{5.16a}$$

$$\nu_i^{(2)} = \mathring{\nu}_i^{(2)} \frac{|E|^{2s}}{E_i^{2s}}. \tag{5.16b}$$

The applicability of these relations is connected to conditions of comparatively high and low temperatures (see (2.59a) and (2.60a)), respectively; here $s = U_i/U_*$. The expressions (5.16) are used later in the description of the discharge maintenance in a recombination-controlled regime.

In the more general case of nonlocal heating and with direct and step ionization treated separately, the dependences of the frequency of direct ionization (2.57) and the rate coefficient (2.62a) of step ionization on the maintenance field intensity are

$$\nu_i = \overset{\circ}{\nu}_i \left(\frac{\langle |E|^2 \rangle}{E_i^2} \right)^s \quad \text{(case A)}, \tag{5.17a}$$

$$\nu_i = \overset{\circ}{\nu}_i \left(\frac{E_{(r)}^2}{E_i^2} \right)^s \quad \text{(case B)}. \tag{5.17b}$$

$$\varrho_{si} = \overset{\circ}{\varrho}_{si} \left(\frac{\langle |E|^2 \rangle}{E_i^2} \right)^s \quad \text{(case A)}, \tag{5.18a}$$

$$\varrho_{si} = \overset{\circ}{\varrho}_{si} \left(\frac{E_{(r)}^2}{E_i^2} \right)^s \quad \text{(case B)}. \tag{5.18b}$$

The regime of nonlocal Joule heating in the plasma volume is denoted by case A. Equations (5.17a) and (5.18a) are obtained by using (5.13a). Case B is related to nonlocal Joule heating through resonance absorption. Equation (5.15a) is utilized to obtain (5.17b) and (5.18b).

5.3.4 Plasma Density Expressed in Terms of the Maintenance Field Intensity

The relationships of the plasma density to the intensity of the HF field which sustains the discharge and the nonlinear plasma permittivity given here are deduced from the particle balance equation (5.2). They are obtained by using the results for the thermal nonlinearity presented in the previous subsection. Both recombination- and diffusion-controlled regimes are considered [5.17,20,27,28,32–34].

Recombination-Controlled Regime. In this case (5.2) reduces to

$$\nu_i = \varrho_r n, \tag{5.19a}$$

from which, through ν_i, the relationships of the density n to T_e and, through $T_e(|E|^2)$, to $|E|^2$ are obvious. With dissociative recombination, n on the right-hand side of (5.19a) is in the form $n_{i_{(m)}} = \tilde{\xi} n_i$; see the comments on (2.64). With $\tilde{\xi} = 1$ and after taking into account that ϱ_r is a slowly varying function of T_e, (5.19a) yields, after using (5.16),

$$n = \frac{\overset{\circ}{\nu}_i^{(1)}}{\varrho_r} \frac{|E|^2}{E_i^2}, \tag{5.19b}$$

$$n = \frac{\overset{\circ}{\nu}_i^{(2)}}{\varrho_r} \frac{|E|^{2s}}{E_i^{2s}}, \tag{5.19c}$$

for comparatively high and low temperatures, according to (2.59a) and (2.60a).

Diffusion-Controlled Regime. First, by following the steps of the treatments in [5.32,33], the relationships of the density to the field intensity in discharges maintained by Joule heating in the plasma volume are determined, as they stem from the nonlinear terms (recombination losses and step ionization) in the particle-balance equation (5.2). These results cover regions of comparatively high and low plasma densities, respectively, in the density range of the diffusion-controlled regime. Then, by a generalization [5.34] accounting for the saturation [5.23] in the step ionization, the region of intermediate values of density is also covered. The mechanism of Joule heating due to resonance absorption is involved in the generalization of the model, too. With respect to SW-sustained discharges, the two mechanisms of heating are related to the main, usually long, part of the discharge length and to the end of the discharge, respectively.

Case (i): diffusion-controlled discharge with nonlinearity due to recombination losses. Discharge maintenance in a diffusion-controlled regime at a comparatively high plasma density and, correspondingly, comparatively high gas pressure is considered. The nonlinearity in (5.2),

$$\Delta(D_A n) + \nu_i n = \varrho_r n^2 \,, \tag{5.20a}$$

is related to the recombination losses. For $R \ll L_\chi$, i.e. under conditions of nonlocal heating, $T_e(r) = $ const. and therefore D_A, ν_i and ϱ_r do not depend on the radial space coordinates. An exact solution of (5.20a) can easily be obtained in a model of a planar discharge – a slab $(-d < x < d)$ of thickness $2d$. After introducing a dimensionless density $n_1 \equiv n(\varrho_r/\nu_i)$ and a dimensionless transverse coordinate $\xi \equiv x(\nu_i/D_A)^{1/2}$, (5.20a) reads

$$\frac{\mathrm{d}^2 n_1}{\mathrm{d}\xi^2} + n_1 = n_1^2 \,. \tag{5.20b}$$

Its solution can be expressed in terms of Jacobian elliptic functions. The phase portrait of (5.20b) is depicted in Fig. 5.1; $\dot{n}_1 = \mathrm{d}n_1/\mathrm{d}\xi$. The phase trajectories are given by

$$\dot{n}_1^2 = n_{1_0}^2 \left(1 - \frac{2}{3}n_{1_0}\right) - n_1^2 \left(1 - \frac{2}{3}n_1\right) \,, \tag{5.21a}$$

where $n_{1_0} \equiv n_1(\xi = 0) = \varrho_r n(x = 0)/\nu_i$ is the dimensionless density at the discharge axis; $n_1 = 0$ and $n_1 = 1$ are the equilibrium points. The equation of the separatrix corresponds to $n_{1_0} = 1$ and has the form

$$\dot{n}_1^2 = \frac{2}{3} \left(n_1 + \frac{1}{2}\right) (n_1 - 1)^2 \,. \tag{5.21b}$$

The boundary conditions $n(x = \pm d) = n_w$ determine, as physically meaningful solutions, only those sections of the curves which are inside the closed part of the separatrix, i.e. the region $1 \geq n_1 \geq n_{1_w}$; here n_{1_w} is the dimensionless electron density at the discharge walls.

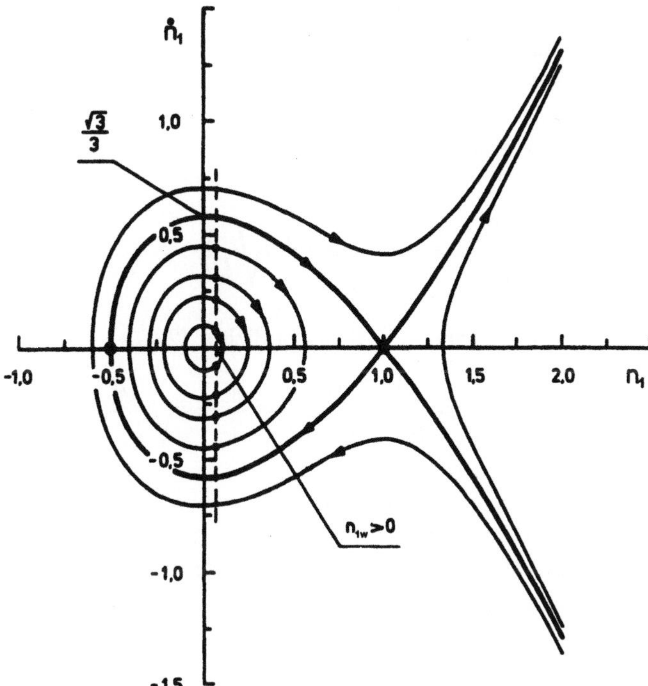

Figure 5.1. Phase portrait of (5.20b). The separatrix is also marked (*bold curve*) and the case of a plasma density at the wall (n_{1_w}) not equal to zero is also included (*dashed line*) ([5.32], Fig. 1)

The phase trajectories in the vicinity of the origin of the $(n_1, \dot{n}_1)$ space, correspond to a "pure" diffusion regime, whereas those very close to the separatrix describe a "pure" recombination regime. In fact, the diffusion length L_D (2.66a) and the density $n_{\max} = \nu_i/\varrho_r$ in the recombination regime, i.e. those quantities which characterize the diffusion and recombination regimes, respectively, are the quantities used to obtain the normalized equation (5.20b). The phase portrait shows that the plateau (i.e. the region of small $|\dot{n}_1|$ values), which is formed close to the discharge axis, spreads over a larger area when $n_{1_0} \to 1$, i.e. when the phase trajectories are close to the separatrix. In this case the dimensionless density gradients at the walls

$$\dot{n}_1\big|_{r=\pm d(\nu_i/D_A)^{1/2}} = n_{1_0}\sqrt{1 - \frac{2}{3}n_{1_0}} \tag{5.21c}$$

tend to the limiting value $|\dot{n}_1|_{\max} = \sqrt{3}/3$.

A second integration of (5.20b), after applying the boundary condition $n_1(\xi = d\sqrt{\nu_i/D_A}) = n_{1_w}$, yields

$$\int_{n_{1_w}}^{n_{1_0}} \frac{\mathrm{d}n_1}{\sqrt{\frac{2}{3}n_1^3 - n_1^2 + n_{1_0}^2\left(1 - \frac{2}{3}n_{1_0}\right)}} = d\sqrt{\frac{\nu_i}{D_A}}, \tag{5.22a}$$

i.e. a relation which has the meaning of a condition for discharge maintenance. It contains the effect of nonlinearity: relating T_e (through ν_i and D_A) to n_{1_0} (the density at the discharge axis), it relates – because of the dependence of T_e on $|E|^2$ – the density of the plasma to the intensity $|E|^2$ of the field sustaining the discharge. In the case of $n_{1_w} = 0$, an assumption made for simplicity, (5.22a) reduces to

$$I(n_{1_0}) \equiv \int_0^{n_{1_0}} \frac{dn_1}{\sqrt{\frac{2}{3}n_1^3 - n_1^2 + n_{1_0}^2\left(1 - \frac{2}{3}n_{1_0}\right)}} = d\sqrt{\frac{\nu_i}{D_A}}. \tag{5.22b}$$

For $n_{1_0} \ll 1$ (5.22b) transforms into the Schottky condition (2.67a)

$$I(n_{1_0} \to 0) \equiv \int_0^{n_{1_0}} \frac{dn_1}{\sqrt{n_{1_0}^2 - n_1^2}} \equiv \int_0^1 \frac{dy}{\sqrt{1 - y^2}} \equiv \frac{\pi}{2} = d\sqrt{\frac{\nu_i}{D_A}}, \tag{5.22c}$$

and the linkage between n_{1_0} and $|E|^2$ disappears.

The integral on the left-hand side of (5.22b) can be represented exactly, in terms of an elliptical integral of the first kind. However, over a wide range of n_{1_0}, this integral can be approximated by a simpler one, which corresponds to the solution of the equation

$$\frac{d^2 n_1}{d\xi^2} + (1 - n_{1_0})n_1 = 0, \tag{5.23a}$$

obtained by the replacement $n_1^2 \to n_1 n_{1_0}$. Such a replacement accounts for the importance of the nonlinearity in the vicinity of the discharge axis. The solution of (5.23a) is

$$n_1(\xi) = n_{1_0} \cos\left(\sqrt{1 - n_{1_0}}\,\xi\right). \tag{5.23b}$$

The conditon for discharge maintenance

$$\tilde{I}(n_{1_0}) \equiv \frac{1}{\sqrt{1 - n_{1_0}}} \int_0^{n_{1_0}} \frac{dn_1}{\sqrt{n_{1_0}^2 - n_1^2}} \equiv \frac{\pi}{2} \frac{1}{\sqrt{1 - n_{1_0}}} = d\sqrt{\frac{\nu_i}{D_A}}, \tag{5.23c}$$

which corresponds to (5.23a), accounts for the linkage between n_{1_0} and $T_e(|E|^2)$, i.e. for the nonlinearity.

The integrals in (5.22b) and (5.23c), denoted by $I(n_{1_0})$ and $\tilde{I}(n_{1_0})$, respectively, are compared in Fig. 5.2, which illustrates the good accuracy of the approximation made and its validity over a wide range of n_{1_0}.

Turning back to the original variables, the electron density at the discharge axis and its transverse profile are

$$n(x = 0) = \frac{1}{\varrho_r}\left[\nu_i - \left(\frac{\pi}{2d}\right)^2 D_A\right], \tag{5.24a}$$

$$n(x) = \frac{1}{\varrho_r}\left[\nu_i - \left(\frac{\pi}{2d}\right)^2 D_A\right]\cos\left(\frac{\pi}{2d}x\right), \tag{5.24b}$$

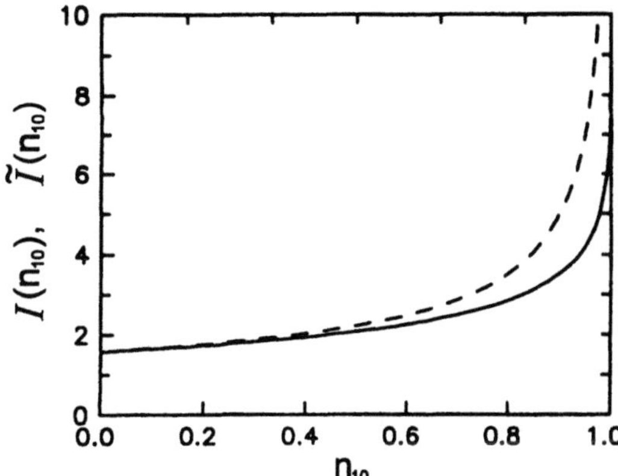

Figure 5.2. Check of the accuracy of the approximation $n_1^2 \simeq n_1 n_{1_0}$. The dependences on n_{1_0} of the integral (*solid curve*) in (5.22b), denoted by $I(n_{1_0})$, and of the solution (*dashed curve*) of the integral in (5.23c), denoted by $\tilde{I}(n_{1_0})$, are compared ([5.32], Fig. 2)

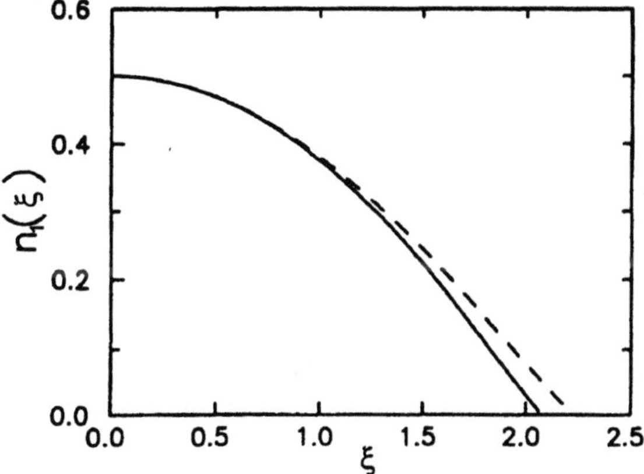

Figure 5.3. Comparison of the density profiles (in normalized quantities) obtained from the exact (*solid curve*) and approximate (*dashed curve*) solutions of (5.20b) given by (5.25) and (5.23b), respectively, at $n_{0_1} = 0.5$ ([5.32], Fig. 3)

respectively. These expressions show that, in contrast to a "pure" diffusion regime ($\varrho_r = 0$ in (5.20a) and the condition (5.22c) for sustaining the discharge), in which the density value at the axis is not specified by the particle balance equation, accounting for a finite value of ϱ_r leads to a determination of $n(x = 0)$. Because the very small value of ϱ_r, even a small deviation from equality of the two terms in (5.24a) ensures values of $n(x = 0)$ that are reasonable for the diffusion-controlled regime.

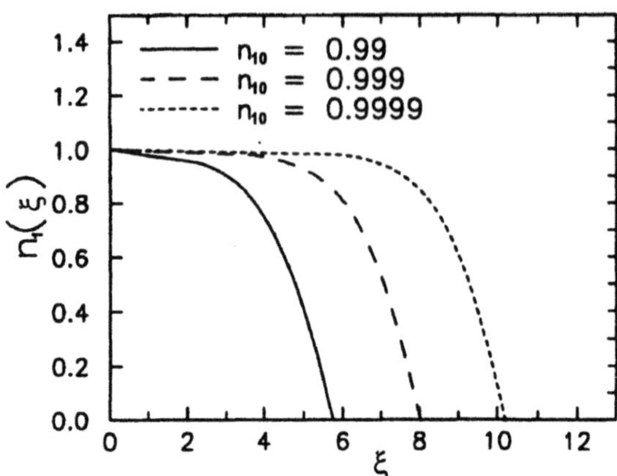

Figure 5.4. Density profiles (in normalized quantities) obtained from the exact solution (5.25) at values of n_{1_0} close to unity ([5.32], Fig. 4)

The validity of the approximation $n_1^2 \to n_1 n_{1_0}$ in the nonlinear term of (5.20a) can be checked also by comparing (Fig. 5.3) the transverse density profile $n_1(\xi)$ determined by the exact solution of (5.20b),

$$\int_{n_1(\xi)}^{n_{1_0}} \frac{dn_1}{\sqrt{\frac{2}{3}n_1^3 - n_1^2 + n_{1_0}^2\left(1 - \frac{2}{3}n_{1_0}\right)}} = \xi, \tag{5.25}$$

with the profile (5.23b) given by the approximate solution. Obviously, the approximation with a cosine function profile is good enough over a wide range of n_{1_0}, excluding the values $n_{1_0} \lesssim 1$ where a transition to the recombination regime shows up via an almost constant value of the density over most of the cross-section of the discharge (Fig. 5.4). The transverse profile possesses a flat, wide plateau with only a narrow region of steep descent towards the wall.

Obviously, the form of the transverse density profile depends on the geometry of the discharge. Assuming that in cylindrical geometry the replacement $n_1^2 \to n_1 n_{1_0}$ gives the same accuracy as for planar geometry, the solution of (5.20a) can be written as

$$n(r) = \frac{1}{\varrho_r}\left[\nu_i - \left(\frac{2.4}{R}\right)^2 D_A\right] J_0\left(\frac{2.4}{R}r\right). \tag{5.26a}$$

After introducing the effect of the thermal nonlinearity (5.6b) and (5.17a), the dependences of the plasma density at the axis and of the complete density profile on the field intensity come out straightforwardly as

$$n(r = 0, |E|^2) = N_{1_0}\left(\frac{\langle E^2\rangle^s}{E_{\text{th}}^{2s}} - 1\right), \tag{5.26b}$$

$$n(r) = N_{1_0} \left(\frac{\langle E^2 \rangle^s}{E_{\text{th}}^{2s}} - 1 \right) J_0 \left(\frac{2.4}{R} r \right) .$$
(5.26c)

Here

$$N_{1_0} = \left(\frac{2.4}{R} \right)^2 \left(\frac{D_A}{\varrho_r} \right)$$
(5.26d)

has the dimensions of a density and

$$E_{\text{th}} = E_i \left(\frac{2.4}{R} \sqrt{\frac{D_A}{\mathring{\nu}_{i_{1,2}}}} \right)^{(1/s)}$$
(5.27)

is a normalizing field, defined as the field amplitude for discharge maintenance by direct ionization in which the losses are associated only with ambipolar diffusion (a definition according to the Schottky condition). In a planar discharge, the factor $(2.4/R)$ in (5.27) is replaced by $(\pi/2d)$.

As (5.26) shows, discharge maintenance in a diffusion-controlled regime at comparatively high pressure and electron density – a case in which recombination contributes to the charged-particle losses – requires a field intensity slightly above the Schottky threshold field E_{th}. In the limit of a field intensity which strongly exceeds the normalizing field intensity, (5.26b) tends to the result (5.19c) for a recombination-controlled regime.

The density averaged over the discharge cross-section obtained from (5.24) and (5.26) is

$$\bar{n} = \frac{2}{\pi} n(x = 0)$$
(5.28a)

and

$$\bar{n} \simeq \frac{1}{2} n(r = 0) ,$$
(5.28b)

in planar and cylindrical discharges, respectively.

The nonlinear plasma permittivity is directly obtained after inserting (5.24b) and (5.26b) into (2.8a):

$$\varepsilon(|E|^2) = 1 - \frac{n(|E|^2)}{n_c} .$$
(5.29)

Case (ii): diffusion-controlled regime with a nonlinearity due to step ionization. This is the case of discharge maintenance in a diffusion-controlled regime at a comparatively low plasma density. Under such conditions the step ionization [5.36,37] is the nonlinear process in the particle balance. Equation (5.2) becomes

$$\Delta(D_A n) + \nu_i n + \varrho_{\text{si}} n^2 = 0 .$$
(5.30a)

The difference between this case and that before, with recombination, (5.20a), lies not only in the sign of the nonlinear term: whereas ϱ_r is a slowly varying function of the temperature, ϱ_{si} (2.62), the rate of step ionization, depends strongly on T_{e}.

Following the pattern of the treatment of case (i), the validity of an approximate solution of (5.30a), obtained for a planar discharge, is checked by comparing it with the exact solution. Then the same approximation is used for a discharge with a cylindrical geometry. However, in the present case of a nonlinearity associated with step ionization, the application of an approximation proved to be valid in the planar case to a cylindrical case is additionally checked by comparison with the numerical results given in [5.37].

In dimensionless variables $n_2 = n\varrho_{\mathrm{si}}/\nu_{\mathrm{i}}$ and $\xi = x\sqrt{\nu_{\mathrm{i}}/D_{\mathrm{A}}}$, (5.30a) takes the form

$$\frac{\mathrm{d}^2 n_2}{\mathrm{d}\xi^2} + n_2 + n_2^2 = 0 \,. \tag{5.30b}$$

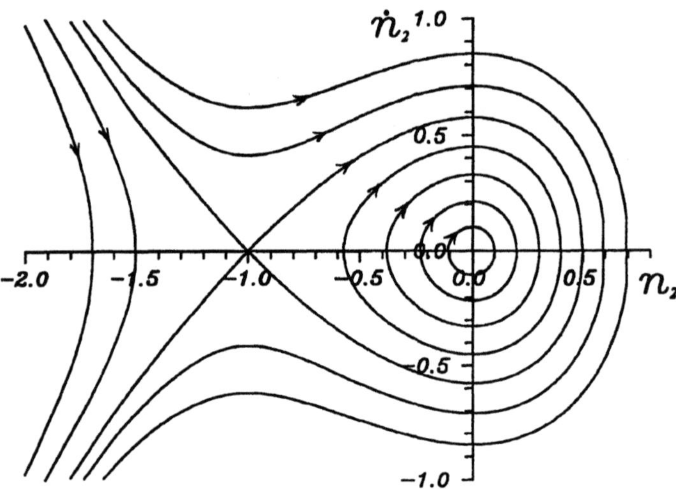

Figure 5.5. Phase portrait of (5.30b) ([5.33], Fig. 1)

A first integration of (5.30b), taking into account the requirement for a maximum value of the density $n_{2_0} \equiv n_2(\xi = 0)$ at the axis, leads to

$$\dot{n}_2^2 = n_{2_0}^2 \left(1 + \frac{2}{3} n_{2_0}\right) - n_2^2 \left(1 + \frac{2}{3} n_2\right) \,, \tag{5.31a}$$

presented in Fig. 5.5. The phase portrait (Fig. 5.5) is a mirror image, in the $\dot{n}_2$ axis, of that of (5.20b) (Fig. 5.1). The physically reasonable region is specified by $n_2 > 1$. Similarly to case (i), the region close to the origin corresponds to a diffusion regime, with direct ionization being the only process for charged-particle gain. A value of $n_{2_0} = 1$ corresponds to equal contributions from the processes of direct and step ionization at the discharge axis. Increasing n_2 means increasing the contribution of the step ionization. Compared with the recombination–diffusion regime (case i), there is no limitation associated with the separatrix.

A second integration of (5.30b) results in the transverse density profile

$$\int_{n_2(\xi)}^{n_{2_0}} \frac{\mathrm{d}n_2}{\sqrt{n_{2_0}^2 \left(1 + \frac{2}{3}n_{2_0}\right) - n_2^2 \left(1 + \frac{2}{3}n_2\right)}} = \xi .$$

(5.31b)

With a vanishing electron density $n_2(\xi = d\sqrt{\nu_i/D_A}) = 0$ at the discharge wall assumed for simplicity, (5.31b) yields a relation between the normalized density at the axis, the electron temperature and the normalized slab width

$$I(n_{2_0}) = \int_0^{n_{2_0}} \frac{\mathrm{d}n_2}{\sqrt{n_{2_0}^2 \left(1 + \frac{2}{3}n_{2_0}\right) - n_2^2 \left(1 + \frac{2}{3}n_2\right)}} \equiv d\sqrt{\frac{\nu_i}{D_A}} .$$

(5.31c)

This is the condition for discharge maintenance. For $\varrho_{si} \to 0$, $n_{2_0} \ll 1$ and (5.31c) reduces to the Schottky approximation (2.67a).

A replacement $n_2^2 \to n_2 n_{2_0}$ in the nonlinear term in (5.30b) transforms (5.31c) into

$$\tilde{I}(n_{2_0}) = \frac{\pi}{2} \frac{1}{\sqrt{1 + n_{2_0}}} \equiv d\sqrt{\frac{\nu_i}{D_A}}$$

(5.32a)

and leads to the normalized density profile

$$n_2(\xi) = n_{2_0} \cos\left(\sqrt{1 + n_{2_0}}\; \xi\right) ,$$

(5.32b)

which, in the original variables, is of the form

$$n(x) = \frac{\nu_i}{\varrho_{si}} \left[\left(\frac{\pi}{2d}\right)^2 \frac{D_A}{\nu_i} - 1\right] \cos\left(\frac{\pi}{2d} x\right) .$$

(5.32c)

A comparison of the exact (5.31c) and approximate (5.32a) solutions for the condition of discharge maintenance and for the transverse density profile ((5.31b) and (5.32c)) shows the applicability of the approximate solutions even at large values of n_{2_0} (Figs. 5.6 and 5.7).

The approximation $n_2^2 \to n_2 n_{2_0}$ applied to a cylindrical discharge (a plasma column of radius R), together with a zero value of the density at the walls, yields a density profile analogous to (5.32b)

$$n(r) = n_0 J_0\left(\frac{2.4}{R} r\right) ,$$

(5.33a)

where the density at the discharge axis

$$n_0 \equiv n(r = 0) = \frac{\nu_i}{\varrho_{si}} \left[\left(\frac{2.4}{R}\right)^2 \frac{D_A}{\nu_i} - 1\right]$$

(5.33b)

is expressed in terms of T_e. Introducing (5.17a) and (5.18a) into (5.33b) leads straigtforwardly to the final result for the effect of the nonlinearity, the plasma density at the axis:

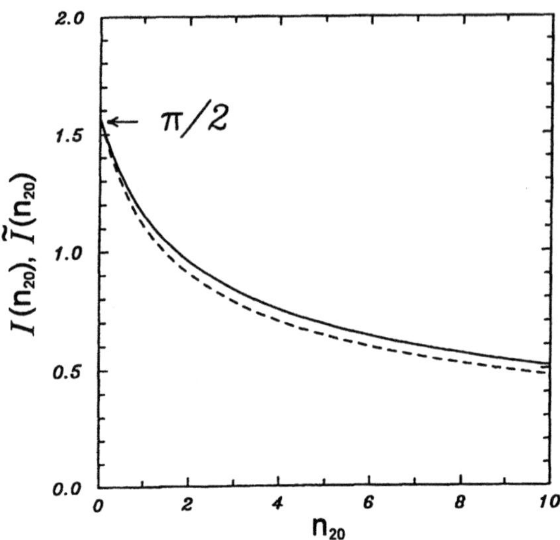

Figure 5.6. Check of the accuracy of the approximation $n_2^2 \simeq n_2 n_{2_0}$. Dependences on n_{2_0} of the integral (*solid curve*) in (5.31c), denoted by $I(n_{2_0})$, and of the expression (*dashed curve*) in (5.32a), denoted by $\tilde{I}(n_{2_0})$ ([5.33], Fig. 2)

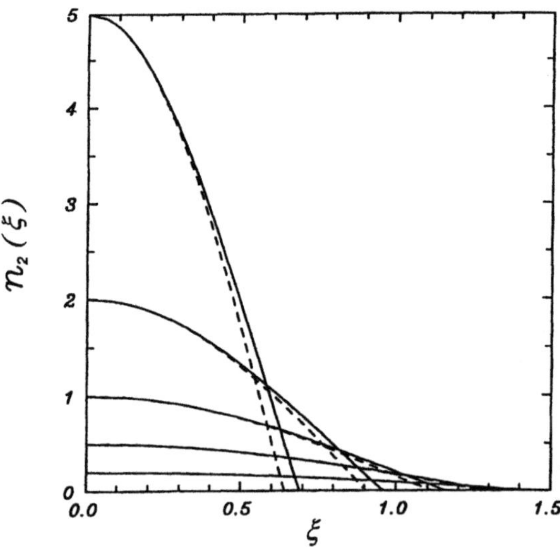

Figure 5.7. Normalized density profiles from the exact (*solid curves*) (5.31b) and approximate (*dashed curves*) solution (5.32c) ([5.33], Fig. 3)

$$n(r = 0) = N_{2_0} \left[\left(\frac{\langle E^2 \rangle}{E_{\text{th}}^2} \right)^{-s} - 1 \right] , \tag{5.34a}$$

expressed in terms of the maintenance field intensity; here

$$N_{2_0} = \frac{\overset{\circ}{\nu}_i}{\overset{\circ}{\varrho}_{\text{si}}} . \tag{5.34b}$$

The normalizing field E_{th} is the same (5.27) as in the case of nonlinearity associated with recombination losses. Equation (5.34a) shows that in the case of a nonlinearity associated with step ionization, which contributes to charged-particle production by direct ionization, a field intensity lower than the normalizing field determined by the mechanism of direct ionization maintains the discharge.

The quantity

$$\frac{n_0}{N_{2_0}} = \left(\frac{\langle E^2 \rangle}{E_{\text{th}}^2} \right)^{-s} - 1 \tag{5.34c}$$

is the ratio of the frequencies of step ionization and direct ionization at the discharge axis: $n_0/N_{2_0} < 1$ corresponds to a direct-ionization-dominated regime, whereas $n_0/N_{2_0} \gg 1$ presents a regime where step ionization is strongly dominant.

The nonlinear plasma permittivity is obtained after inserting (5.33a) and (5.34a) in (5.29).

When the alternative expression (2.62b) for the rate coefficient of step ionization ϱ_{si} is used instead of (2.62a), the result is

$$n(r = 0) = N_{2_0}' \left(\frac{\langle E^2 \rangle}{E_{\text{th}}^2} \right)^{s-1} \left[\left(\frac{\langle E^2 \rangle}{E_{\text{th}}^2} \right)^{-s} - 1 \right] , \tag{5.35a}$$

where

$$N_{2_0}' = \left[\left(\frac{2.4}{R} \right)^2 \frac{D_{\text{A}}}{\overset{\circ}{\nu}_i} \right]^{1-s^{-1}} \frac{\overset{\circ}{\nu}_i}{\overset{\circ}{\varrho}_{\text{si}}'} . \tag{5.35b}$$

Case (iii): generalized procedure with respect to mechanisms of nonlinearity and regimes of plasma heating. The particle balance equation (5.2), written in the form

$$\Delta(D_{\text{A}} n) + \nu_i n + \varrho_{\text{NL}} n^2 = 0 , \tag{5.36a}$$

takes care now, through

$$\varrho_{\text{NL}} = \varrho_{\text{si}} - \varrho_{\text{r}} , \tag{5.36b}$$

of both step ionization and volume recombination.

For a planar discharge, in the dimensionless variables $\xi = x\sqrt{\nu_i/D_{\text{A}}}$ and $n_3 = n(x)/n(x = 0)$, (5.36a) has the form

$$\frac{d^2 n_3}{d\xi^2} + n_3 + \gamma_{NL} n_3^2 = 0 , \tag{5.37a}$$

with $\gamma_{NL} = \varrho_{NL} n(\xi = 0)/\nu_i$. In contrast to cases (i) and (ii), the coefficient ϱ_{NL} is left out in the normalization of the density, because it now accounts for different processes. The density is normalized to its value at the axis.

The coefficient γ_{NL} in (5.37a) represents the ratio, at the discharge axis, of the contribution of the nonlinear term to the contribution of the process of direct ionization. With $\gamma_{NL} > 0$, the step ionization dominates over the recombination, whereas for $\gamma_{NL} < 0$ the recombination losses dominate over the charged-particle production by step ionization at the discharge axis. For $\gamma_{NL} = -1$, which is a limiting value, the volume recombination compensates completely the charged-particle production by both direct and step ionization. In this case, i.e. a "'pure"' recombination regime, the diffusion losses should tend to zero. For $\gamma_{NL} = 0$ the nonlinear contributions of step ionization and recombination are equal and (5.36a) is linear. However, this does not mean that the problem is then linear. Longitudinal diffusion or another nonlinear mechanism which is weaker than the step ionization and recombination should then be included in (5.36a) or the radial electron-temperature inhomogeneity should be treated more accurately and taken into account in the frequency of direct ionization. Moreover, a strictly quadratic dependence of the step ionization rate on the electron density, as so far assumed, also represents only an approximation.

A first integration of (5.37a) yields the equation

$$\dot{n}_3 = \pm\sqrt{C - U(n_3)} \tag{5.37b}$$

of the trajectories on the phase portrait $(n_3, \dot{n}_3)$ with γ_{NL} being a parameter. Using an analogy from mechanics, C is a constant which corresponds to the "total energy of the system" and the function

$$U(n_3) = n_3^2 \left(1 + \frac{2}{3}\gamma_{NL} n_3\right) , \tag{5.37c}$$

having the meaning of a "potential energy", determines the type of trajectory in the phase portrait.

Figure 5.8 exhibits the phase trajectories of (5.37a), which satisfy – for given γ_{NL} values – the requirement for a maximum density at the discharge axis (i.e. $\dot{n}_3 = 0$ at $n_3 = 1$). This requirement relates the value of C to γ_{NL}:

$$C = 1 + \frac{2}{3}\gamma_{NL}. \tag{5.37d}$$

The physically meaningful parts of the trajectories are for $n_3 > 0$. At $\gamma_{NL} = 0$, $C = 1$ and the trajectory is a circle. Increasing the γ_{NL} values ($\gamma_{NL} > 0$) correspond to increasing the value of C, related to an increased contribution of the step ionization, and the transverse density profile becomes steeper. For $\gamma_{NL} < 0$, the value of C decreases because of the losses by volume recombination, and the transverse density profile becomes flatter. $C = 1/3$ is a limiting value obtained at $\gamma_{NL} = -1$.

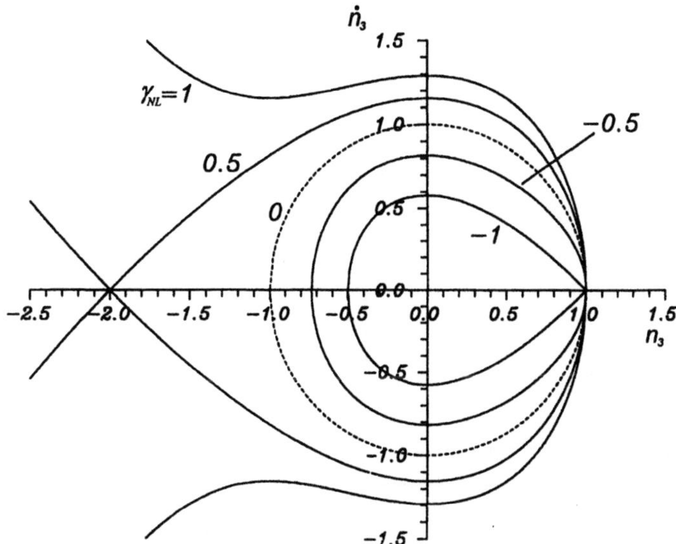

Figure 5.8. Phase trajectories of (5.37a) which, for a given value of $\gamma_{\rm NL}$, satisfy the condition of a maximum of the density at the discharge axis ($\dot{n}_3 = 0$ at $n_3 = 1$) ([5.34], Fig. 1)

The second integration of (5.37a), with the boundary condition $n_3 = 0$ at $\xi = d\sqrt{\nu_{\rm i}/D_{\rm A}}$, i.e. at the discharge walls, represents the condition for discharge maintenance:

$$I(\gamma_{\rm NL}) \equiv \int_0^1 \frac{{\rm d}n_3}{\sqrt{1 - n_3^2 + \frac{2}{3}\gamma_{\rm NL}(1 - n_3^3)}} = d\sqrt{\frac{\nu_{\rm i}}{D_{\rm A}}}. \tag{5.37e}$$

This relates $T_{\rm e}$ (through $\nu_{\rm i}$ and $D_{\rm A}$) to $n(x = 0)$ (through $\gamma_{\rm NL}$) and, therefore, $|E|^2$ (through $T_{\rm e}$) to $n(x = 0)$. When the nonlinearity is ignored ($\gamma_{\rm NL} = 0$), (5.37e) reduces to the Schottky condition (2.67a).

The replacement $n_{1,2}^2 \to n_{1,2}n_{1_0,2_0}$ in the nonlinear terms of cases (i) and (ii) corresponds here to the replacement $n_3^2 \to n_3$ in the nonlinear term of (5.37a). Such an approximation replaces (5.37d) with

$$C^* = 1 + \gamma_{\rm NL}, \tag{5.38a}$$

and transforms (5.37e) into

$$\tilde{I}(\gamma_{\rm NL}) \equiv \frac{\pi}{2}\frac{1}{\sqrt{1 + \gamma_{\rm NL}}} = d\sqrt{\frac{\nu_{\rm i}}{D_{\rm A}}}, \tag{5.38b}$$

keeping the relation between $T_{\rm e}$ and $n(x = 0)$.

Here, as well as before, the replacement $n_3^2 \to n_3$ is a good enough approximation (excluding the region of $\gamma_{\rm NL} \to -1$, which corresponds to a "'pure'" recombination régime). This is depicted in Fig. 5.9, where (5.37e)

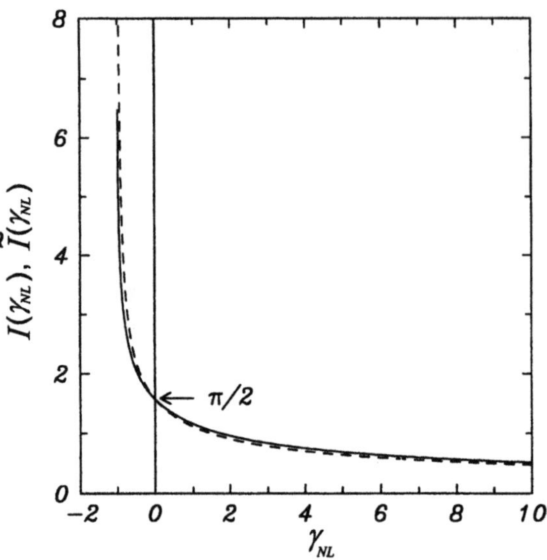

Figure 5.9. Plots of the integral (*solid curve*) in (5.37e), denoted by $I(\gamma_{\mathrm{NL}})$, and of the expression (*dashed curve*) in (5.38b), denoted by $\tilde{I}(\gamma_{\mathrm{NL}})$ ([5.34], Fig. 2)

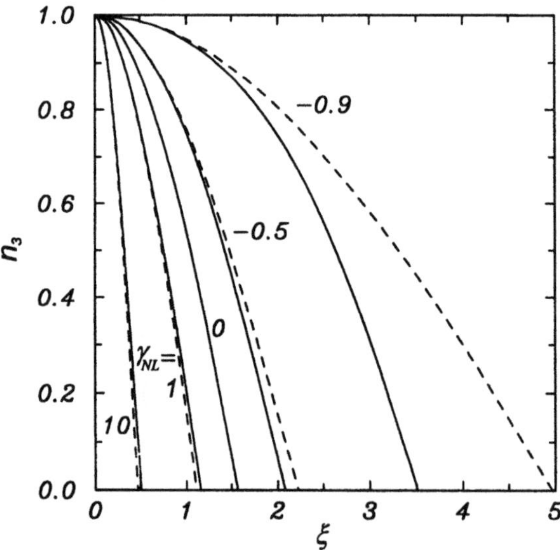

Figure 5.10. Transverse density profiles at different γ_{NL} values obtained as an exact solution of (5.37a) ((5.39a), *solid curves*) and those given by an approximation ((5.39b), *dashed curves*) ([5.34], Fig. 3)

and (5.38b) are compared, and also in Fig. 5.10, where the transverse density profiles obtained from the exact solution

$$\int_{n_3(\xi)}^{1} \frac{\mathrm{d}n_3}{\sqrt{1 - n_3^2 + \frac{2}{3}\gamma_{\mathrm{NL}}(1 - n_3^3)}} = \xi \tag{5.39a}$$

and from the approximate solution

$$n_3(\xi) = \cos\left[\sqrt{1 + \gamma_{\mathrm{NL}}}\ \xi\right] \tag{5.39b}$$

are presented.

For a discharge with cylindrical geometry, the solution of (5.37a) obtained by using the replacement $n_3^2 \to n_3$ in the nonlinear term is

$$n_3(\xi) = \mathrm{J}_0(\sqrt{1 + \gamma_{\mathrm{NL}}}\ \xi), \tag{5.40a}$$

with $\xi = r\sqrt{\nu_\mathrm{i}/D_\mathrm{A}}$. In the original variables it reads

$$n(r) = \frac{\nu_\mathrm{i}}{\varrho_{\mathrm{NL}}}\left[\left(\frac{2.4}{R}\right)^2 \frac{D_\mathrm{A}}{\nu_\mathrm{i}} - 1\right] \mathrm{J}_0\left(\frac{2.4}{R}r\right), \tag{5.40b}$$

where the density at the discharge axis is expressed in terms of the electron temperature:

$$n_0 \equiv n(r = 0) = \frac{\nu_\mathrm{i}}{\varrho_{\mathrm{NL}}}\left[\left(\frac{2.4}{R}\right)^2 \frac{D_\mathrm{A}}{\nu_\mathrm{i}} - 1\right]. \tag{5.40c}$$

The introduction of (5.17) and (5.18) into (5.40c) gives the plasma density in terms of the maintenance field intensity. Moreover, the linkage between n and $|E|^2$ can be obtained not only in the case of Joule heating in the plasma volume (the situation commented on in cases (i) and (ii), which corresponds to case A in (5.17) and (5.18)), but also in the case of Joule heating through resoance absorption of SWs (case B there).

Equation (5.40c) explains the meaning of the parameter γ_{NL}. For example, in the case of Joule heating in the plasma volume (i.e. case A in (5.17) and (5.18)), the plasma density at the axis is of the form

$$n_0 = N_{3_0}\left[\left(\frac{E_{\mathrm{th}}^2}{\langle E^2\rangle}\right)^s - 1\right], \tag{5.41a}$$

where $N_{3_0} = \nu_\mathrm{i}/\varrho_{\mathrm{NL}}$ and E_{th} is the normalizing field (5.27). As has been commented on before (in cases (i) and (ii)), E_{th}^2, determined by the Schottky condition (2.67), has the role of either an upper or a lower limit of the maintenance field intensity $\langle E^2\rangle$ actually required. Representing the parameter γ_{NL} in the form

$$\gamma_{\mathrm{NL}} \equiv \frac{\varrho_{\mathrm{NL}}}{\nu_\mathrm{i}} n_0 = \left(\frac{E_{\mathrm{th}}^2}{\langle E^2\rangle}\right)^s - 1 \tag{5.41b}$$

reveals that its value is fixed by the deviation of the maintenance field intensity from the Schottky field intensity. The relations (5.41) hold (with $\langle E^2 \rangle$ replaced by $E_{(r)}^2$) in the case when the discharge is produced predominantly by heating in the regions of resonance absorption. For $\gamma_{NL} > 0$, i.e. with step ionization as a nonlinear mechanism, the maintenance field intensity is less than E_{th}^2, whereas when recombination is the nonlinear mechanism, it is above E_{th}^2.

The final results for the ionization nonlinearity, i.e. the expressions for the density at the discharge axis in terms of the maintenance field intensity, according to (5.40c) with (5.17) and (5.18) for the two regimes of electron heating are as follows:

- Case A: predominant Joule heating in the plasma volume:

$$n_0 = \frac{\mathring{\nu}_i}{\mathring{\varrho}_{si} - \varrho_r \beta_s (E_{th}^2 / \langle E^2 \rangle)^s} \left[\left(\frac{E_{th}^2}{\langle E^2 \rangle} \right)^s - 1 \right]. \tag{5.42a}$$

- Case B: predominant Joule heating in the regions of resonance absorption:

$$n_0 = \frac{\mathring{\nu}_i}{\mathring{\varrho}_{si} - \varrho_r \beta_s (E_{th}^2 / E_{(r)}^2)^s} \left[\left(\frac{E_{th}^2}{E_{(r)}^2} \right)^s - 1 \right]. \tag{5.42b}$$

Here $\beta_s = (R/2.4)^2 (\mathring{\nu}_i / D_A)$.

The radial density profile is described by (5.40b), (5.28b) gives the average value (over the cross-section) of the density and the nonlinear plasma permittivity stems directly from (5.29) with (5.40a) and (5.42) introduced into it.

As has already been commented on, the quadratic dependence on the density of the charged-particle gain through step ionization is applicable to regimes with a comparatively low density, whereas an effect of nonlinearity related to the recombination losses should be expected for comparatively high densities of a diffusion-controlled discharge. Covering the range of intermediate densities requires the two nonlinear mechanisms to be taken into account simultaneously, and a reduction of the relative contribution of step ionization (2.63) with increasing density must be included in the treatment. Replacement of $\varrho_{si} n^2$ in (5.36a) by (2.63) transforms (5.37a) into

$$\frac{d^2 n_3}{d\xi^2} + n_3 + \left(\frac{\gamma_{si}}{1 + \tilde{\mu}_0 n_3} - \gamma_r \right) n_3^2 = 0, \tag{5.43}$$

where $\gamma_{si} = \varrho_{si} n(\xi = 0)/\nu_i$, $\gamma_r = \varrho_r n(\xi = 0)/\nu_i$ and $\tilde{\mu}_0 = \tilde{\mu} n(\xi = 0)$. Then the transverse density profile $n_3(\xi)$ can be written as

$$\int_{n_3(\xi)}^1 \frac{dn_3}{\dot{n}_3} = \xi, \tag{5.44a}$$

with

$$\dot{n}_3 = \left[\left(1 + \frac{\gamma_{\text{si}}}{\tilde{\mu}_0} \right) (1 - n_3^2) - \frac{2}{3}\gamma_{\text{r}}(1 - n_3^3) \right.$$

$$\left. - 2\frac{\gamma_{\text{si}}}{\tilde{\mu}_0^2}(1 - n_3) + 2\frac{\gamma_{\text{si}}}{\tilde{\mu}_0^3}\ln\left(\frac{1 + \tilde{\mu}_0}{1 + \tilde{\mu}_0 n_3} \right) \right]^{1/2}, \tag{5.44b}$$

which reduces to (5.37b) when the saturation effect is neglected (i.e. for $\tilde{\mu}_0 \to$ 0). The gradual influence of changing $\tilde{\mu}_0$ is visualized by the phase trajectories shown in Fig. 5.11.

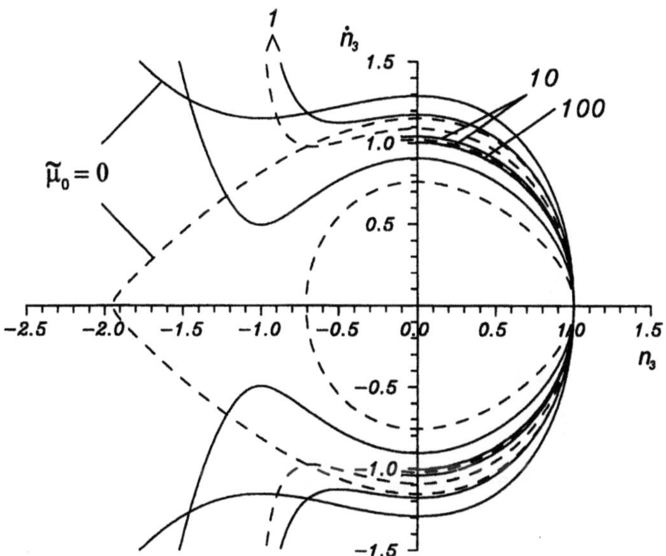

Figure 5.11. Phase trajectories of (5.43) similar to those in Fig. 5.8, for two values of $\gamma_{\text{si}} = 0.5$ (*dashed curves*) and 1 (*solid curves*) and various values of $\tilde{\mu}_0$. γ_{r} is chosen to be 5×10^{-4}, except for the regions closest to the origin, where γ_{r} is chosen as $\gamma_{\text{r}} = 1$ with $\tilde{\mu}_0 = 0.5$ to show the transition to the recombination regime at high electron densities ([5.34], Fig. 4)

An analytical approximation of the resultant maintenance condition corresponding to (5.37e) with the inclusion of the saturation effect is cumbersome, but the solution can be carried out numerically instead. Equation (5.44) can be well approximated (assuming a negligible density at the wall) by profiles of the type

$$n_3(\xi) = a(\xi)\cos\left(\frac{\xi}{\xi_{\text{w}}} \right), \tag{5.45a}$$

where $a(\xi)$ is a slowly varying function over the scale ξ_{w}, the ξ value at the wall. The good agreement between the exact solutions according to (5.44) and those obtained from (5.45a), neglecting the small variation of $a(\xi)$, have been commented on before for $\tilde{\mu} = 0$. Therefore in the slab situation the transverse density profiles may be well approximated by

$$n(x) = n_0(T_\mathrm{e}) \cos\left(\frac{\pi}{2d} x\right) , \tag{5.45b}$$

with $n_0(T_\mathrm{e})$ depending on the maintenance field intensity through T_e.

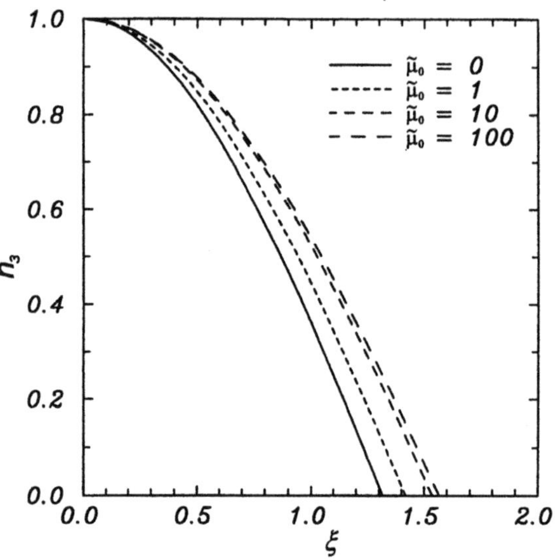

Figure 5.12. Transverse density profiles for $\gamma_\mathrm{si} = 0.5$ and varying $\tilde{\mu}_0$. $\gamma_\mathrm{r} = 5 \times 10^{-4}$ as in Fig. 5.11 ([5.34], Fig. 5)

Transferring the results to the cylindrical case replaces the cosine function with the J_0 Bessel function. The influence of the saturation in the step ionization is shown in Fig. 5.12, where the radial density profiles display the limited changes introduced into the profiles in comparison with those shown in Fig. 5.10. An increase of $\tilde{\mu}_0$ reduces the effects introduced by step ionization.

However, the dependence of the maintenance field intensity on the density

$$\frac{\langle E^2 \rangle}{E_\mathrm{th}^2} = \left(\frac{1 + n_0 \dfrac{\varrho_\mathrm{r}}{\mathring{\nu}_i} \beta_\mathrm{s}}{1 + \dfrac{n_0 \varrho_\mathrm{si}/\mathring{\nu}_i}{1 + \tilde{\mu} n_0}} \right)^{\frac{1}{s}} \tag{5.46}$$

is of the greatest importance. Equation (5.46) is a generalization of the inversion of (5.42a). For case B of predominant resonance-absorption heating, the same expression holds, except that the place of $\langle E^2 \rangle$ is taken by $E_{(\mathrm{r})}^2$.

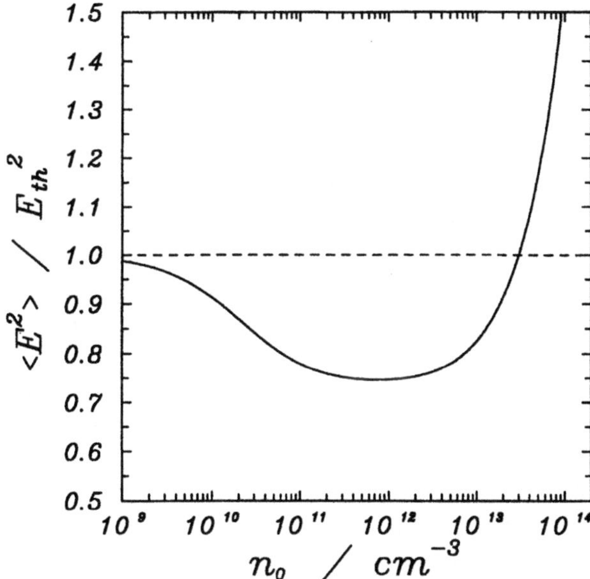

Figure 5.13. Normalized maintenance field strength $\langle E^2 \rangle / E_{\text{th}}^2$ versus the electron density at discharge centre n_0 (*solid curve*) according to (5.46) for $(\varrho_r \beta_s / \mathring{\nu}_i) n_n = 5 \times 10^{-4}$, $\mathring{\varrho}_{si} \beta_s / \mathring{\nu}_i) n_n = 0.5$, $\tilde{\mu} n_n = 1$ with a normalizing value $n_n = 3 \times 10^{10}\,\text{cm}^{-3}$ and $s = 1.3$. The Schottky approximation $\langle E^2 \rangle = E_{\text{th}}^2$ is also marked (*dashed line*) ([5.34], Fig. 6)

To give an overview of the trends to be expected, in Fig. 5.13 an example of $\langle E^2 \rangle / E_{\text{th}}^2$ versus n_0 is plotted over a large range of n_0. Evidently, for low n_0 the value of $\langle E^2 \rangle$ is close to the Schottky value E_{th}^2. An increased efficiency of ionization due to stepwise processes with increasing n_0 leads to a diminishing of $\langle E^2 \rangle$, and E_{th}^2 acts here in a role of an upper bound. As a consequence of saturation in the metastable densities described by a growing $\tilde{\mu} n_0$, $\langle E^2 \rangle$ levels off towards increased n_0, thus keeping the deviations from E_{th}^2 within limits. Then, with further increase of n_0, the effect of recombination becomes more and more noticeable, leading to a minimum in $\langle E^2 \rangle$ and finally resulting in a growing $\langle E^2 \rangle$ towards a rather high n_0. In the latter region E_{th}^2 acquires more the role of a lower bound.

In depicting $n_0(\langle E^2 \rangle)$, Fig. 5.13 gives access to the nonlinear plasma permittivity, including the range of saturated metastable densities.

The left-hand side of (5.46) can be written as $\Theta / \Theta_{\text{th}}$, where Θ as before, is the power absorbed on average per electron, and Θ_{th} is the corresponding threshold (Schottky) value.

The simplifying assumption of no variation in Θ, which has often been used until recently (e.g. [5.38]), is obviously avoided here, not only to obtain a functional description of Θ, but also for the very reason of preserving full self-consistency stressed at the outset.

5.4 The Electrodynamic Part of the Problem of Maintenance of a Waveguided Discharge

So far, one of the relationships between plasma density and maintenance field intensity has been obtained. Representing the description of an ionization nonlinearity, it was arrived at as a solution of (5.2) and (5.3), the equations involved in the gas-discharge part of the problem, by excluding T_e as an unknown quantity. Now, the next step necessary for the description of waveguided discharges will be made. This is to obtain a second equation which relates the same quantities n and $|E|^2$, i.e. one which stems from the electrodynamic part of the problem for discharge maintenance in the field of a propagating SW. Thus, it is required to deal with (5.4) and (5.5) and to reduce them to one equation [5.17,20,27,28,32–34] by excluding k as an unknown quantity. In fact, for the case of discharge maintenance by slow EM SWs under the conditions of small σ values ($\sigma < 0.3$) – the case which exists in almost all experimental arrangements for SW-produced discharges – this has already been done in Chap. 3 by giving (3.70b) there for the wave energy flux and excluding β from it by introducing the slowly varying (along the phase diagrams) function $\bar{f}$ (3.73). In a way, the representation of the electrodynamic part of the problem of discharge maintenance has already been prepared for its analytical treatment.

When the electrons gain energy by Joule heating in the plasma volume, the Joule losses, Q of the wave, which enter into (5.4), are given by (3.72). In this case the electrodynamic part of the problem, (5.4) and (5.5), reduces to

$$\frac{\mathrm{d}}{\mathrm{d}z}\left(\bar{n}^2|E_z(r=R)|^2\right) = -l\bar{n}|E_z(r=R)|^2,\tag{5.47}$$

where $l = (\nu/\omega R)(1/\langle \bar{f}\rangle)n_c$.

When the electrons gain energy by Joule heating in the regions of resonance absorption, Q in (5.4) is given by (4.73). If the latter is expressed through $E^2_{(r)}$, which is the maintenance field intensity (5.15b), (5.4) and (5.5) reduce to [5.34]

$$\frac{\mathrm{d}}{\mathrm{d}z}\left(\bar{n}^2|E_z(r=R)|^2\right) = -l\bar{n}|E^2_{(r)}|.\tag{5.48a}$$

The wave number β is excluded from (5.15b) for $E^2_{(r)}$ by transforming it to

$$E^2_{(r)} = \frac{\pi}{2}g^2\left(\frac{n_c}{\bar{n}}\right)\frac{L_n^{(r)}}{R}\frac{\omega}{\nu}|E_z(r=R)|^2.\tag{5.48b}$$

It is taken into account that σ is small, and

$$g = kR|\bar{\varepsilon}|\tag{5.48c}$$

is introduced as a slowly varying function in the region of the phase diagram where this mechanism of heating could be important, i.e. the region of comparatively large $\omega/\overline{\omega}_{\mathrm{p}}$ values. The characteristic length $L_n^{(\mathrm{r})}$ can still be a function of $\bar{n}$, to be considered below.

Equations (5.47) and (5.48) hold over the main, longer part of the discharge length and over the end of the plasma column, respectively. The value of $\bar{n}$ where the solutions of (5.47) and (5.48) should be matched [5.35] is given by

$$P_{\mathrm{r}}\frac{n_{\mathrm{c}}}{\bar{n}} \simeq 1\,, \tag{5.49a}$$

where P_{r}, as given by (4.73c), can be approximated as

$$P_{\mathrm{r}} \simeq \frac{\pi}{2}\,(\beta R)^2|\bar{\varepsilon}|^2\frac{L_n^{(\mathrm{r})}}{R}\,\frac{\omega}{\nu}\,. \tag{5.49b}$$

5.5 Self-Consistent Axial Structure

The axial structure of the discharge describes the self-consistent plasma–wave behaviour along the discharge length. It appears as a solution of a set of two equations: (5.42a) and (5.47) or (5.42b) and (5.48a). These equations relate the plasma density $n(z)$ and the maintenance field intensity $|E(z)|^2$ (or $\langle|E(z)|^2\rangle$ or $E_{(\mathrm{r})}^2(z)$) under the various regimes of charged-particle losses and plasma heating considered in Sects. 5.3 and 5.4.

5.5.1 Recombination-Controlled Regime

Equations (5.19) and (5.47) make up the set of equations which describes the axial structure of a discharge in the recombination-controlled regime [5.17,20,27,28]. In the case of comparatively high temperatures (2.59a), i.e. when (5.19b) holds, the solution is

$$\frac{\mathrm{d}\bar{n}}{\mathrm{d}z} = -\frac{1}{3}\frac{\nu\omega}{R}\frac{m}{4\pi e^2}\frac{1}{\langle\bar{f}\rangle}\,, \tag{5.50a}$$

$$|E(z)| = E_0\left(1 - \frac{z}{N(z=0)L_{\mathrm{eff}}}\right)^{1/2}\,. \tag{5.50b}$$

Here $E_0 = |E_z(r = R, z = 0)|$ and $N(z = 0) \equiv n(z = 0)/n_{\mathrm{c}}$ are the maintenance field intensity and the normalized density, respectively, at the onset of the range of maintenance of a discharge by slow SWs;

$$L_{\mathrm{eff}} = 3\frac{R\omega}{\nu}\,\langle\bar{f}\rangle \tag{5.50c}$$

is a dimensionless effective length of the discharge.

In the case of comparatively low temperatures (2.60c), i.e. when (5.19b) holds, the solution is

$$\frac{\mathrm{d}\bar{n}}{\mathrm{d}z} = -\frac{1}{2+s^{-1}}\,\frac{\nu\omega}{R}\,\frac{m}{4\pi e^2}\,\frac{1}{\langle\bar{f}\rangle}\,, \tag{5.51a}$$

$$|E(z)| = E_0\left(1 - \frac{z}{N(z=0)L_{\mathrm{eff}}}\right)^{1/2s}, \tag{5.51b}$$

with the effective length of the discharge

$$L_{\mathrm{eff}} = \left(2 + \frac{1}{s}\right)\frac{R\omega}{\nu}\,\langle\bar{f}\rangle\,. \tag{5.51c}$$

A linear density decrease accompanied by a linear reduction of the maintenance field intensity along the discharge length makes up the self-consistent axial structure of a discharge in a recombination-controlled regime.

5.5.2 Diffusion-Controlled Regime

The results for the axial structure of the discharge are given in the same sequence as in Sect. 5.3.4b.

Case (i): diffusion-controlled regime under conditions of Joule heating in the plasma volume and nonlinearity due to recombination losses. The set of equations (5.26), (5.28b) and (5.47) is reduced to the following equation for the normalized intensity $u(z) \equiv |E_z|^2/E_{\mathrm{th}}^2$ of the maintenance field [5.32]:

$$\frac{\mathrm{d}}{\mathrm{d}\zeta}\left[u(u^s - 1)^2 u\right] = -(u^s - 1)u\,, \tag{5.52}$$

where a dimensionless axial variable $\zeta = (z/R)(n_{\mathrm{c}}/N_{1_0})(2\nu/\omega\langle\bar{f}\rangle)$ has been introduced.

Its solution is

$$\frac{2s+1}{s}(u^s - u_1^s) - \ln\frac{u}{u_1} = -(\zeta - \zeta_1)\,, \tag{5.53}$$

where u_1 is the dimensionless intensity at $\zeta = \zeta_1$.

In a diffusion-controlled regime (Sect. 5.3.4) the deviation of the maintenance field intensity from the Schottky field intensity is small. Thus, the normalized intensity $u(z)$ can be written as $u = 1 + \Delta_n$ with $\Delta_n \ll 1$. A decreasing intensity of the SW field along the discharge axis

$$u = 1 + \Delta_{n_1} - \frac{1}{2s}(\zeta - \zeta_1) \tag{5.54a}$$

results from (5.53). Here $\Delta_{n_1} = \Delta_n(\zeta - \zeta_1)$, and ζ_1 corresponds to the onset of discharge maintenance by slow SWs. The self-consistent axial variation of the plasma density obtained by substituting (5.54a) into (5.26b) and (5.28b) is

$$\bar{n}(\zeta) = \frac{1}{2}N_{1_0}\left\{\left[1 + \Delta_{n_1} - \frac{1}{2s}(\zeta - \zeta_1)\right]^s - 1\right\} \tag{5.54b}$$

or, equivalently (with $s = 1$),

$$\bar{n}(\zeta) = \frac{1}{2} N_{1_0} \left[\Delta_{n_1} - \frac{1}{2}(\zeta - \zeta_1) \right] . \tag{5.54c}$$

Therefore, a linear axial density profile with the gradient

$$\frac{d\bar{n}}{dz} = -\frac{1}{2} \frac{\nu\omega}{R} \frac{1}{\langle \bar{f} \rangle} \frac{m}{4\pi e^2} \tag{5.55a}$$

is self-consistently determined by a linear axial variation of $|E_z|^2$, the intensity of the field which sustains the discharge. The gradient of this field intensity,

$$\frac{d|E_z|^2}{dz} = -0.2 \frac{\nu\omega}{\langle \bar{f} \rangle} \frac{m}{4\pi e^2} \frac{R\varrho_r}{D_A} E_{th}^2, \tag{5.55b}$$

is much smaller than that of the density, because of the small value of the recombination coefficient.

Therefore, in a diffusion-controlled discharge with a relatively high electron density (a regime which could be reached at relatively high pressures), and where the recombination losses govern the nonlinearity in the particle balance equation, a linear decrease of the plasma density is self-consistently connected to a slight linear decrease of the maintenance field intensity along the discharge length. At the end of the column the maintenance field intensity tends to the value of E_{th}, which is related to the Schottky condition.

Case (ii): diffusion-controlled regime under conditions of Joule heating in the plasma volume and nonlinearity due to step ionization. The set of equations (5.34a), (5.28b) and (5.47) reduces to the following equation for the normalized intensity $u(z) = |E_z|^2/E_{th}^2$ [5.33]:

$$\frac{d}{d\zeta} \left[(u^{-s} - 1)^s u \right] = -(u^{-s} - 1)u, \tag{5.56a}$$

with $\zeta = 2lz/N_{2_0}$. The solution is

$$\frac{2s-1}{s}(u^{-s} - u_1^{-s}) - \ln \frac{u}{u_1} = -(\zeta - \zeta_1), \tag{5.56b}$$

where $u_1 = u(\zeta = \zeta_1)$.

With u close to 1 ($u = 1 - \Delta_n$ and $\Delta_n \ll 1$), (5.56b) yields

$$u = 1 - \Delta_{n_1} + \frac{1}{2s}(\zeta - \zeta_1), \tag{5.57a}$$

and the axial profile of the plasma density, self-consistently associated with the axial profile (5.57a) of the field intensity, is

$$\bar{n}(z) = \frac{1}{2} N_{2_0} \left[s\Delta_{n_1} - \frac{1}{2}(\zeta - \zeta_1) \right] . \tag{5.57b}$$

Starting in the region close to the SW launcher, which has a higher density value than elsewhere with a maintenance field intensity which is below the value of the normalizing field E_{th} for discharge maintenance by direct ionization, the field intensity increases along the discharge length because of the decreased contribution (at lower densities) of step ionization. The variations in the longitudinal direction of the field intensity and electron density are linear, with gradients

$$\frac{du}{d\zeta} \simeq \frac{1}{2s} \tag{5.58a}$$

and

$$\frac{d\bar{n}}{d\zeta} \simeq -\frac{1}{4}N_{2_0}, \tag{5.58b}$$

which, in terms of the original variables, means

$$\frac{d|E_z|^2}{dz} = \frac{1}{s}\frac{\nu}{\omega R}\frac{1}{\langle\bar{f}\rangle}\frac{\mathring{\varrho}_{si}n_c}{\mathring{\nu}_i}E_{th}^2 \equiv \frac{1}{s}\frac{\nu\omega}{R}\frac{1}{\langle\bar{f}\rangle}\frac{\mathring{\varrho}_{si}}{\mathring{\nu}_i}\frac{m}{4\pi e^2}E_{th}^2, \tag{5.59a}$$

$$\frac{dn}{dz} = -\frac{1}{2}\frac{\nu}{\omega R}\frac{1}{\langle\bar{f}\rangle}n_c \equiv -\frac{1}{2}\frac{\nu\omega}{R}\frac{1}{\langle\bar{f}\rangle}\frac{m}{4\pi e^2}. \tag{5.59b}$$

Therefore, in a diffusion-controlled discharge with a relatively low electron density, when step ionization is the predominant nonlinear mechanism in the particle balance equation, a linear decrease of the plasma density is self-consistently connected to a linear increase of the maintenance field intensity along the discharge length. Similarly to the case of recombination losses, here also – because of the small values of $\mathring{\varrho}_{si}/\mathring{\nu}_i$ (corresponding to small values of $\mathring{\varrho}_{si}n_c/\nu_i$) – the axial variation of the field intensity is much smoother than that of the plasma density.

When the coefficient of step ionization is given by (2.62b) and, consequently, (5.35) governs the relation of the plasma density to the maintenance-field intensity, the axial gradient of the density is the same (5.59b). However, the axial gradient of the field intensity

$$\frac{d|E_z|^2}{dz} = \frac{1}{s}\frac{\nu\omega}{R}\frac{1}{\langle\bar{f}\rangle}\frac{\mathring{\varrho}_{si}}{\mathring{\nu}_i}\left[\left(\frac{R}{2.4}\right)^2\frac{\mathring{\nu}_i}{D_A}\right]^{1-s^{-1}}\frac{m}{4\pi e^2}E_{th}^2 \tag{5.60}$$

is a little higher than that determined by (5.59a). This can be associated with the increased contribution of step ionization when its rate coefficient is given by (2.62b).

Case (iii): diffusion-controlled regime under conditions of Joule heating in the volume and in regions of resonance absorption. This case describes the self-consistent axial profiles of the plasma density and maintenance field intensity along the complete length of a plasma column: in the main, long, part of the discharge where plasma heating occurs through collisions in the volume, as well as in the region close to the discharge end where the linear transformation of SWs into volume plasmons (Chap. 4) is the predominant mechanism of damping and, therefore, the plasma heating occurs mainly through resonance absorption. The results for the axial structure of the discharge [5.34] presented here are based on case (iii) in Sect. 5.3.4b where the two nonlinear mechanisms in the particle balance – recombination losses and step ionization – have been taken into account simultaneously. With respect to the dependence presented in Fig. 5.13 these mechanisms concern discharges with relatively low ($n < 10^{10}\,\mathrm{cm}^{-3}$) and relatively high $n > 3 \times 10^{13}\,\mathrm{cm}^{-3}$) densities.

The case where the thermal nonlinearity is related to Joule heating in the plasma volume is given first. The axial structure of the discharge (axial profiles of plasma density and field intensity) is obtained by coupling (5.42a) and (5.47) into a set. Written in the dimensionless variables $\bar{N} = \bar{n}/n_{\mathrm{c}}$, $u = |E_z|^2/E_{\mathrm{th}}^2$ and $\zeta = lz/n_{\mathrm{c}}$, this set is

$$\frac{1}{\bar{N}u}\frac{\mathrm{d}}{\mathrm{d}\zeta}(\bar{N}^2 u) = -1\,, \tag{5.61a}$$

$$\bar{N} = \frac{1}{2n_{\mathrm{c}}}\frac{\mathring{\nu}_{\mathrm{i}}}{\varrho_{\mathrm{si}} - \varrho_{\mathrm{r}}\beta_{\mathrm{s}}u^{-s}}\left(u^{-s} - 1\right)\,. \tag{5.61b}$$

Its solution

$$\frac{\mathrm{d}|E_z|^2}{\mathrm{d}z} = \frac{1}{s}\left(\frac{\mathring{\varrho}_{\mathrm{si}}}{\mathring{\nu}_{\mathrm{i}}} - \frac{\varrho_{\mathrm{r}}}{\mathring{\nu}_{\mathrm{i}}}\beta_{\mathrm{s}}\right)\frac{\nu}{\omega R}\frac{n_{\mathrm{c}}}{\langle\bar{f}\rangle}E_{\mathrm{th}}^2\,, \tag{5.62a}$$

$$\frac{\mathrm{d}\bar{n}}{\mathrm{d}z} = -\frac{1}{2}\frac{\nu}{\omega R}\frac{n_{\mathrm{c}}}{\langle\bar{f}\rangle} \tag{5.62b}$$

combines the results given separately for cases (i) and (ii).

Within the limits of logarithmic accuracy used here, the density decrease along the discharge is not influenced by the mechanism of nonlinearity. However, the variations of the field intensity are completely determined by it. If step ionization is the predominant nonlinear mechanism, the maintenance field intensity, starting at the launcher side with a value *below* the (Schottky) normalizing field intensity, *increases* along the discharge length. Indications of such an increase have also been obtained in kinetic modelling of discharges in helium gas [5.39,40]. If recombination is the predominant nonlinear mechanism, the maintenance field intensity, starting with a value *above* the Schottky field intensity, *decreases* along the discharge length.

These conclusions are illustrated in Figs. 5.14 and 5.15, where the numerical solutions of the set of equations (5.61) are presented ($\Sigma = \mathring{\varrho}_{\mathrm{si}}n_{\mathrm{c}}/\mathring{\nu}_{\mathrm{i}}$ and $\Pi = \varrho_{\mathrm{r}}\beta_{\mathrm{s}}n_{\mathrm{c}}/\mathring{\nu}_{\mathrm{i}}$ are the parameters which characterize the nonlinear effects of step ionization and recombination losses, respectively). These solutions show effects beyond the logarithmic-accuracy approximation within which the analytical results (5.62) have been obtained. These are the slight deviations (Fig. 5.14) of the $u(\zeta)$ dependence from a linear relation and the rather weak dependence of the slopes of the density profiles (Fig. 5.15) on the mechanism of nonlinearity. With an increasing role of step ionization the density profiles become steeper, whereas an increasing contribution from recombination leads to a lower steepness.

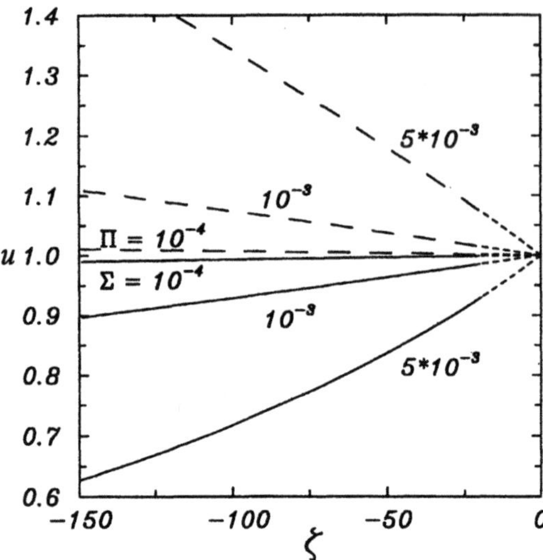

Figure 5.14. Normalized field intensity u versus normalized axial position ζ (measured from the discharge end $\zeta = 0$) for the range of volume Joule heating under conditions of predominant step ionization (*solid curves* in the *lower* part of the figure, i.e. $u < 1$, various values of Σ as indicated, $\Pi = 10^{-5}$) and predominant recombination (*dashed curves* in the *upper* part of the figure, i.e. $u > 1$, various values of Π as indicated, $\Sigma = 10^{-5}$); $s = 1.3$. The *dotted portions* of the curves mark the regions where Joule heating is not predominant ([5.34], Fig. 7)

A variation of the maintenance field intensity along the discharge length means, according to (5.6b) and (5.13a), an axial variation of the electron temperature as well. An increasing field intensity along the discharge length when step ionization is the predominant nonlinear mechanism means an increase of the electron temperature from the launcher towards the discharge end. When recombination losses are predominant, the decrease of the maintenance field intensity leads to a decrease of the temperature. However, because of the slow (logarithmic) dependence of the temperature on the maintenance field intensity, the axial variation of T_e is even weaker than the weak variation of the maintenance field intensity. The axial variation of the maintenance field intensity means an axial variation of Θ. In the case of step ionization Θ increases along the discharge column, whereas for recombination losses it decreases.

The procedure just used in solving the set of equations (5.61) can be directly applied to the case of Joule heating in regions of resonance absorption. This type of heating determines the self-consistent wave–plasma behaviour at the end of the plasma column. Now, the quantity $E^2_{(r)}$ as defined by (5.15b) plays the role of a maintenance field intensity. In addition, (5.42) shows that the relation between plasma density and maintenance field intensity $E^2_{(r)}$ under conditions of Joule heating in regions of resonance absorption is the same

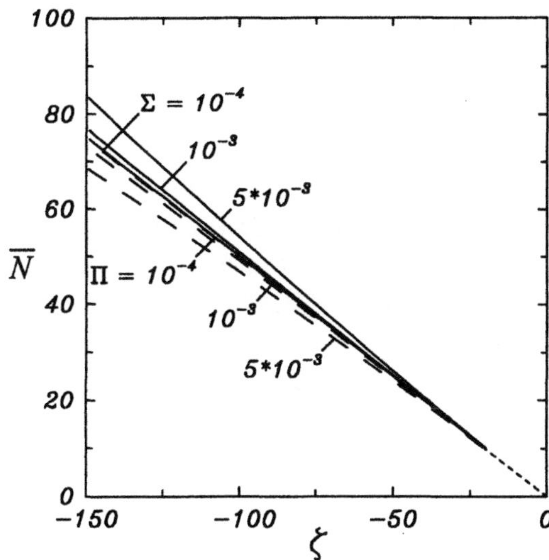

Figure 5.15. Normalized averaged plasma density $\bar{N}$ versus normalized axial position ζ (with $\zeta = 0$ at discharge end) for the range of volume Joule heating under conditions of predominant step ionization (*solid curves* with various values of Σ as indicated, $\Pi = 10^{-5}$) and predominant recombination (*dashed curves* with various vaues of Π as indicated, $\Sigma = 10^{-5}$); $s = 1.3$. The *dotted portion* of the curves mark the region where Joule heating is not predominant ([5.34], Fig. 8)

as that between n and $\langle E^2 \rangle$ under conditions of Joule heating in the plasma volume. After the axial variations of n and $E^2_{(r)}$ have been determined, the axial variation of the SW field $(|E_z(z, r = R)^2|)$ can also be obtained from (5.15b). However, there is an important difference which distinguishes the cases of Joule heating in the plasma volume and Joule heating in the regions of resonance absorption. In the latter case, $E^2_{(r)}$ – the maintenance field intensity – depends on n. This means that the main cause of the self-consistent wave–plasma structure stems from the mechanism of thermal nonlinearity.

Equations (5.42b) and (5.48) complete the set which should be solved in this case. $L_n^{(r)}/R \approx n_c/\bar{n}$ results from the expression for the radial profile (5.40b). Introducing the dimensionless variables $\bar{N} = \bar{n}/n_c$, $w = E^2_{(r)}/E^2_{\text{th}}$ and $\zeta = (\pi g^2 z)/(2R \langle \bar{f} \rangle)$, (5.42b) and (5.48) become

$$\frac{1}{\bar{N}w} \frac{\mathrm{d}}{\mathrm{d}\zeta}(\bar{N}^4 w) = -1 \,, \tag{5.63a}$$

$$\bar{N} = \frac{1}{2n_c} \frac{\mathring{\nu}_i}{\mathring{\varrho}_{si} - \varrho_r \beta_s w^{-s}} (w^{-s} - 1) \,. \tag{5.63b}$$

The solution of the first equation, within the limits of logarithmic accuracy, is

$$\bar{n} = \bar{n}_{\mathrm{r}} \sqrt[3]{1 - \frac{3}{8} \frac{n_{\mathrm{c}}^3}{\bar{n}_{\mathrm{r}}^3} \frac{g^2}{\langle \bar{f} \rangle} \frac{\pi}{R} (z - z_{\mathrm{r}})}, \tag{5.64}$$

where $\bar{n}_{\mathrm{r}}$ is the density at $z = z_{\mathrm{r}}$, i.e. at the point where the region of predominant Joule heating through resonance absorption starts, and (5.64) matches the linear density profile (5.62b). According to (5.64), at the end of the discharge, the density drops faster than over the main (long) part of the discharge. The type of density decrease is changed because of the change in the mechanism of energy transfer. Similarly to (5.62b), (5.64) does not show any influence of the mechanism of nonlinearity.

The maintenance field intensity is

$$w \equiv \frac{E_{(\mathrm{r})}^2}{E_{\mathrm{th}}^2} = 1 - \Delta_{\mathrm{r}} \sqrt[3]{1 - \frac{3}{8} \frac{n_{\mathrm{c}}^3}{\bar{n}_{\mathrm{r}}^3} \frac{g^2}{\langle \bar{f} \rangle} \frac{\pi}{R} (z - z_{\mathrm{r}})}, \tag{5.65a}$$

where

$$\Delta_{\mathrm{r}} = \frac{2}{s} \left(\frac{\mathring{\varrho}_{\mathrm{si}}}{\mathring{\nu}_{\mathrm{i}}} - \frac{\varrho_{\mathrm{r}}}{\mathring{\nu}_{\mathrm{i}}} \beta_{\mathrm{s}} \right) \bar{n}_{\mathrm{r}}. \tag{5.65b}$$

This relation (5.65a), giving the axial structure of the maintenance field intensity, shows also the axial variation of the electron temperature, according to (5.15a). For step ionization as the dominating nonlinear mechansim, $E_{(\mathrm{r})}^2$ increases towards the discharge end, leading to an increase of T_{e} as well. Since the parameter Θ is proportional to $E_{(\mathrm{r})}^2$, an increase of $E_{(\mathrm{r})}^2$ means an increase of Θ towards the discharge end. When recombination losses control the nonlinearity of the discharge, the axial decrease of the maintenance field intensity at the end of the discharge results in a decrease of T_{e} and Θ. As in the case of Joule heating in the plasma volume, the axial variations of T_{e} are slower than that of the maintenance field intensity.

With (5.15b), the relation of $E_{(\mathrm{r})}^2$ to the field intensity $E_z(r = R)|^2$, (5.65a) finally gives the following solution for the axial variation – in original variables – of the SW field intensity at the discharge end:

$$u \equiv \frac{|E_z|^2}{E_{\mathrm{th}}^2} = \frac{2}{\pi g^2} \frac{\nu}{\omega} \frac{\bar{n}^2(z)}{n_{\mathrm{c}}^2} \left[1 - \Delta_{\mathrm{r}} \sqrt[3]{1 - \frac{3}{8} \frac{n_{\mathrm{c}}^3}{\bar{n}_{\mathrm{r}}^3} \frac{g^2}{\langle \bar{f} \rangle} \frac{\pi}{R} (z - z_{\mathrm{r}})} \right], \tag{5.66}$$

where Δ_{r} is given by (5.65b) and $\bar{n}(z)$ by (5.64).

Numerical solutions of (5.63) are plotted in Figs. 5.16 and 5.17. With the values of Σ and Π chosen, the results for the density profiles (Fig. 5.17), similarly to the analytical solution, are not influenced by the nonlinear mechanism. With step ionization as the nonlinear mechanism (Fig. 5.16), the wave field intensity, reaching the region of Joule heating due to resonance absorption with values below the threshold intensity, decreases again. When the recombination losses are the predominant nonlinear mechanism, the SW field intensity, approaching the region of Joule heating due to resonance absorption with values above the normalizing field intensity, continue to decrease.

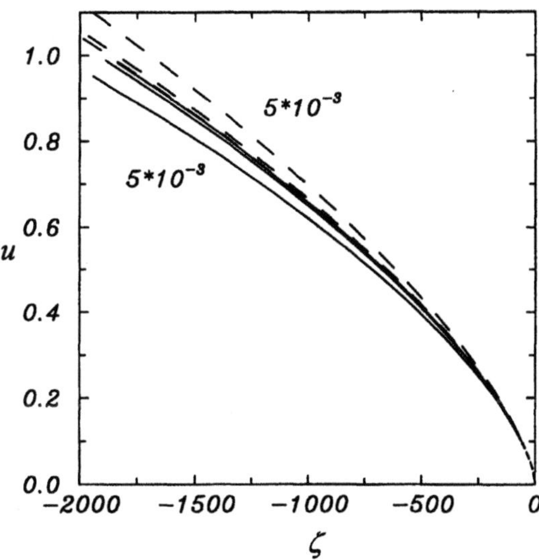

Figure 5.16. Normalized field intensity u versus normalized axial position ζ for the range of resonance absorption under conditions of predominant step ionization (*solid curves*: $\Sigma = 10^{-4}$, 10^{-3} and 5×10^{-3} with $\Pi = 10^{-5}$) and predominant recombination (*dashed curves*: $\Pi = 10^{-4}$, 10^{-3} and 5×10^{-3} with $\Sigma = 10^{-5}$); $s = 1.3$. $\zeta = 0$ is at the discharge end. The *dotted portion* of the curves marks the region outside the range of applicability of the approximations used ([5.34], Fig. 9)

If it is assumed that $L_n^{(\mathrm{r})}/R$, which characterizes the radial density gradient, is simply a constant, the solutions of (5.63) are similar, i.e.

$$\bar{n} = \bar{n}_{\mathrm{r}}\,\wp(z)\,, \tag{5.67a}$$

$$w \equiv \frac{E_{(\mathrm{r})}^2}{E_{th}^2} = 1 - \Delta_{\mathrm{r}}\,\wp(z)\,, \tag{5.67b}$$

$$u = \frac{|E_z|^2}{E_{th}^2} = \frac{2}{\pi g^2}\,\frac{R}{L_n^{(\mathrm{r})}}\,\frac{\nu}{\omega}\,\frac{\bar{n}(z)}{n_{\mathrm{c}}}\,[1 - \Delta_{\mathrm{r}}\,\wp(z)]\,. \tag{5.67c}$$

Here

$$\wp(z) = \sqrt{1 - \frac{1}{3}\,\frac{n_{\mathrm{c}}^2}{\bar{n}_{\mathrm{r}}^2}\,\frac{g^2}{\langle\bar{f}\rangle}\,\frac{\pi L_n^{(\mathrm{r})}}{R^2}\,(z - z_{\mathrm{r}})}\,.$$

The density drops with a square-root dependence on the axial coordinate when $L_n^{(\mathrm{r})}$ is constant. Proportionality between the axial variations of the plasma density and the SW field intensity $|E_z|^2$ is the basic relation between them. This stems from the nonlocality of the electron heating when the latter occurs through resonance absorption. In addition, the mechanisms of nonlinearity in the particle balance influence (through the quantity Δ_{r}) the axial profile, but this effect is weak. With step ionization as the predominant

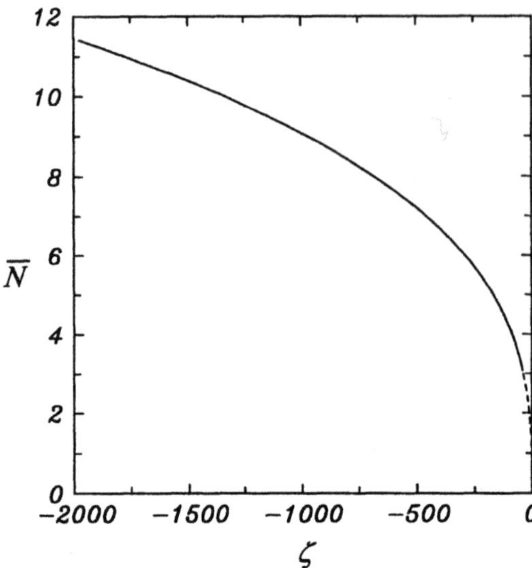

Figure 5.17. Normalized averaged plasma density $\bar{N}$ versus normalized axial position ζ for the range of resonant absorption under conditions of predominant step ionization ($\Sigma = 10^{-4}$ to 5×10^{-3} with $\Pi = 10^{-5}$) and predominant recombination ($\Pi = 10^{-4}$ to 5×10^{-3} with $\Sigma = 10^{-5}$); $s = 1.3$. The *dotted portion* of the curves marks the region outside the range of applicability of the approximations used ([5.34], Fig. 10)

mechanism of nonlinearity (i.e. with $\Delta_{\mathrm{r}} > 0$), the small deviations from proportionality between $\bar{n}(z)$ and $|E_z(z)^2|$ are in the direction of flattening of the axial profile of the field intensity towards the discharge end. In contrast, when the recombination losses control the nonlinearity in a discharge sustained in a diffusion-controlled regime, the deviations from proportionality between $\bar{n}(z)$ and $|E_z(z)|^2$ are in the direction of steepening of the axial profile of the field intensity. This is in accordance with the numerical solutions (Fig. 5.16).

5.5.3 Comparison of the Axial Structures of Discharges in Diffusion- and Recombination-Controlled Regimes

A comparison [5.27,28] of the axial gradients for the density profiles obtained for discharges sustained by Joule heating in the plasma volume in diffusion-controlled (5.62b) and recombination-controlled ((5.50a) and (5.51a)) regimes shows that the results can be unified in the form

$$\frac{\mathrm{d}\bar{n}}{\mathrm{d}z} = -\frac{1}{\kappa} \frac{\nu\omega}{R} \frac{m}{4\pi e^2} \frac{1}{\langle \bar{f} \rangle} . \tag{5.68}$$

Here, the value of the numerical factor κ accounts for the discharge regime: $\kappa = 2$ for a diffusion-controlled regime and $\kappa = 3$, $\kappa = 2 + s^{-1}$ (which is a value between 2 and 3) for recombination-controlled regimes under conditions specified by (2.59a) and (2.60a), respectively. The influence of the gas discharge conditions occurs through the factor $(\nu\omega/R)$. The value of $\langle \bar{f} \rangle$ specifies the particular features of the wave dispersion behaviour. The ratio of the density gradients in the diffusion- and recombination-controlled regimes (when the latter is specified by (2.59a)) is $3:2 = 1.5$.

Equation (5.68) combines the influence of the various factors: the regime of the discharge (in which the type of nonlinearity is reflected), the gas-discharge conditions and the wave dispersion behaviour. All analytical dependences are exact. The only approximation is the replacement of the slowly varying (over the phase diagrams) function $\bar{f}$ by its averaged value $\langle \bar{f} \rangle$.

The factor $\nu\omega/R$ is related to the SW space damping rate (3.74b) at the column end [5.28]:

$$\alpha_L = \frac{\nu}{\omega R} \, . \tag{5.69}$$

In dimensionless variables (5.68) looks like

$$\frac{\mathrm{d}\bar{N}}{\mathrm{d}\zeta} = -1 \, , \tag{5.70a}$$

with $\zeta = z/L_{\mathrm{eff}}$, and assumes the form of an electrodynamic similarity law for SW-sustained discharges. L_{eff}, having the meaning of an effective length of the discharge, is related to the space damping rate at the column end through

$$L_{\mathrm{eff}} = \kappa \langle \bar{f} \rangle \alpha_L^{-1} \equiv \kappa \frac{\omega R}{\nu} \langle \bar{f} \rangle \, , \tag{5.70b}$$

with $\kappa = 2$, 3 or $2 + s^{-1}$, as specified before.

Whereas the axial variation of the field intensity in diffusion-controlled regimes is weak ((5.55b) and (5.59a)) in recombination-controlled regimes it is strong ((5.50b) and (5.51b)), following that of the density.

In conclusion, the results in this section yield the self-consistent axial structure of waveguide discharges sustained by propagating SWs. In order to obtain these results, the model of HF discharges presented in Sect. 5.3 had to be coupled with a description of the wave propagation behaviour. With the two mechanisms of heating included – Joule heating in the volume and in regions of resonance absorption – and with their predominant role along the discharge length, the results present a complete picture of the axial structure of the discharge. The wave–plasma self-consistency along the discharge length is the basic pattern of this type of discharge. This is exactly the result the model gives, showing the mutually connected axial variations of the parameters of the plasma produced by the wave and of the characteristics of the wave which produces the plasma. The ionization nonlinearity, which is the basis of discharge production in general, ensures the self-consistency in the axial structure of SW-sustained discharges.

5.6 Axial Density Profiles

In this section, the results for diffusion-controlled regimes are extended beyond the limits of discharge maintenance by slow SW propagation in the thin-cylinder approximation (i.e. at small σ values: $\sigma < 0.3$) given above. However, this extension concerns only the axial density profile. It is valid within the limits of logarithmic accuracy discussed in Sect. 5.5. The presentation here aims to show how the changes in wave dispersion behaviour over the complete EM range of wave existence for weak collisions ($\omega \gg \nu$) and for strong collisions ($\nu \gg \omega$) influence the axial density profile. The considerations also extend the discussions in the previous section on discharge maintenance by Joule heating in regions of resonance absorption. Conditions of weak collisions ($\nu \ll \omega$) are considered in this case. Similarly to the case of Joule heating in the plasma volume, this extension also aims at covering the complete EM range of the waves. A planar discharge is treated first and then the case of a cylindrical waveguide is treated, too. The validity of these results is also within the limits of logarithmic accuracy (Sect. 5.5).

The discussions start from some generalized form of the electrodynamic part of the problem of discharge maintenance [5.30,41], in which the gas discharge part is introduced, first in a general manner and then specifcally for cases of diffusion and recombination regimes.

5.6.1 General Relations

Weakly damped EM waveguided modes, in particular SWs, are considered. The wave energy $W(z,t)$ and the wave energy flux $P(z,t)$ obtained after integration (2.32) over the transverse cross-section of the waveguided structure are related to each other through

$$P = v_{\mathrm{gr}} W. \tag{5.71a}$$

The group velocity v_{gr} has already been given in Chap. 3. According to the phase diagrams of the waves, it is large in the regions of higher-density plasma, and decreases with the density. It should be mentioned that the wave energy can be used as a quantity characterizing the behaviour of the waves not only for weak collisions ($\omega \gg \nu$), but also for strong collisons ($\nu \gg \omega$), provided the waves are weakly damped, i.e. $\omega \ll \gamma$, where γ is the time damping rate in

$$\partial W / \partial t = 2\gamma W. \tag{5.71b}$$

The rate of wave-energy losses is defined through the work of the wave field on electrons (see (2.29) and (2.32)):

$$Q(z,t) = \frac{1}{2} \mathrm{Re} \left\{ \int_{S_\perp} (\boldsymbol{E}^* \cdot \boldsymbol{j}) \, \mathrm{d}S_\perp \right\}. \tag{5.72}$$

Thus (5.71b) and (5.72) relate the wave energy to Q:

$$W = -\frac{Q}{2\gamma}.$$ (5.73)

The stationary form (5.4) of the wave-energy balance equation (2.32a), after (5.73) and (5.71a) have been used, transforms into

$$\frac{\mathrm{d}}{\mathrm{d}z}\left(\frac{v_{\mathrm{gr}}}{\gamma}\,Q\right) = 2Q$$ (5.74a)

or, equivalently, into

$$\frac{\mathrm{d}}{\mathrm{d}z}\left(\frac{Q}{\alpha}\right) = -2Q\,,$$ (5.74b)

where α is the space damping rate of the wave.

In (5.74) the wave power losses Q are functions of the density n and of the electric-field intensity, and, through it, of the electron temperature T_{e}. These three quantities are connected through the electron transport equations (5.2) and (5.3). With these relations, (5.74) can be reduced to an equation for the longitudinal profile of, for instance, the plasma density. The other quantities involved in (5.74) are the time and space damping rates (γ and α), which can directly be found from the dispersion relation.

Equations (5.74), deduced from basic electrodynamic relations, are the equations which describe the axial structure of stationary waveguided discharges. They are valid regardless of the type of guided mode and geometry of the guiding structure, the type of wave damping and the mechanism of plasma heating by the wave, the regime of discharge maintenance and the ratio between ω and ν. The only requirement for their validity is weakly damped waves ($\gamma < \omega$ or $\alpha < \beta$). The relations (5.74) are general in the sense that they can be used for obtaining any of the quantities n, T_{e} and $|E|^2$ which, in combination, determine the axial structure of the discharge. On introducing the quantity

$$\Theta(\bar{n}) = \frac{Q}{\bar{n}},$$ (5.75)

i.e. the mean power absorbed per electron, and assuming that the variation of α is due to density changes, $\alpha = \alpha(\bar{n})$, (5.74b) reduces to the well-known [5.16,18,19,42–44] and widely used [5.45–57] formula for the axial profile of the plasma density

$$\frac{\mathrm{d}\bar{n}(z)}{\mathrm{d}z} = -2\alpha\bar{n}(z)\left/\left(1 - \frac{\mathrm{d}\alpha}{\mathrm{d}\bar{n}}\frac{\bar{n}}{\alpha} + \frac{\mathrm{d}\Theta}{\mathrm{d}\bar{n}}\frac{\bar{n}}{\Theta}\right)\right. .$$ (5.76)

This equation results directly from (5.4) when, in accordance with the geometrical-optics approximation, its left-hand side is written in terms of the space damping rate $\alpha(\bar{n})$:

$$\frac{\mathrm{d}P(\bar{n})}{\mathrm{d}z} = -2\alpha(\bar{n})P(\bar{n}).$$ (5.77a)

The decrease of P brought about by changing $\bar{n}$ in an axially inhomogeneous plasma is, to a good approximation, the same as that encountered in a homogeneous plasma (Sect. 4.3). Of course, in the self-consistent case this decrease has to be included in the plasma maintenance via $Q(\bar{n})$. Using thus $P = Q/2\alpha$ in (5.4) and recalling (5.75), one ontains

$$\frac{\mathrm{d}}{\mathrm{d}z}\left(\frac{\Theta\bar{n}}{2\alpha}\right) = -\Theta\bar{n} \tag{5.77b}$$

and, thus, (5.76).

Equation (5.76) has usually been applied with the assumption that in a diffusion-controlled regime and with Joule heating in the plasma volume, $\Theta = $ const. and, thus, $\mathrm{d}\Theta/\mathrm{d}\bar{n} = 0$. Although it is an approximation which replaces the gas-discharge part of the problem of discharge maintenance, this assumption makes (5.76) unique in its simplicity: the determination of the axial density profile is reduced, in a way, to knowing the space damping rate of the wave. The latter can easily be obtained as a numerical solution of the dispersion relation even when the configuration of the waveguide is complicated.

However, as has been stressed before, a self-consistent description of the axial structure requires Θ to be expressed in terms of the maintenance field intensity and, through it, the density. Though small, the modification which the step ionization and recombination introduce into the diffusion-controlled regime are connected to the term $\mathrm{d}\Theta/\mathrm{d}\bar{n}$ in (5.76). As a consequence they involve the nonlinear permittivity, i.e. the dependence of $\bar{n}$ on $\langle E^2 \rangle$, constituting in a way the physical basis of the discharge production.

As discussed in Sect. 5.3, the electron energy balance equation (5.3) has, in the more general case of nonlocal heating, the form

$$Q = 3\pi U_* \int_0^R rn\nu_* \, \mathrm{d}r. \tag{5.78a}$$

After integration over the cross-section, the particle balance equation (5.2), with a nonlinearity due to recombination losses included in it, reduces to

$$-\left(rD_A\frac{\partial n}{\partial r}\right)\bigg|_{r=R} + \int_0^R r\varrho_r n^2 \, \mathrm{d}r = \int_0^R rn\nu_i \, \mathrm{d}r. \tag{5.78b}$$

With $\nu_* \simeq \nu_i$ which is valid at least under the conditions of (2.59a), (5.78a) and (5.78b) can be combined into

$$Q = 3\pi U_*\left[-R\left(D_A\frac{\partial n}{\partial r}\right)\bigg|_{r=R} + \int_0^R r\varrho_r n^2 \, \mathrm{d}r\right]. \tag{5.79}$$

With an electron density represented in the form

$$n(r,z) = n(r=0,z)g(r), \tag{5.80}$$

i.e. supposing invariability of the radial density profile in the axial direction, (5.79) gives

$$\Theta = -3\pi U_* R D_{\mathrm{A}}(r = R, z)\overline{g(r)}^{-1} \left.\frac{\mathrm{d}g(r)}{\mathrm{d}r}\right|_{r=R}$$

$$+ 3\pi\bar{n}U_*\overline{g(r)}^{-2} \int_0^R r\varrho_{\mathrm{r}}g^2(r)\,\mathrm{d}r\,, \tag{5.81}$$

where the bar denotes averaging over r.

In the diffusion-controlled regime, the second term on the right-hand side of (5.81) is small compared with the first one, which means weak variations of Θ and $|E|^2$, as obtained in Sect. 5.5, and the longitudinal density profile ((5.74b), (5.76) and (5.77b)) is determined by the axial variation of the space damping rate α:

$$\frac{\mathrm{d}}{\mathrm{d}z}\left(\frac{\bar{n}(z)}{\alpha(z)}\right) = -2\bar{n} \tag{5.82a}$$

or, equivalently,

$$\frac{\mathrm{d}\bar{n}}{\mathrm{d}z} = -\frac{2\alpha\bar{n}}{1 - \dfrac{\bar{n}}{\alpha}\dfrac{\mathrm{d}\alpha}{\mathrm{d}\bar{n}}}\,. \tag{5.82b}$$

Neglecting the second term in the denominator of (5.76) is equivalent to obtaining results for the axial density profile which are within the limits of the logarithmic accuracy commented on in Sect. 5.5. When $\alpha(n)n = \mathrm{const.}$, which is a good approximation over a wide range of parameters (see (3.69a)), $(\mathrm{d}\alpha/\mathrm{d}\bar{n})(\bar{n}/\alpha) \approx -1$ and (5.82b) reduces to $(\mathrm{d}\bar{n}/\mathrm{d}z) \approx -\alpha\bar{n} \approx \mathrm{const.}$, resulting directly in a linear axial profile of the plasma density. Neglecting the $\Theta(z)$ variation is equivalent to neglecting the axial variation of the maintenance field intensity $|E_z(r = R, z)|^2$ in (5.47). Within this approximation (5.47) also directly results in a linear axial density profile. Therefore, this linear profile of the plasma density, which is always stressed in describing SW-sustained discharges, is due to the dispersion properties of the wave (energy flux $P \propto \bar{n}^2$, according to (3.70b) and (3.73), and/or space damping rate $\alpha \propto (1/\bar{n})$, according to (3.69a) and (3.74b)) acting together with Joule heating by collisions in the plasma volume.

At comparatively high pressures, when the gas discharge is in a recombination-controlled regime, the second term in (5.81) is the larger one, and then Θ is proportional to $\bar{n}$. Such a conclusion is in agreement with (5.19), which also shows proportionality of n and $|E|^2$. With $\Theta \propto \bar{n}$, $(\mathrm{d}\Theta/\mathrm{d}\bar{n})(\bar{n}/\Theta)$ approaches 1, and (5.76) reduces to

$$\frac{\mathrm{d}\bar{n}}{\mathrm{d}z} = -\frac{2\alpha\bar{n}}{2 - \dfrac{\bar{n}}{\alpha}\dfrac{\mathrm{d}\alpha}{\mathrm{d}\bar{n}}}\,. \tag{5.82c}$$

A comparison of (5.82b) and (5.82c) shows that a numerical factor constitutes the difference between the results for the axial gradient in the diffusion-

and recombination-controlled regimes. This is in agreement with the comments of Sect. 5.5.3 about the value of 3/2, which appears as the ratio between the gradients of the axial profiles in these two regimes.

5.6.2 Discharges Maintained in a Diffusion-Controlled Regime by Joule Heating in the Volume

Results for the axial profile of the plasma density in discharges sustained in a diffusion-controlled regime by Joule heating in the volume are given for two cases: weak ($\nu \ll \omega$) [5.41] and strong ($\nu \gg \omega$) [5.30] collisions. The treatment is analytical and it covers the complete EM range of the existence of the waves.

Conditions of Weak Collisions: Electrodynamic Similarity Law.
Since – for $\nu^2 \ll \omega^2$ – $\mathrm{d}|\varepsilon|/\mathrm{d}z = \mathrm{d}\bar{N}/\mathrm{d}z$ with $\bar{N} = \bar{n}/n_{\mathrm{c}}$, (5.82b) for the axial density profile is equivalent to

$$\frac{\mathrm{d}|\varepsilon|}{\mathrm{d}z} = -\frac{2(|\varepsilon|+1)\alpha}{1 - [(|\varepsilon|+1)/\alpha](\mathrm{d}\alpha/\mathrm{d}|\varepsilon|)} \,. \tag{5.83}$$

For

$$\alpha \propto \frac{1}{|\varepsilon|+1} \,, \tag{5.84a}$$

i.e. $\alpha \propto 1/\bar{n}$, one obtains

$$H(|\varepsilon|) \equiv \frac{1}{\alpha}(|\varepsilon|+1)\left(\frac{\mathrm{d}\alpha}{\mathrm{d}|\varepsilon|}\right) \simeq \text{const.} \tag{5.84b}$$

The statement that $H(|\varepsilon|)$ is a constant or, at least, a slowly varying function of the density is equivalent to $(\bar{n}/\alpha)(\mathrm{d}\alpha/\mathrm{d}\bar{n}) \approx$ const. As discussed in connection with (5.82b), $\alpha\bar{n} = $ const., i.e. $\alpha \propto (1/\bar{n})$, means a linear profile of the plasma density, as it has been obtained over a long length of the discharge in all experiments on SW-sustained discharges at comparatively low pressure [5.15,56,58].

The determination of the axial density profile from (5.83) reduces to analysis of the dependence of the space damping rate α on the density n. The results on the dispersion behaviour of the waves in cylindrical waveguides (see (3.66)–(3.68)) and the slowly varying function (3.69b) extracted from the $\alpha(\bar{n})$ dependence can be used directly.

Results for the axial gradient of the plasma density in discharges maintained under conditions of different σ values are plotted in Figs. 5.18 and 5.19, obtained with (3.66d), which gives $\alpha(\bar{n})$ in the complete EM range of SW existence. The normalizations of $(\mathrm{d}|\varepsilon|/\mathrm{d}z)$ in Figs. 5.18 and 5.19 are those suitable for EM waves and for quasi-static ones. The slow variation of $\bar{F}$ (3.69b) over the phase diagrams, which correlates with a linear density profile, is illustrated in Fig. 5.20, where the function $H(|\varepsilon|)$ (5.84b) is represented. A comparison of Figs. 5.18 and 5.19 with Fig. 5.20 shows that

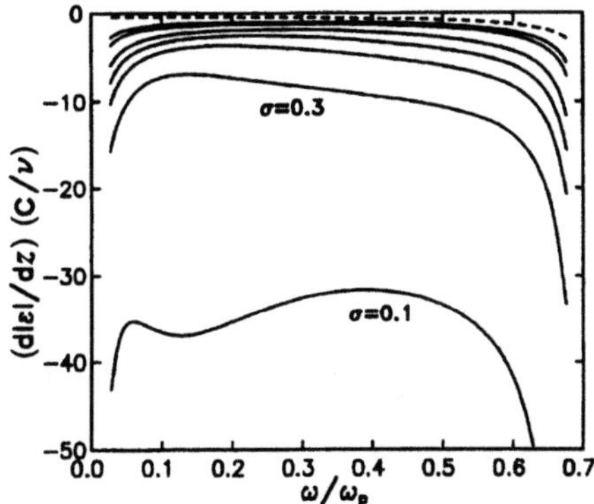

Figure 5.18. Normalized axial density gradient $d\bar{N}/dz \equiv d|\varepsilon|/dz$ in cylindrical discharges (*solid curves*) compared with that for a single-interface configuration (*dashed curve*); σ values as in Fig. 3.16 ([5.41], Fig. 7)

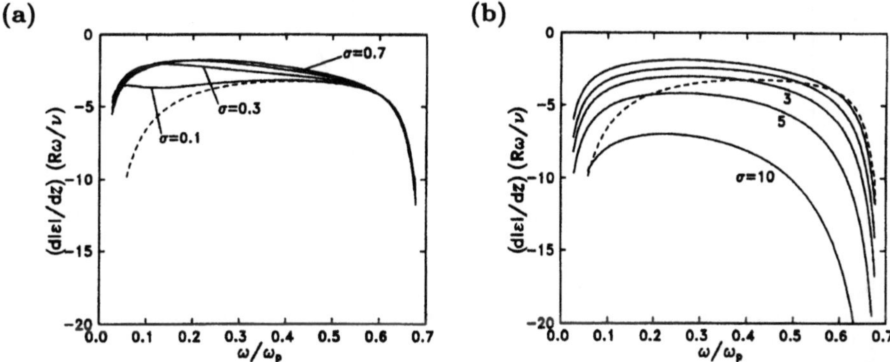

Figure 5.19. Normalized axial density gradient $d\bar{N}/dz \equiv d|\varepsilon|/dz$ in cylindrical discharges (*solid curves*) compared with that for the $\sigma \to 0$ limit, quasi-static approximation (*dashed curves*). (**a**) $\sigma = 0.1$, 0.3, 0.5 and 0.7; (**b**) $\sigma = 1, 2, 3, 5$ and 10 ([5.41], Fig. 8)

the domain of the linear profile (Figs. 5.18 and 5.19) extends over a range of $(\omega/\omega_{\rm p})$ larger than that given by $H(|\varepsilon|) \simeq -1$.

The approximate formulae (3.67d) and (3.68d) for the space damping rate, and $H(|\varepsilon|) \simeq -1$ lead to the following analytical result for the normalized density gradient:

$$\frac{d|\bar{N}|}{dz} = -\frac{1}{2} b \frac{1}{\bar{F}} \, .$$

(5.85)

This equation contains (5.55a) and (5.59b), and keeping the meaning of the different factors discussed in connection with (5.68) extends the validity of (5.55a) and (5.59b) over the complete EM range of wave existence. The value of the constant $1/\bar{F}$ should be taken from Fig. 3.20: $\bar{F} = \bar{F}_1$ (3.67) for $\sigma < 0.3$

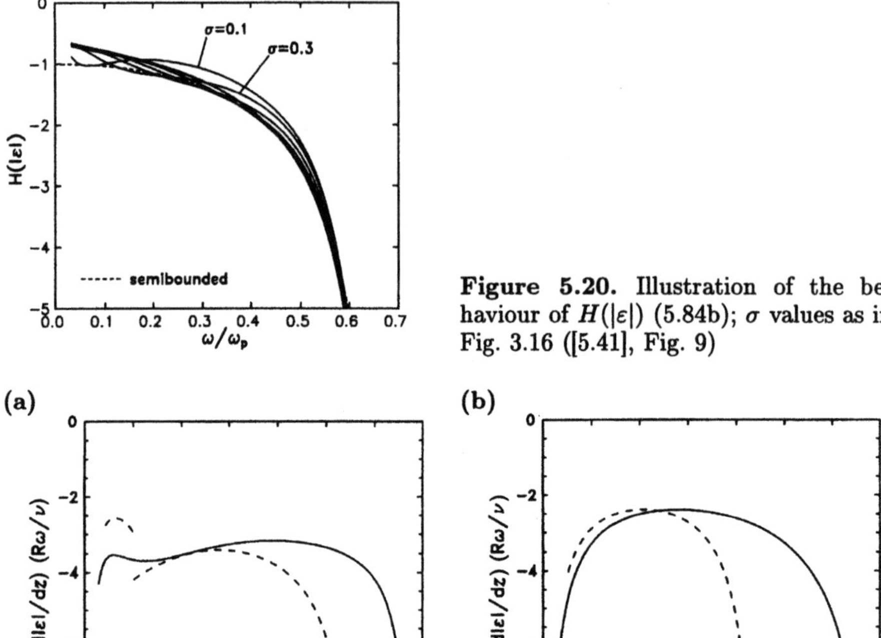

Figure 5.20. Illustration of the behaviour of $H(|\varepsilon|)$ (5.84b); σ values as in Fig. 3.16 ([5.41], Fig. 9)

Figure 5.21. Comparison of the approximate formula (5.85) (*dashed curve*) with the numerical solution of (5.83) (*solid curve*) for $\sigma = 0.1$ (a) and $\sigma = 2.0$ (b) ([5.41], Figs. 11a,c)

(region I in Fig. 3.16), $\bar{F} = \bar{F}_2$ (3.68) for $0.3 < \sigma < 3$ (region II in Fig. 3.16) and $\bar{F} = \sigma\bar{F}_2$ for $\sigma > 3$ (regions II and III in Fig. 3.16). The coefficient $1/2$ indicates that the discharge is in a diffusion-controlled regime. The role of the quantity b can be interpreted as an electrodynamic similarity law. For small ($\sigma < 0.3$) and intermediate ($0.3 \leq \sigma \leq 3$) σ values, $b = \nu/\omega R \equiv (\nu/\omega)(k_{\mathrm{v}}/\sigma)$, i.e. the combination of the vacuum wave number k_{v} and the parameter σ results in the normalizing parameter $1/R$ for quasi-static waves. For comparatively large σ values ($\sigma > 3$), $b = (\nu/\omega)(\omega/c) \equiv (\nu/\omega)(\sigma/R)$, i.e. the combination of σ and the normalizing parameter for quasi-static waves $1/R$ results in the vacuum wave number $k_{\mathrm{v}} = \omega/c$, which is the EM wave normalizing parameter. These conclusions are compiled in Table 5.1. The applicability of the formula (5.85) is shown in Fig. 5.21, where it is compared with the numerical solution of (5.83). Equation (5.85) extends the validity of the result (5.68) over the complete EM region of SW existence. Equation (5.82b) has been employed here for obtaining the density profile when the waveguide configuration is a plasma column surrounded by a vacuum. However, it is also

Table 5.1. Specification of the quantities included in (5.85)

		Equation	$\tilde{G}$	$\bar{F}$	b
$\tilde{x} < 1$	$\sigma < 0.3$	(3.67)	$\tilde{A}$	$\bar{F}_1$	$\dfrac{\nu}{\omega}\dfrac{1}{R}$
$\tilde{x} > 1$	$0.3 \leq \sigma \leq 3$	(3.68)	$2\tilde{C}$	$\bar{F}_2$	$\dfrac{\nu}{\omega}\dfrac{1}{R}$
$\tilde{x} > 1$	$\sigma > 3$	(3.68)	$2\tilde{C}$	$\sigma\bar{F}_2$	$\dfrac{\nu}{\omega}k_{\mathrm{v}}$

useful when α is calculated for cases of cylindrical plasmas taking also into account (dielectric) glass walls and, conceivably, enclosing metallic shields. The modifications of α introduced then usually correspond to slight modifications of the slowly varying function $\bar{F}$ or $\langle \bar{f} \rangle$ in (5.85) or (5.68).

Discharge Maintenance by Fast SWs at Strong Collisions ($\nu \gg \omega$). Discharges with strongly collisional plasmas ($\nu \gg \omega$) sustained by SWs which have a phase velocity close to the speed of light and the field of which penetrates weakly into the plasma are considered now. The density along the complete length is relatively high.

In order to obtain the axial density profile of a discharge sustained in a diffusion-controlled regime from (5.82), the quantity $(\bar{n}/\alpha)\,(\mathrm{d}\alpha/\mathrm{d}\bar{n})$ should be calculated first. As (3.75b) for the space damping rate of SWs in strongly collisional plasmas shows, the dependence of α on $\bar{n}$ is, within the limits of logarithmic accuracy, of the type $\alpha \propto \sqrt{n_{\mathrm{c}}/\bar{n}}$. This means that

$$\frac{\bar{n}}{\alpha}\frac{\mathrm{d}\alpha}{\mathrm{d}\bar{n}} = -\frac{1}{2} \tag{5.86a}$$

and, thus, (5.82b) reduces to

$$\frac{\mathrm{d}\bar{n}}{\mathrm{d}z} = -\frac{4}{3}\alpha\bar{n}. \tag{5.86b}$$

At high densities and with the space damping rate given by (3.75b), (5.86b) results in

$$\frac{\mathrm{d}}{\mathrm{d}\zeta}\sqrt{\frac{\bar{n}}{n_{\mathrm{c}}}} = \frac{-1}{\mu_* \ln\left(\mu_*\sqrt{\bar{n}/n_{\mathrm{c}}}\right)}, \tag{5.87a}$$

where

$$\mu_* = \frac{2c}{R\omega\Gamma^2}\sqrt{\frac{\omega}{\nu}} \tag{5.87b}$$

and

$$\zeta = \frac{1}{3}\frac{z}{R}\frac{2c}{R\omega\Gamma^2}\sqrt{2} \tag{5.87c}$$

is a dimensionless longitudinal coordinate.

Within the limits of logarithmic accuracy, integration of (5.87a) yields quadratic law

$$\bar{n} \approx \bar{n}_0 \left(1 - \sqrt{\frac{n_{\rm c}}{\bar{n}_0}} \frac{\zeta}{\mu_* \ln \left(\mu_* \sqrt{\bar{n}_0/n_{\rm c}} \right)} \right)^2 \tag{5.88a}$$

for the axial density profile, where $\bar{n}_0 = \bar{n}(z = 0)$ is a constant. In the original variables (5.88a) reads

$$\frac{\bar{n}}{\bar{n}_0} = \left\{ 1 - \frac{\sqrt{2}}{3} \sqrt{\frac{\nu}{\omega}} \sqrt{\frac{n_{\rm c}}{\bar{n}_0}} \left[\ln \left(\frac{2c}{R\omega\Gamma^2} \sqrt{\frac{\omega}{\nu}} \sqrt{\frac{\bar{n}_0}{n_{\rm c}}} \right) \right]^{-1} \frac{z}{R} \right\}^2 . \tag{5.88b}$$

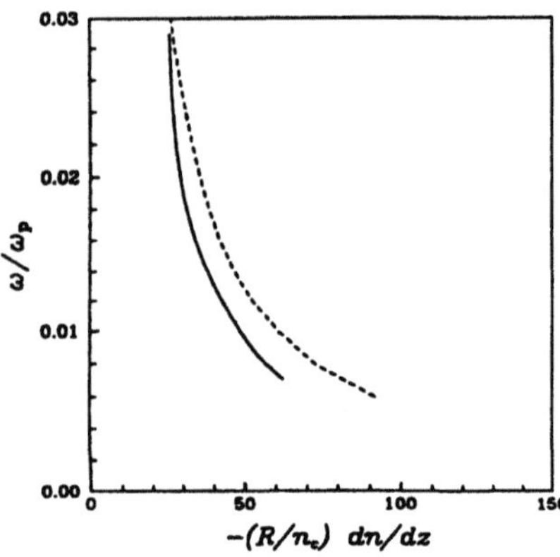

Figure 5.22. Normalized density gradient of the electron density versus $\omega/\overline{\omega_{\rm p}}$, i.e. along the plasma column, for strong collisions, $\nu/\omega = 10$. The result (*dashed curve*) according to (5.88b) is compared with the numerical result (*solid curve*) obtained from (5.82b) ([5.30], Fig. 4)

Figure 5.22 presents the gradient of the density profile obtained from (5.86a) using the analytical solution (3.75b) for α, and the results obtained from (5.82b) by using the numerical solution for α (Fig. 3.23).

For the recombination-controlled regime (5.82c) one obtains the same result as (5.88b), except that the factor 3 in the denominator is replaced by a factor 5.

The profile of the density is smoothly inhomogeneous along the discharge and steepens with increasing density. This dependence is in agreement with the numerical results obtained in [5.52].

5.6.3 Discharges Maintained by Joule Heating in the Plasma Volume and in Regions of Resonance Absorption

In this subsection results for different mechanisms of wave damping are presented which, with respect to discharge production, mean different mechanisms of plasma heating. The complete longitudinal profile of the plasma density along the discharge length is obtained.

The first part of this subsection contains treatments based on a model of a planar discharge in a diffusion-controlled regime [5.59,60]. This is a model in which the geometrical effects are taken out and only the physically meaningful effects associated with the transverse dimension of the waveguide are present. In general, the planar model retains the same functional dependences as the cylindrical geometry provides and drops geometrical effects which are related to the slowly varying (along the phase diagrams) functions introduced in Chap. 3, for example in the case of a plasma column surrounded by vacuum: $\bar{f}$ (3.73) and $\bar{F}$ (3.69b). Recently the interest in planar SW-sustained discharges has increased with their realization in experiments [5.61].

In the second part of this subsection, a cylindrical discharge is treated [5.35]. The two mechanisms of plasma heating – Joule heating in the volume and in regions of resonance absorption – are involved in the description of the complete longitudinal structure of discharges in diffusion- and recombination-controlled regimes.

Longitudinal Profile of the Plasma Density in a Planar Discharge in a Diffusion-Controlled Regime. Similarly to the previous subsection, the treatment here – a planar discharge in a diffusion-controlled regime – is based on an analysis of (5.82a). This means that the results for the space damping rate of an SW (Chaps. 3 and 4), in which account is taken of both space damping through collisions and resonance absorption, can be used directly.

In order to obtain the complete profile of the plasma density, including the regions of high density at the discharge onset, it is assumed that a high enough power for discharge production is applied. Thus, the skin depth (2.26a) at the beginning of the discharge is much smaller than the discharge scale in the transverse direction and the behaviour of SWs in this region of the discharge is that of SWs in a semi-infinite plasma (i.e. the dispersion relation (4.38a) and space damping rate (4.38b)).

If the plasma density at the beginning of the discharge is high enough, the inequality

$$\frac{\pi L_n^{(\mathrm{r})}}{\lambda_{\mathrm{sk}}(z)} \gg \frac{\nu}{2\omega} \tag{5.89a}$$

holds. The dissipation of the SW energy is due to the losses in the resonance layer close to the discharge wall. In this case (4.38b) reduces to

$$\alpha(z) = \frac{\omega^3}{c^2 \omega_{\mathrm{p}}(z)} \pi L_n^{(\mathrm{r})}, \tag{5.89b}$$

and (5.82a) becomes

$$\frac{\mathrm{d}}{\mathrm{d}z}\left(\frac{\omega_{\mathrm{p}}(z)}{\omega}\right) = -\frac{2}{3}\frac{\omega^2}{c^2}\pi L_n^{(\mathrm{r})}. \tag{5.89c}$$

This leads to a quadratic law for the density decrease in the axial direction

$$\bar{n}(z) = \bar{n}_0 \left[1 - \sqrt{\frac{n_\mathrm{c}}{\bar{n}_0}} \frac{2\pi}{3} \frac{\omega^2}{c^2} L_n^{(\mathrm{r})}(z - z_0) \right]^2 , \tag{5.89d}$$

where $\bar{n}_0$ is the density at $z = z_0$. As the plasma density decreases along the discharge length, the SW field penetrates more deeply into the plasma and the losses of wave power in the plasma volume increase. When the condition

$$\frac{\nu}{2\omega} \gg \frac{\pi L_n^{(\mathrm{r})}}{\lambda_\mathrm{sk}(z)} , \tag{5.90a}$$

opposite to (5.89a), starts to hold, the SW attenuation is totally determined by Joule losses in the plasma volume, and (4.38b) reduces to

$$\alpha(z) = \frac{\nu}{2c} \frac{\omega^2}{\omega_\mathrm{p}^2(z)} . \tag{5.90b}$$

Equation (5.82a) takes the form

$$\frac{\mathrm{d}}{\mathrm{d}z} \left(\frac{\omega_\mathrm{p}^2(z)}{\omega^2} \right) = -\frac{\nu}{2c} \tag{5.90c}$$

and gives a linear profile

$$\bar{n}(z) = \bar{n}_0 \left[1 - \frac{n_\mathrm{c}}{\bar{n}_0} \frac{\nu}{2c}(z - z_0) \right] \tag{5.90d}$$

for the axial variation of the plasma density. The two profiles (5.89d) and (5.90d) can appear consecutively only if the collision frequency is high enough $((\nu/2\omega) > (\pi L_n^{(\mathrm{r})}/d))$. Otherwise the fast EM waves in a thick slab (i.e. semi-infinte plasma) sustain only the quadratic longitudinal profile (5.89d).

With a further decrease of the density, the skin depth $\lambda_\mathrm{sk}(z)$ exceeds the transverse dimension d of the slab, and the approximation of a semi-infinite plasma is not valid any longer. Under such conditions (i.e. for $\lambda_\mathrm{sk}(z) \gg d$), the approximation of a thin slab starts to hold and the dispersion relation is now (4.41).

At the onset of the thin-slab regime, the SW is still fast (Sect. 3.3.1), with a space damping rate given by (4.42). According to (3.55) and (3.56a), a region of fast waves ($v_\mathrm{ph} \simeq c$) in a thin slab ($\varkappa_\mathrm{p}d < 1$) exists if the plasma density fulfils the inequalities

$$\frac{c}{\omega d} \ll \frac{\omega_\mathrm{p}^2(z)}{\omega^2} \ll \frac{c^2}{\omega^2 d^2} \tag{5.91a}$$

and the slab is suffciently thin:

$$\frac{c}{\omega d} \gg 1, \quad \text{i.e.} \quad \sigma < 1 . \tag{5.91b}$$

Under the condition

$$\frac{c}{\omega d} \ll 1, \quad \text{i.e.} \quad \sigma > 1, \tag{5.92}$$

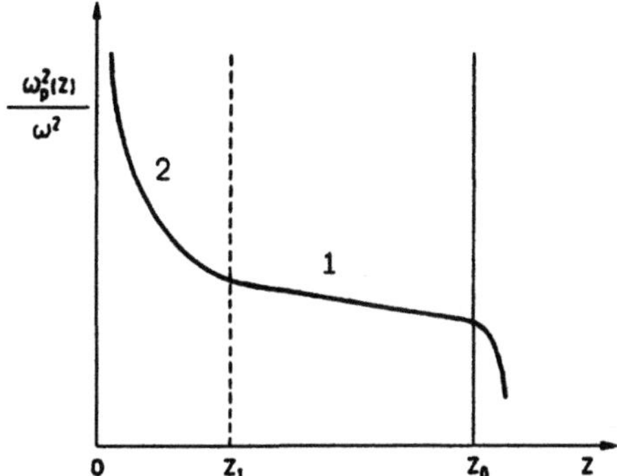

Figure 5.23. Schematic illustration of the normalized density profile $\bar{N}(z)$ in a discharge sustained in the thick-slab regime ($\sigma \gg 1$). The notation "2" and "1" indicates the type of law obeyed by the axial density dependence (2 – quadratic law (5.89d); 1 – linear law (5.90d)) ([5.60], Fig. 1)

the thick-slab regime ($\varkappa_\mathrm{p} d \gg 1$, semi-infinite plasma) governs the total length of the SW discharge: the density profile starts (region $0 < z < z_1$ in Fig. 5.23) with the quadratic dependence (5.89d) and then transforms (region $z_1 < z < z_0$ in Fig. 5.23) into the linear dependence (5.90d). In this case, the vicinity of the quasi-static resonance (3.1) of the SW (point $z = z_0$), where $\varepsilon(z_0)+1 = 0$, is the region of the discharge end. With respect to the phase diagrams given in Fig. 3.16 for SWs in a cylindrical waveguide, this is a longitudinal density profile of a discharge sustained at comparatively large σ ($\sigma > 0.3$), a case in which the phase diagrams lie in regions II and III .

Returning to the conditions fixed by (5.91b), two cases may be realized in the range of densities given by (5.91a). In the first case, when the collisions are sufficiently strong (5.90a), the regime of collisional dissipation is realized in the whole region of densities (5.91a) and, according to (4.42), the space damping is

$$\alpha(z) = \frac{\omega^3}{c^3} \frac{\lambda_\mathrm{sk}^4(z)}{d^2} \frac{\nu}{\omega} \ . \tag{5.93a}$$

Equation (5.82a) reduces to

$$\frac{\mathrm{d}}{\mathrm{d}z} \left(\frac{\omega_\mathrm{p}^2(z)}{\omega^2} \right)^2 = -\frac{4}{3} \frac{c\nu}{\omega^2 d^2} \tag{5.93b}$$

and, after integration, results in a square-root dependence for the density decrease in the axial direction:

$$\bar{n}(z) = \bar{n}_0 \sqrt{1 - \left(\frac{n_\mathrm{c}}{\bar{n}_0} \right)^2 \frac{4c\nu}{3\omega^2 d^2} (z - z_0)} \ . \tag{5.93c}$$

In the second case, of sufficiently weak collisions $(\nu/2\omega < \pi L_n^{(\mathrm{r})}/d)$, the damping of the fast SWs in the thin-slab regime at density values determined by

$$\frac{\omega_{\mathrm{p}}^2(z)}{\omega^2} > \frac{\nu}{\omega}\frac{c^2}{\omega^2}\frac{1}{\pi L_n^{(\mathrm{r})}d} \tag{5.94a}$$

is governed by the mechanism of resonant absorption. Then, (4.42) gives the space damping rate

$$\alpha(z) = \frac{\omega^3}{c^3}\lambda_{\mathrm{sk}}^2(z)\frac{\pi L_n^{(\mathrm{r})}}{d}\,, \tag{5.94b}$$

and (4.82a) takes the form

$$\frac{\mathrm{d}}{\mathrm{d}z}\left(\frac{\omega_{\mathrm{p}}^2(z)}{\omega^2}\right) = -\frac{\pi L_n^{(\mathrm{r})}\omega}{cd}\,. \tag{5.94c}$$

It results in a linear profile of the density

$$\bar{n}(z) = \bar{n}_0\left[1 - \frac{n_{\mathrm{c}}}{\bar{n}_0}\frac{\pi L_n^{(\mathrm{r})}\omega}{cd}(z - z_0)\right]\,. \tag{5.94d}$$

If the condition

$$\frac{\nu}{\omega} < \frac{\pi L_n^{(\mathrm{r})}\omega}{c} \tag{5.95}$$

holds, the linear profile (5.94d) covers the whole range (5.91a) of plasma densities produced by the propagation of fast waves.

With further decrease of the density (see (3.57a)),

$$\frac{\omega_{\mathrm{p}}^2(z)}{\omega^2} \ll \frac{c}{\omega d}\,, \tag{5.96}$$

the thin-slab regime $(\varkappa_{\mathrm{p}}d \ll 1)$ ensures the transformation of the SW from a fast one into a slow one with a space damping rate determined by (4.43c).

Under the condition (5.95) the resonance absorption governs the formation of the density profile over the whole region of (5.96). According to (4.43c) the space damping rate is

$$\alpha(z) = \frac{\omega^4}{\omega_{\mathrm{p}}^4(z)}\frac{\pi L_n^{(\mathrm{r})}}{d^2}\,. \tag{5.97a}$$

Equation (5.82a) becomes

$$\frac{\mathrm{d}}{\mathrm{d}z}\left(\frac{\omega_{\mathrm{p}}^2(z)}{\omega^2}\right)^2 = -\frac{4}{3}\frac{\pi L_n^{(\mathrm{r})}}{d^2} \tag{5.97b}$$

and gives a square-root dependence for the axial distribution of the density:

$$\bar{n}(z) = \bar{n}_0 \sqrt{1 - \left(\frac{n_c}{\bar{n}_0}\right)^2 \frac{4}{3} \frac{\pi L_n^{(r)}}{d^2}(z - z_0)} \ . \tag{5.97c}$$

Under the conditions

$$\frac{\pi L_n^{(r)}}{d} > \frac{\nu}{\omega} > \frac{\pi L_n^{(r)} \omega}{c} \tag{5.98a}$$

and in the density range

$$\frac{\omega_p^2(z)}{\omega^2} \gg \frac{\pi L_n^{(r)}}{d} \frac{\omega}{\nu} \ , \tag{5.98b}$$

the dissipation of the wave energy is due to collisions in the plasma volume. According to (4.43c) the space damping rate is

$$\alpha(z) = \frac{\nu}{\omega} \frac{\omega^2}{\omega_p^2(z)d} \ . \tag{5.99a}$$

Equation (5.82a) takes the form

$$\frac{d}{dz}\left(\frac{\omega_p^2(z)}{\omega^2}\right) = -\frac{\nu}{\omega d} \tag{5.99b}$$

and, after integration, gives the linear density profile

$$\bar{n}(z) = \bar{n}_0 \left[1 - \frac{n_c}{\bar{n}_0} \frac{\nu}{\omega d}(z - z_0)\right] \ . \tag{5.99c}$$

With a further decrease of the density (5.98b) is not valid any longer, and this changes the type of SW damping. In the range of density values

$$\frac{\omega_p^2(z)}{\omega^2} \ll \frac{\pi L_n^{(r)}}{d} \frac{\omega}{\nu} \ , \tag{5.100}$$

the mechanism of resonance absorption determines the SW damping, and the law for the density decrease is that given by (5.97c).

The rapid growth of the space damping rate with this density decrease leads to a strong resonance absorption of the SW in the region $z \simeq z_0$, determined by the condition $\alpha(z_0) \simeq \beta(z_0)$:

$$\frac{\omega_p^2(z = z_0)}{\omega^2} = \frac{\pi L_n^{(r)}}{d} \ . \tag{5.101}$$

The point $z = z_0$ may be considered as the point of the discharge end.

Figure 5.24 illustrates the variation of the axial density distribution under conditions of relatively weak collisions (5.95). The regime of resonance absorption of the SW is the predominant one for plasma heating. It determines the density profile over the complete length of the discharge ($0 < z < z_0$). At the onset of the discharge ($0 < z < z^*$) the fast wave in the thick slab shows up in the quadratic law (5.89d), then ($z^* < z < z_0$) the fast wave in the thin slab appears with a transition to the linear law (5.94d). At the discharge

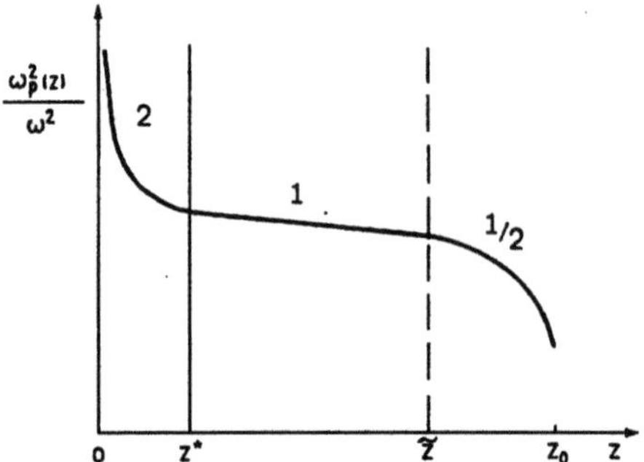

Figure 5.24. Schematic illustration of the normalized density profile $\bar{N}(z)$ in a discharge sustained under conditions of predominant resonance absorption, inequality (5.95): 2 – quadratic law (5.89d) governed by the thick–slab regime; 1 – linear law (5.94d) determined by fast wave propagation in the thin-slab regime; 1/2 – square-root law (5.97c) determined by slow wave propagation in the thin-slab regime ([5.60], Fig. 2)

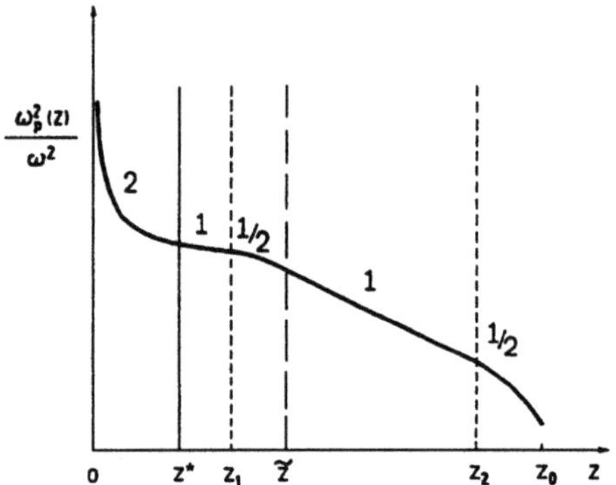

Figure 5.25. Schematic illustration of the normalized density profile $\bar{N}(z)$ in a discharge sustained under conditions determined by inequalities (5.98a) ([5.60], Fig. 3)

end the slow wave in the thin slab determines a square-root density decrease (5.97c).

Figure 5.25 presents schematically the axial density dependence when the collision frequency ν satisfies the inequalities (5.98a). In the thick-slab regime ($0 < z < z^*$) the resonance absorption leads to the quadratic density profile (5.89d). Thereafter ($z^* < z < \tilde{z}$) two regimes of fast waves in a thin slab appear: resonance absorption ($z^* > z > z_1$) with the linear law (5.94d), and collisional damping in the plasma volume ($z_1 < z < \tilde{z}$) with the square-root dependence (5.93c). At the end, the two regimes of slow waves in the thin slab show up consecutively: damping through collisions in the plasma volume ($\tilde{z} < z < z_2$) with the linear profile (5.99c), and damping through resonance absorption ($z > z_2$) with the square-root dependence (5.97c). Compared with a cylindrical waveguide, this behaviour corresponds to phase diagrams with small σ ($\sigma < 0.3$, Fig. 3.16).

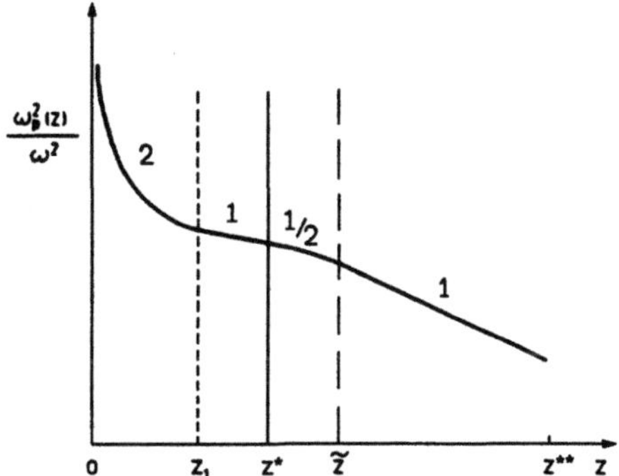

Figure 5.26. Schematic illustration of the normalized density profile $\bar{N}(z)$ in a discharge sustained under conditions of strong collisions (5.102) ([5.60], Fig. 4)

Figure 5.26 depicts the axial density profile under the condition of strong collisions:

$$\frac{\nu}{\omega} > \frac{\pi L_n^{(\mathrm{r})}}{d} . \tag{5.102}$$

Fast waves in a thick slab trigger the beginning of the discharge ($0 < z < z_1$) via resonance absorption (the quadratic law (5.89d)) and Joule heating in the plasma volume ($z_1 < z < z^*$, linear law (5.90d)), consecutively. In the thin-slab regime the plasma heating is by Joule heating in the plasma volume: at first ($z^* < z < \tilde{z}$) the fast waves lead to the square-root dependence (5.93c) and then ($\tilde{z} < z < z^{**}$) slow waves lead to the linear profile (5.99c). Transferred to the phase diagrams of a cylindrical waveguide, such behaviour again corresponds to the case of small σ ($\sigma < 0.3$, Fig. 3.16). In the vicinity of point z^{**} (Fig. 5.26), $\beta d \approx 1$, and the SWs transform from slow waves in a thin slab into slow waves in a thick slab. (According to the phase diagram (Fig. 3.16) for a cylindrical waveguide, this is region III.) The dispersion behaviour (4.39) is again given by the model of a semi-infinite plasma. Sufficiently far before resonance, the condition

$$\frac{\nu}{\omega} \gg \frac{\pi L_n^{(\mathrm{r})}}{2} \frac{\omega}{c} \left(\frac{\omega_p^2(z)}{\omega^2} - 2 \right)^{-1/2} \tag{5.103}$$

reduces (4.39c) to

$$\alpha(z) = \frac{\nu}{\omega} \left(\frac{\omega_p^2(z)}{\omega^2} - 2 \right)^{-3/2} . \tag{5.104a}$$

Equation (5.82a) becomes

$$\frac{\mathrm{d}}{\mathrm{d}z}\left(\frac{\omega_\mathrm{p}^2(z)}{\omega^2}-2\right)^{3/2}=-\frac{2\nu}{c} \tag{5.104b}$$

and leads to a power law (with a power of 2/3) for the density profile. In the vicinity of the SW resonance, if the inequality opposite to (5.103) is valid, the absorption is resonant and (4.39c) yields

$$\alpha(z)=\frac{\pi\omega^2}{2c^2}L_n^{(\mathrm{r})}\left(\frac{\omega_\mathrm{p}^2(z)}{\omega^2}-2\right)^{-2}. \tag{5.105a}$$

Equation (5.82a), in the form

$$\frac{\mathrm{d}}{\mathrm{d}z}\left(\frac{\omega_\mathrm{p}^2(z)}{\omega^2}-2\right)^2=-\frac{\pi\omega^2}{c^2}L_n^{(\mathrm{r})}, \tag{5.105b}$$

gives a square-root law for describing the way in which the density approaches its value at the quasi-static resonance point.

It must be stressed that the treatment above gives only a qualitative description of the region close to resonance. In this region the assumptions and simplifications of the model may be questionable. According to Sect. 5.5 a treatment solely of the density profile is acceptable only when the changes in the longitudinal direction of the maintenance field intensity and electron temperature are weak. The assumption of a constant value of $L_n^{(\mathrm{r})}$ along the discharge length is also a strong simplification.

Longitudinal Profile of the Plasma Density in Cylindrical Discharges Sustained in Different Regimes. The longitudinal structure of diffusion- and recombination-controlled cylindrical discharges sustained by Joule heating in the plasma volume and in regimes of resonance absorption is considered now. Again a planar model is utilized first, for stressing that in the case of Joule heating via resonance absorption the self-consistency of the plasma–wave behaviour is ensured by the mechanisms of heating. Then, the treatment is extended to cylindrical waveguide discharges. In both cases (5.74a) is analysed and solved together with the electron energy balance equation (5.10).

As has been commented on in Sect. 5.5.2, case (iii), as well as in Sect. 5.3.1b and Sect. 5.3.4b (case (iii)), the field $E_{(\mathrm{r})}$ which sustains a discharge through Joule heating in regions of resonance absorption is determined not only by the wave field but also by the plasma density. Because of this, as (5.65)–(5.67) show, the interrelation between plasma density and field intensity which builds the self-consistent axial structure of the discharge stems directly from the mechanism of heating. Thus, the basic relation between these two characteristics is maintained even when an assumption of a constant electron temperature $(T_\mathrm{e}(z) = \mathrm{const.})$ along the discharge length is made. Therefore, disregarding the nonlinear terms in the particle balance equation (5.2) leaves the weak variation of $T_\mathrm{e}(z)$ undetermined because the changes of the maintenance field $E_{(r)}^2$ are not specified, but the basic relation between plasma density and field intensity, and their simultaneous variation along the discharge, are determined as they are related to the mechanisms of heating.

In the case of a planar discharge sustained by slow EM waves in a thin inhomogeneous slab, the quantities involved in (5.74a) – the time damping rate γ and Joule losses Q – are as given by (4.43b) and (4.47). Therefore, (5.74a) becomes

$$\frac{d\omega}{\nu}\frac{1}{n_{\rm c}}\frac{\rm d}{{\rm d}z}\left(\bar{n}^2|E_z|^2\right) = -2\bar{n}|E_z|^2\left(1 + \pi\frac{L_n^{(\rm r)}}{d}\frac{\omega}{\nu}n_{\rm c}\right) \tag{5.106a}$$

and takes into account the two mechanisms of heating (Joule heating in the plasma volume and in regions of resonance absorption). Equation (5.10) yields

$$\bar{n}\nu_*U_* = \frac{e^2\nu}{3m\omega^2}|E_z|^2\left(\bar{n} + \pi\frac{L_n^{(\rm r)}}{d}\frac{\omega}{\nu}n_{\rm c}\right). \tag{5.106b}$$

In the discharge region where the density $\bar{n}$ is comparatively high, i.e. for

$$\frac{\bar{n}}{n_{\rm c}} \gg \pi\frac{\omega}{\nu}\frac{L_n^{(\rm r)}}{d}, \tag{5.107a}$$

which is a condition for Joule heating in the plasma volume, (5.106) results in a linear axial density profile:

$$\bar{n} = \bar{n}_0 - n_{\rm c}\kappa\frac{\nu}{d\omega}(z - z_0). \tag{5.107b}$$

Here $z \geq z_0$, with z_0 denoting the beginning of a discharge sustained by slow waves in a thin-slab approach, $\bar{n}_0 = \bar{n}(z = z_0)$ and κ takes the values $\kappa = 1$ and $2/3$ or $2/(2 + s^{-1})$ for the diffusion regime and for the volume recombination regimes under conditions fixed by (2.59) and (2.60), respectively.

Equation (5.107b) is equivalent to (5.68), discussed before for a cylindrical discharge. Their comparison shows that the quantity $(2\langle\bar{f}\rangle)^{-1}$ constitutes the difference between planar and cylindrical discharges. For that reason the appearance of $\bar{f}$ in the case of a cylindrical discharge can be connected to geometrical effects specifying the behaviour of the phase diagrams of cylindrical waveguides.

In recombination-controlled discharges the profiles of the maintenance field intensity which correspond to the density profile (5.107b) are as given by (5.19): $\bar{n} \propto |E_z|^2$ and $\bar{n} \propto |E_z|^{2s}$ under the conditions specified by (2.59) and (2.60), respectively.

With a density decrease along the discharge length so that

$$\frac{\bar{n}}{n_{\rm c}} \ll \pi\frac{\omega}{\nu}\frac{L_n^{(\rm r)}}{d}, \tag{5.108a}$$

damping through resonance absorption and, correspondingly, Joule heating in the regions of resonance absorption show up. The set of equations (5.106) leads to

$$\frac{d\omega}{\nu}\frac{1}{n_c}\frac{d}{dz}\left(\bar{n}^2|E_z|^2\right) = -2\pi\frac{L_n^{(r)}}{d}\frac{\omega}{\nu}n_c|E_z|^2 ,$$
(5.108b)

$$\bar{n}\nu_*U_* = \frac{e^2\nu}{3m\omega^2}\pi\frac{L_n^{(r)}}{d}\frac{\omega}{\nu}n_c|E_z|^2 .$$
(5.108c)

In a diffusion-controlled regime (5.108c) reduces to

$$T_e = -\frac{U_*}{\ln\left(|E_z|^2/\beta_r\bar{n}\right)} ,$$
(5.109a)

with

$$\beta_r^{-1} = \frac{1}{U_*\mathring{\nu}_*}\frac{e^2\nu}{3m\omega^2}\pi\frac{L_n^{(r)}}{d}\frac{\omega}{\nu}n_c .$$
(5.109b)

It shows that even when the nonlinear terms in the particle balance equation are disregarded (which corresponds to the assumption of $T_e(z) = $ const.) the self-consistency of the longitudinal variation of the plasma density and field intensity is preserved. This appears in the proportionality between them $(\bar{n} \propto |E_z|^2)$.

In a recombination-controlled regime, for which the solutions obtained are completely self-consistent, the result for the electron temperature is

$$T_e = -\frac{U_{*,i} + U_*}{\ln\left(|E_z|^2/E_{n(r)}^2\right)} ,$$
(5.110a)

where

$$E_{n(r)} = \left(\frac{3m\omega^2\mathring{\nu}_{i(1,2)}\mathring{\nu}_*U_*}{e^2\nu\varrho_r\pi(L_n^{(r)}/d)(\omega/\nu)n_c}\right)^{1/2}$$
(5.110b)

is a normalizing field. If the conditions specified by (2.59) hold, the dependence of the density on the field intensity is $\bar{n} \propto |E_z|$, whereas with (2.60) it is $\bar{n} \propto |E_z|^{2/(1+s^{-1})}$. Therefore, the regime of electron heating – Joule heating in the plasma volume and in regions of resonance absorption – determines, in an essential manner, the interrelation between plasma density and field intensity.

The axial density profile

$$\bar{n} = \bar{n}_1\left[1 - (z - z_1)\kappa\pi\frac{L_n^{(r)}}{d}\left(\frac{n_c}{\bar{n}_1}\right)^2\right]^{1/2}$$
(5.111a)

is obtained from the relations between $\bar{n}$ and $|E_z|$ introduced in (5.108b). Here $\kappa = 4/3$ for a discharge in a diffusion regime (see (5.97c)) and $\kappa = 1$ and

$\kappa = 4/(3 + s^{-1})$ for recombination regimes as specified by (2.59) and (2.60), respectively. In (5.111a) $z \geq z_1$ and z_1 is the position along the discharge length where the linear profile (5.107b), which covers the main, long part of the discharge, transforms into a faster decrease of the density (5.111a), at the end of the discharge. After matching (5.107b) and (5.111a), the latter finally becomes

$$\bar{n} = n_{\mathrm{c}} \pi \frac{\omega}{\nu} \frac{L_n^{(\mathrm{r})}}{d} \sqrt{1 - (z - z_1)\frac{\kappa}{\pi} \frac{1}{L_n^{(\mathrm{r})}} \left(\frac{\nu}{\omega}\right)^2} \ . \tag{5.111b}$$

When laws of density variation other than linear ones are obtained, discussions of the conditions of their applicability are necessary. In general, each deviation from a linear profile obtained from equations in which longitudinal diffusion and thermal conductivity are neglected requires comment. In the case of (5.111), longitudinal diffusion and thermal conductivity may be neglected if

$$\frac{3}{2} L_n^{(\mathrm{r})} \left(\frac{\omega}{\nu}\right)^2 - (z - z_1) \gg \sqrt{\frac{5}{3} \frac{D_{\mathrm{e}}}{\nu_*}} \tag{5.112a}$$

and

$$\frac{3}{4} L_n^{(\mathrm{r})} \left(\frac{\omega}{\nu}\right)^2 - (z - z_1) \gg \frac{d}{\pi} \tag{5.112b}$$

hold. These are, in fact, requirements for weak collisions, which in any case should hold in order to ensure conditions for the mechanism of resonance absorption.

As an extension of the treatment to cylindrical discharges [5.35], the complete EM range of wave existence (see the phase diagrams in Fig. 3.16) will now be covered. Moreover, the discussions will concern both diffusion- and recombination-controlled regimes. Again (5.74a) is the starting point. After introducing (4.72) and (4.73b) for γ and Q into (5.74a), the latter becomes

$$\frac{\mathrm{d}}{\mathrm{d}z} \left[\frac{2c}{\nu} \sigma(\tilde{y}^2 + \sigma^2)^{1/2} \frac{|\bar{\varepsilon}|}{|\bar{\varepsilon}| + 1} \frac{2}{\tilde{x}} \frac{\mathrm{I}_1(\tilde{x})}{\mathrm{I}_0(\tilde{x})} \tilde{T}\bar{n}|E_z(r = R)|^2 \right]$$
$$= -\frac{2}{\tilde{x}} \frac{\mathrm{I}_1(\tilde{x})}{\mathrm{I}_0(\tilde{x})} \bar{n}|E_z(r = R)|^2 \left(1 - \frac{\sigma^2}{\tilde{x}^2} |\bar{\varepsilon}|\tilde{M} + P_{\mathrm{r}}\frac{n_{\mathrm{c}}}{\bar{n}}\right) \ . \tag{5.113}$$

Since the regime of Joule heating in the plasma volume has already been treated (Sect. 5.6.2), attention is directed here to those regions along the discharge length where the heating occurs through resonance absorption. This mechanism of heating is predominant for

$$P_{\mathrm{r}}\frac{n_{\mathrm{c}}}{\bar{n}} \gg 1 - \frac{\sigma^2}{\tilde{x}^2} |\bar{\varepsilon}|\tilde{M} \ . \tag{5.114}$$

For $\tilde{x} \ll 1$, i.e. slow SWs in the thin-cylinder approximation, the left-hand side of (5.114) is proportional to $|\bar{\varepsilon}|^{-1}$, whereas for $\tilde{x} \gg 1$ (fast SWs in the thick-cylinder approximation), it is proportional to $|\bar{\varepsilon}|^{1/2}$. This means that in the first case ($\tilde{x} \ll 1$) the contribution of the resonance absorption to the plasma heating increases with decreasing density, i.e. towards the end of the discharge. On the other hand, if the discharge is long enough and the plasma density in the region close to the SW launcher is high enough to ensure SW behaviour in the thick-cylinder approximation ($\tilde{x} \gg 1$), the mechanism of resonance absorption might also act at the onset of the discharge (see earlier in Sect. 5.6.3). Thus, both the onset and the end of the discharge can be governed by the mechanism of resonance absorption, whereas the main (long) part of the discharge between these two regions is sustained by Joule heating in the plasma volume performed by slow EM SW propagation under the conditions of the thin-cylinder approximation.

For $\tilde{x} \ll 1$ the right-hand side of (5.114) is approximately equal to 1, whereas for ($\tilde{x} \gg 1$) it approaches 1/2. Therefore, (5.114) reduces to

$$\frac{1}{2}\,\pi(\beta R)^2 |\bar{\varepsilon}| \frac{L_n^{(\mathrm{r})}}{R}\,\frac{\omega}{\nu} \gg 1 \quad \text{for } \tilde{x} \ll 1, \tag{5.115a}$$

$$2\pi\sigma\sqrt{|\bar{\varepsilon}|}\,\frac{L_n^{(\mathrm{r})}}{R}\,\frac{\omega}{\nu} \gg 1 \quad \text{for } \tilde{x} \gg 1. \tag{5.115b}$$

The result obtained for the longitudinal density profile at the end of a diffusion-controlled discharge where the maintenance is by means of resonance absorption:

$$\bar{n} = \bar{n}_1 \sqrt{1 - \frac{1}{3}\pi \frac{L_n^{(\mathrm{r})}}{R^2}\,\frac{n_{\mathrm{c}}^2}{\bar{n}_1^2}\,\frac{1}{\bar{F}_3}(z - z_1)}\,, \tag{5.116a}$$

coincides with (5.67) derived under the same conditions. Here

$$\bar{F}_3 = \frac{2[1 - (\tilde{y}^2 |\bar{\varepsilon}|/4) + |\varepsilon|^{-1}]}{\tilde{y}^2 \beta R |\bar{\varepsilon}|(|\bar{\varepsilon}| + 1)^2} \tag{5.116b}$$

is a slowly varying function over the region of the phase diagram considered.

For $\tilde{x} \gg 1$, i.e. in the region close to the SW launcher, the slowly varying function $\bar{F}_3$ can be written as

$$\bar{F}_3 = \frac{\bar{F}_3'}{|\bar{\varepsilon}|^{3/2}}, \tag{5.117a}$$

where

$$\bar{F}_3' = \frac{\sigma}{\beta R}\,\frac{\tilde{L}}{\tilde{y}^2 |\bar{\varepsilon}|} \tag{5.117b}$$

is a slowly varying function over this region of the phase diagram. The integration of (5.113) gives

$$\bar{n} = \bar{n}_2 \left[1 - \frac{1}{6}\pi\frac{L_n^{(\mathrm{r})}}{R^2}\,\frac{1}{\bar{F}_3'}\sqrt{\frac{n_{\mathrm{c}}}{\bar{n}_2}}(z - z_2)\right]^2, \tag{5.118}$$

where $\bar{n}_2 = \bar{n}(z = z_2)$ is the density at $z = z_2$, the point where the linear density profile determined by the collisional volume Joule heating over the main part of the discharge starts. The behaviour of the axial density profile demonstrated in Fig. 5.25 for the planar case shows up also in a cylindrical discharge: on both sides of the linear density profile obtained under conditions of predominant Joule heating in the plasma volume, regions with a steeper density gradient appear close to the discharge end and at the discharge onset, where the plasma heating is due to the mechanism of resonance absorption in a radially inhomogeneous plasma.

A density profile at the end of the column determined by the mechanism of Joule heating in regions of resonance absorption can also be discussed for a discharge in a recombination-controlled regime. The electron temperature obtained from the local approximation of the electron energy balance equation (5.3), with Joule heating, as given by (4.73b), is

$$T_{\mathrm{e}} = -\frac{U_{(*,\mathrm{i})} + U_*}{\ln\left[2\varUpsilon|E_z(r{=}R)|^2/E^2_{\mathrm{i}_{(1,2)}}\right]}\,, \tag{5.119a}$$

where

$$\varUpsilon = |\bar{\varepsilon}|^2\frac{(\beta R)^2}{\tilde{x}^2}\frac{\mathrm{I}_1(\tilde{x})}{\mathrm{I}_0(\tilde{x})}\,, \tag{5.119b}$$

$$E^2_{\mathrm{i}_{(1,2)}} = \frac{3m\omega^2\mathring{\nu}_{\mathrm{i}_{(1,2)}}\mathring{\nu}_*U_*}{e^2\nu\varrho_{\mathrm{r}}A}\,, \tag{5.119c}$$

$$A = \pi n_{\mathrm{c}}\frac{L_n^{(\mathrm{r})}}{R}\frac{\omega}{\nu}\,. \tag{5.119d}$$

The following expressions determine the relation between the plasma density and the SW field amplitude:

$$\bar{n} = \frac{\mathring{\nu}_{\mathrm{i}_{(1)}}}{\varrho_{\mathrm{r}}}\sqrt{2\varUpsilon}\frac{|E_z(r = R)|}{E_{\mathrm{i}_{(1)}}}\,, \tag{5.120a}$$

$$\bar{n} = \frac{\mathring{\nu}_{\mathrm{i}_{(2)}}}{\varrho_{\mathrm{r}}}\left(\sqrt{2\varUpsilon}\frac{|E_z(r = R)|}{E_{\mathrm{i}_{(2)}}}\right)^{2/(1+s^{-1})}\,. \tag{5.120b}$$

These expressions are the results for ionization nonlinearity under conditions of Joule heating in regions of resonance absorption, in the recombination regimes specified by (2.59) and (2.60), respectively. After inserting (5.120) into the wave energy balance equation (5.113), one obtains

$$\frac{2c\sigma}{\nu n_{\mathrm{c}}}\frac{\mathrm{d}}{\mathrm{d}z}(\bar{F}_3\bar{n}^4) = -A\bar{n}^2\,, \tag{5.121a}$$

$$\frac{2c\sigma}{\nu n_{\mathrm{c}}}\frac{\mathrm{d}}{\mathrm{d}z}(\bar{F}_3\bar{n}^{3+s^{-1}}) = -A\bar{n}^{1+s^{-1}}\,, \tag{5.121b}$$

which govern the axial variation of the plasma density at the end of the discharge in the recombination regimes. The density

$$\bar{n}(z) = \bar{n}_1 \sqrt{1 - \kappa \frac{\pi}{\bar{F}_3} \frac{L_n^{(\mathrm{r})}}{R^2} \frac{n_c^2}{\bar{n}_1^2} (z - z_1)} \tag{5.122}$$

at the end of the discharge decreases according to a square-root law. Here $\kappa = 1/4$ and $\kappa = 1/(3 + s^{-1})$ specify the cases corresponding to (2.59) and (2.60), respectively.

When *comparing* the results above with those in Sect. 5.5, it should be remembered that in that section the situations treated were limited to the σ range $\sigma \leq 0.3$ and excluded extremely high electron densities at the discharge onset (in the fast-wave regime), and thus were limited to the situations encountered in the majority of experiments so far. The situations studied in Chap. 6 will essentially be of this type, though with some extension in the range of σ. In the above Sect. 5.6 aimed to substantially enlarge the range of conditions to encompass all situations that could conceivably be encountered, including cases of strong collisions. The results for the axial density profile obtained in the preceding specific approach of Sect. 5.5, namely a virtually linear density profile modified by resonance absorption to a square-root-type dependence towards the end of the discharge, are, of course, contained in the general approach above. They appear, in particular, in Sect. 5.6.3 concerning plasma cylinders via (5.107b) and (5.111b).

The axial structure of the discharge, as discussed here in some detail, is a typical feature of the self-consistent interplay and coexistence of wave and plasma. This structure is composed of interrelated axial variations of plasma density, electron temperature, maintenance field intensity, SW field intensity and wave number. The density profile obviously turns out always to have a negative sign in the associated spatial derivative. It might be mentioned that this latter property has been pointed out as a requirement for stable operation of SW discharges [5.19], and is indeed encountered in many experiments. Though this consideration can be extended to include the effects studied above, a complete treatment of the stability problem is complicated, calling for consideration of fully time-dependent equations and situations beyond the scope of the present book.

The essential aspect of a radial density inhomogeneity has been included in this chapter by stressing the potential occurence of resonant absorption (due to peaks in $|E_r|$ when $\omega_\mathrm{p}(r) = \omega$). This analysis is extended in the numerical approach of Chap. 6, where the kinetic effect of collisionless damping is also taken into account.

6. Kinetic Numerical Modelling

This chapter introduces the kinetic modelling of SW-sustained discharges. The numerical procedures employed allow one to complement the analytical treatments given in Chap. 4 on the basis of the fluid plasma model. There are three main aspects to this as follows.

- The electron energy distribution function (EEDF) is obtained by solving Boltzmann's equation.
- Numerically, more details of the discharge processes, concerning both plasma and atomic physics, can be taken into account.
- Self-consistency can be achieved by a numerical treatment concerning points where otherwise estimates or assumptions have to be relied on. An example that is important for SW discharges is the self-consistent calculation of radial density profiles. In particular, the occurence of local plasma resonances depends on the existence of suitable radial density profiles.

As has been discussed first by Bernstein and Holstein [6.1] and Rose and Brown [6.2] (and later, with regard to SW-sustained discharges, in [6.3,4]), the approaches to the plasma density inhomogeneity and the ambipolar field related to it, as well as to the problem of locality/nonlocality in the formation of the electron energy distribution, are crucial points in the modelling of discharge maintenance in a diffusion-controlled regime. With regard to this, two approaches have been developed in the kinetic modelling of gas discharges in general [6.5–7] and of SW-sustained discharges, in particular. These are the local approach [6.8–31] and the nonlocal approach [6.1,32–41].

In the presentation here, the approach based on a *nonlocal* approximation is discussed first. It is best for discharges at lower pressures. In pursuing this approach, much emphasis is put on including, as much as possible, cases with a strong radial density nonuniformity. In doing so the atomic-physics basis is kept relatively simple.

A second part of this chapter will report on work performed in the opposite approximation, the *local* one, basically more appropriate for higher pressures. This type of work, as recorded in the literature, often includes more detailed attention to the atomic-physics basis.

In a third part this chapter will turn to estimates of the applicability of the two regimes, the discussion of which will also shed some light on recent developments in formulating a generalized approach that encompasses both

the local and the nonlocal approximations [6.7,42]. The important points of
this approach will be emphasized, and the foreseeable trend to self-consistent
modelling in this more general way will be commented on. It allows one to
simultaneously include detailed attention to the atomic-physics basis and to
the influences of inhomogeneity, and, moreover, to avoid some of the previous
assumptions that limit applicability and accuracy.

There exist a variety of sophisticated numerical procedures which may
even be applied prior to the usual Lorentz two-term approximation (2.34a).
In particular, Monte Carlo simulations, particle-in-cell methods and convec-
tive-scheme methods have been used to numerically interpret Boltzmann's
equation [6.43–54]. Such procedures are computer-intensive, though very important in checking the validity of simplifying assumptions. They may also
be used to address specific aspects, such as aspects of sheaths, particularly
in metal-bounded systems [6.55].

In many cases, however, simplifications may be used to reduce the numerical effort and to provide some transparency to the phenomena involved. For
these reasons the nonlocal and local approximations are stressed here.

Major additions of this chapter over the preceding one are, aside from
allowing non-Maxwellian distribution functions,

- a full numerical treatment of radial inhomogeneity (including the phase
  diagram changes commented on in Chap. 4)
- the inclusion of noncollisional electron heating, as a consequence of the
  radial field inhomogeneity and transformation of SWs into volume plasmons
  [6.56–59].

6.1 Nonlocal Model

SW-sustained discharges – in the microwave region, but also to some extent
in the 100 MHz region – will be modelled kinetically here in pressure ranges
where a nonlocal approximation is reasonable for the calculation of the EEDF
[6.58]. Cases of large energy relaxation length $L_\varepsilon \gg R$ (or at least $L_\varepsilon \simeq R$)
are considered, assuming the estimate (2.76)

$$L_\varepsilon = \lambda_{\mathrm{f.p.}} \sqrt{\frac{\nu}{\nu_\varepsilon}}, \tag{6.1}$$

where $\lambda_{\mathrm{f.p.}}$ is the elastic-scattering mean free path of the electrons, and ν_ε has
been given by (2.35b). This approach aims at taking into account the radial
density inhomogeneity and the associated space charge field E_{A}. Following
the treatment of Bernstein and Holstein [6.1] and that of Tsendin [6.32], the
approach is formulated on the basis of the total energy $\mathcal{W} = u - \Phi$, as defined
in (2.40). As pointed out before, the terminology *nonlocal–local* is used here
to refer to the formation of the isotropic part of the energy distribution.

Helium and argon are taken as typical discharge gases, though results have also been obtained for other inert gases. The assumption of uniform gas temperature used here could be replaced by additional modelling or experimental data. Perfect accuracy is not the main aim, which in any case is not achievable, owing to limitations in the atomic data available even for inert gases; rather, the aim is a model sufficiently detailed to reflect correctly the trend of the basic and interesting features of the self-consistent interrelation of the wave and the plasma.

Of course, the procedure presented below can also be used for other types of discharges in a modified way. Some similarities would remain if high-pressure (and/or large-diameter) discharges were considered – on the basis of a local approach – though some of the interesting features of low-pressure discharges (such as local plasma resonances) might be significantly modified, whereas new problems might arise.

6.1.1 Boltzmann's Equation

As a cornerstone of the complete system of equations required for a self-consistent model, Boltzmann's equation (2.33) is needed. Again, the much-applied two-term approximation given by (2.34a) is relied on. Moreover, the resultant isotropic part of the distribution function, when averaged over the period of the HF field ($T = 2\pi/\omega$), is taken as time-independent and denoted simply by F_0, since the frequencies ω considered for the HF fields used are well above the energy exchange frequency expressed by (2.35). Aiming at a nonlocal scenario for the calculation of the EEDF, a formulation of the problem in the total energy $\mathcal{W} = u - \Phi$, as described in Chap. 2, is used.

Within the frame of the nonlocal approach the isotropic part of the distribution function can be written as

$$F_0(\mathcal{W}, r) = F_0^{(0)}(\mathcal{W}) + F_0^{(1)}(\mathcal{W}, r), \tag{6.2}$$

with the part $F_0^{(1)}(\mathcal{W}, r)$, containing the explicit r dependence, considered to be small: $|F_0^{(1)}| \ll |F_0^{(0)}|$. Thus, Boltzmann's equation (2.41a) can now be rewritten (in cylindrical geometry) as

$$\frac{1}{r}\frac{\partial}{\partial r}\left(\frac{2e}{3m}\frac{u^{3/2}}{\nu}r\frac{\partial F_0^{(1)}}{\partial r}\right) + \frac{2e}{3m}\frac{\partial}{\partial \mathcal{W}}\left(\frac{u^{3/2}}{\nu}E_{\text{eff}}^2\frac{\partial F_0^{(0)}}{\partial \mathcal{W}}\right)$$

$$= S_0\left(F_0^{(0)}(\mathcal{W})\right). \tag{6.3}$$

Equation (6.3) is a reformulation of (2.41) in terms of the total-energy scale $\mathcal{W}$; the effects of the space-charge field $\boldsymbol{E}_A$ being included by the definition $\mathcal{W} = u - \Phi$ (Φ being the space-charge potential). Again $E_{\text{eff}}^2 = (1/2)\,[\nu^2/(\omega^2 + \nu^2)]|\boldsymbol{E}|^2$. Though $F_0^{(0)}(\mathcal{W})$ is independent of r, it should be stressed that $F_0^{(0)}(u)$ is clearly r-dependent through the reverse transformation from the total energy $\mathcal{W}$ to the kinetic energy u. In S_0 the inelastic-collision terms representing the contributions from the transition of electrons from higher to lower energy will be neglected here; see (2.43b) and (2.44b). The inclusion of step processes will be described below.

By integrating (6.3) over the discharge cross-section and neglecting the particle flux on the discharge wall, the equation for $F_0^{(0)}$ is obtained:

$$\frac{\partial}{\partial \mathcal{W}}\left[\bar{D}_\mathcal{W}\,\frac{\partial F_0^{(0)}}{\partial \mathcal{W}} + \bar{V}_\mathcal{W} F_0^{(0)}(\mathcal{W})\right]$$

$$= \sum_k \left[\bar{\nu}_{0_k}(\mathcal{W}) F_0^{(0)}(\mathcal{W}) - \bar{\nu}_{0_k}(\mathcal{W}+U_k) F_0^{(0)}(\mathcal{W}+U_k)\right]$$

$$-\frac{e^2}{4\pi\sqrt{2}\varepsilon_0^2}\left(\frac{e}{m}\right)^{1/2}\ln(\Lambda)\frac{\partial}{\partial \mathcal{W}}\left[\bar{h}(\mathcal{W})F_0^{(0)} + \frac{2}{3}\,\bar{g}(\mathcal{W})\frac{\partial F_0^{(0)}}{\partial \mathcal{W}}\right], \quad (6.4)$$

with the following radially averaged quantities:

$$\bar{D}_\mathcal{W} = \frac{2}{R^2}\int_0^{r^*(\mathcal{W})}\frac{2e}{3m}\frac{u(r)^{3/2}}{\nu(u)}\,E_{\text{eff}}\,r\,\mathrm{d}r\,, \qquad (6.5)$$

$$\bar{V}_\mathcal{W} = \delta\frac{2}{R^2}\int_0^{r^*(\mathcal{W})}u^{3/2}\nu(u)r\,\mathrm{d}r\,, \qquad (6.6)$$

$$\bar{g}(\mathcal{W}) = \frac{2}{R^2}\int_0^{r^*(\mathcal{W})}r\,\mathrm{d}r\left[\int_0^{u(r)}u'^{\,3/2}F_0^{(0)}(u')\,\mathrm{d}u'\right.$$

$$\left.+u^{3/2}(r)\int_{u(r)}^{\infty}F_0^{(0)}(u')\,\mathrm{d}u'\right]\,, \qquad (6.7)$$

$$\bar{h}(\mathcal{W}) = \frac{2}{R^2}\int_0^{r^*(\mathcal{W})}r\,\mathrm{d}r\int_0^{u(r)}u'^{\,1/2}F_0(u')\,\mathrm{d}u'\,, \qquad (6.8)$$

$$\bar{\nu}_{0_k}(\mathcal{W}) = \frac{2}{R^2}\int_0^{r^k(\mathcal{W})}\nu_{0_k}[u(r)]\sqrt{u(r)}\,r\,\mathrm{d}r\,, \qquad (6.9)$$

where $\ln(\Lambda)$ in the electron–electron collision term is given by (2.46). For electrons with a small total energy $\mathcal{W}$, the whole discharge cross-section is not accessible, only a part up to a maximum radius $r^*(\mathcal{W})$, which is the turning point of the electron motion defined by $-\Phi[r^*(\mathcal{W})] = \mathcal{W}$ or $u[r^*(\mathcal{W})] = 0$. $r^k(\mathcal{W})$ is the maximum radius for which the kth inelastic process is possible: $u[r^k(\mathcal{W})] = U_k$. Ionization is one of the inelastic processes.

The last term in (6.4) gives the influence of electron–electron collisions. It should be stressed that their inclusion is important in determining to what extent deviation from a Maxwellian distribution will occur. Whereas the inelastic energy-loss processes obviously tend to lower the tail of the distribution function, the electron–electron collisions have a counterbalancing effect in smoothing out the EEDF between its body and its tail towards a Maxwellian distribution throughout. Since one process is proportional to $N_0 n$ and the other to n^2, the degree of ionization essentially determines the degree of deviation. As calculations in [6.26] show, for the inert gases such as argon and helium mainly addressed here, at degrees of ionization below 10^{-2} deviations become noticeable and more and more pronounced.

As a first addition to the above formulation, the influence of a collisional-radiative model (Sect. 2.2.1b) will be included in an approximate way, in particular the influence of step ionization, by taking into account the most important excited state (denoted below level k). It is assumed that the population N_k (2.49) of this level is determined by the balance between population gain from the ground state, on the one hand, and losses due to diffusion and ionization, on the other hand:

$$N_k = \frac{N_0 n_0 C_0^k}{n_0 S_k + [D_k/(R/2.4)^2]}\,. \tag{6.10}$$

The atomic data were taken from [6.60]. Here N_0 is the ground-state density, C_0^k the rate coefficient for excitation from the ground state, S_k the loss rate by ionization (or excitation in the case of helium, C_k^j), D_k the diffusion coefficient of metastables, and n_0 the electron density in the centre of the discharge. The approximation as given here can at least qualitatively indicate the direction of modifications to be expected due to the inclusion of a more detailed collisional-radiative model. Such a model is expected to quantitatively increase the influence of step processes and might be particularly desirable in the range of frequencies lower than the GHz range preferentially addressed here. If reactions with other constituents are taken into account, of course, further equations are required for determing the concentrations of these constituents.

As a second improvement noncollisional heating will be incorporated, as discussed in Chap. 2. This type of heating appears to be especially important for SW-sustained discharges in the microwave region, where reasonably low values of ν/ω are often encountered, and which are therefore stressed here. Therefore, (2.37a) will be considered and evaluated. Spatially narrow structures in the E_r field in the case of a radial density profile, which may lead to steep peaks in the intensity of this field component owing to the local plasma resonances $\omega_p(r_r) = \omega$ (see Sect. 4.2), are mainly addressed. In contrast to (2.38b) before spatial integration, not a delta function, but rather the actual (self-consistently) calculated $|E_r(r)|^2$, will be used. For the evaluation of these peaks it is assumed that $|E_r(r)|^2$ can be well approximated by a Lorentzian with a resonance half-width Δ. After carrying out the necessary analysis [6.56–58], the quasi-linear term to be added can be written as

$$S_{ql} = \frac{\partial}{\partial u}\left(D_{ql}\frac{\partial F_0}{\partial u}\right) M_{res}(r)\,, \tag{6.11}$$

with the diffusion coefficient for the collisionless (quasi-linear) energy transfer

$$D_{ql} = \frac{\pi e}{2m}\frac{u^{3/2}}{\omega}|E_r(r)|^2 d(\bar{s})\,. \tag{6.12}$$

The function $d(s)$ is defined as [6.56]

$$d(\bar{s}) = \frac{\bar{s}}{2}\left[(1 - \bar{s})e^{-\bar{s}} + \bar{s}^2 \int_0^\infty \frac{e^{-t}}{t}\,dt\right]$$

$$\approx \frac{\bar{s}}{2}e^{-\bar{s}}. \tag{6.13}$$

The argument $\bar{s}$ is given by

$$\bar{s} = \left(\frac{2m\omega^2\Delta^2}{eu}\right)^{1/2} = 4\pi\frac{\tau}{T}. \tag{6.14}$$

Here Δ is the resonance half-width

$$\Delta = L_{n(r)}\left(\frac{\nu_{\text{eff}}}{\omega}\right), \tag{6.15}$$

and $L_{n(r)}$ is the scale length of the radial density inhomogeneity at the resonance point:

$$L_{n(r)} = \left|\frac{d\ln n}{dr}\right|_{r=r_r}^{-1}. \tag{6.16}$$

ν_{eff} is an effective damping frequency, which gives a measure of the absorption in the resonance region (note the similarity with (4.24)). Here ν_{eff} takes into account the energy drain by collisionless heating, yet to be commented on. The definition given by (6.16) is equivalent to the one given by (4.10) in terms of permittivity.

The function M_{res} in (6.11) is defined as the step function

$$M_{\text{res}}(r) = M\left(\frac{|E_r(r)|}{0.2|E_{r_{\max}}|} - 1\right), \tag{6.17}$$

where $M(x) = 0$ for $x \leq 0$ and $M(x) = 1$ for $x > 0$, x being the argument shown on the right-hand side of (6.17). This function is used to limit the influence of the collisionless heating term to the region of the assumed narrow field structure.

Below, Δ is determined computationally, directly from the half-widths of the $|E(r)|^2$ peaks in the plasma resonance case, and from the corresponding width of the sharp increases in the field intensity towards the wall in the nonresonant case (which, as it turns out, usually does not contribute significantly).

In (6.14), $\tau = \Delta/v$ is the transit time for a single pass (of electrons with the energy $u = mv^2/2e$) through the resonance peak, and $T = (\omega/2\pi)^{-1}$ is the wave period. The function $d(\bar{s})$ reaches its maximum near $\bar{s} = 1$. The energy gain remains reasonably high if the transit time is shorter than the field period. It is illustrative to insert (6.13) and (6.14) into (6.12). By introducing the collision frequency $\nu = 1/\tau_c$ one obtains

$$D_{\text{ql}} = \left(\frac{\pi e}{2m}\frac{u^{3/2}}{\nu}|E_r(r)|^2\right)\frac{\tau}{\tau_c}\exp\left(-4\pi\frac{\tau}{T}\right). \tag{6.18a}$$

The first factor, in the large parentheses, is in a form (almost) identical to that of the (velocity/energy space) diffusion coefficient for Joule heating by the E_r field, resembling the case of a DC field ($\nu \gg \omega$; see (6.22) below). However, since the transit time τ is now the relevant time for phase decorrelation rather than the collision time τ_c (the typical time for diffusion in the velocity space due to Joule heating), the expression above is weighted by τ/τ_c. The exponential factor accounts for the fact that actually a time-varying HF field is considered. The energy gain decreases if the electrons spend a time comparable to the wave period in the resonance region, since they may then be accelerated and afterwards decelerated by the reversing field. With decreasing τ there is an exponential decrease.

Thus, in the specific case considered here with relatively sharp radial structures of the E_r field occuring (essentially because of resonances $\omega_p(r_r) = \omega$ due to the radial density inhomogeneity), the final result can be reasonably interpreted as a transit-time mechanism. This case may also be qualitatively considered as mode conversion into plasma waves subjected to Landau damping (see Sect. 3.2.4), leaving open the determination of the relevant k_r spectrum. Therefore, a general basis is Landau-type damping, effective for a (Fourier) decomposition of the radial field structure into "partial" waves with k_r values much larger than the β values associated with the axial SW propagation. This can be visualized by means of an expression for a diffusion coeffcient in the energy space $D_{r_{total}}$ encompassing both collisional and noncollisional damping, which can easily be approximated when the EEDF has a negligible radial variation (weak influence of space-charge effects):

$$D_{r_{total}} = \frac{2}{3}\frac{e}{m}u^{3/2}\,\mathrm{Im}\left\{\frac{E_r(r)}{2}\int\frac{E_r(k_r)e^{-i\,k_r r}\,dk_r}{\omega - i\nu - k_r v_r(u)}\right\}, \tag{6.18b}$$

where $E_r(k_r)$ is the Fourier transform of the field component $E_r(r)$. When ν vanishes, Landau damping obviously enters (for a sufficiently high velocity in radial direction v_r) and an expression of the type (6.18a) results for the specific case of relatively sharp peaks in $|E_r|^2$, whereas for finite ν and small v_r the usual expression relevant to collisional damping is obtained. The estimates of the damping of the E_r field given below are made for the limits valid in the collisional and noncollisional limits. Their addition is considered a reasonably good approximation in the transition region, which is relatively small in the cases under numerical study.

The incorporation of the quasi-linear term into Boltzmann's equation (6.4) for the nonlocal model is now achieved by redefining the *energy* diffusion coefficient (6.5) through the extension

$$\overline{D}_\epsilon = \frac{2}{R^2}\int_0^{r^*(W)} D_\epsilon r\,dr, \tag{6.19}$$

with

$$D_\epsilon = D_z + D_r + D_{ql}. \tag{6.20}$$

D_z and D_r account for Joule heating via the axial and radial components of the SW field, respectively, and D_{ql} for the collisionless (quasi-linear) heating. Explicitly, the definitions are:

$$D_z = \frac{2e}{3m} u^{3/2}(r) \frac{|E_z(r)|^2}{2\nu(u)} \frac{\nu^2(u)}{\nu^2(u) + \omega^2} , \tag{6.21}$$

$$D_r = \frac{2e}{3m} u^{3/2}(r) \frac{|E_r(r)|^2}{2\nu(u)} \frac{\nu^2(u)}{\nu^2(u) + \omega^2} , \tag{6.22}$$

$$D_{\mathrm{ql}} = \frac{\pi e}{m} u^{3/2}(r) \frac{|E_r(r)|^2}{2\nu(u)} \frac{\tau(u)}{\tau_{\mathrm{c}}(u)} \exp\left(-4\pi \frac{\tau}{T}\right) . \tag{6.23}$$

6.1.2 Complementary Relations

To obtain a complete set of equations that allows self-consistent modelling, of course, sufficient additional relations are required to bring the total number up to the number of unknown quantities to be determined simultaneously.

Boltzmann's equation (6.4) – with $\overline{D}_{\mathcal{W}}$ redefined by (6.19)–(6.23) to include noncollisional heating and supplemented by (6.10) to account for step ionization – calls for complementary equations which take into account the presence of plasma ions. This can be accomplished by using the full set of fluid equations, similarly to that described by Self and Ewald [6.61]. Quasi-neutrality ($n_{\mathrm{e}} = n_{\mathrm{i}} = n$) is considered. v_{ir} is the ion drift velocity, ν_{in} is the ion–neutral collision frequency and Φ is again the space-charge potential, i.e.

$$\frac{1}{r} \frac{\mathrm{d}}{\mathrm{d}r}(rn v_{\mathrm{ir}}) = \nu_i n , \tag{6.24}$$

$$v_{\mathrm{ir}} \frac{\mathrm{d}v_{\mathrm{ir}}}{\mathrm{d}r} = -\frac{e}{M} \frac{\mathrm{d}\Phi}{\mathrm{d}r} - (\nu_i + \nu_{\mathrm{in}}) v_{\mathrm{ir}} . \tag{6.25}$$

The mean ion energy is assumed to be small compared with that of the electrons. For simplicity, an infinitely thin sheath may be considered, leading to the (Bohm) sheath criterion

$$\langle v_{\mathrm{i}} \rangle_{r=R} = \sqrt{\frac{2e}{3M}} \, \bar{u} , \tag{6.26}$$

with $\bar{u}$ being a mean electron energy [6.62]. For the ion–neutral collision frequency the following is specifically used [6.63,64]

$$\nu_{\mathrm{in}}(v_{\mathrm{ir}}) = \nu_{\mathrm{in}_0} \left(1 + 0.182 \frac{M v_{\mathrm{ir}}^2}{T_{\mathrm{g}}}\right)^{1/2} , \tag{6.27}$$

owing to symmetric charge exchange collisions; ν_{in_0} is its value for small v_{ir}.

The presence of the ambipolar potential Φ makes necessary an additonal equation which can, however, be provided by the normalizing condition linking $F_0^{(0)}(\mathcal{W})$ and $\Phi(r)$:

$$n[\Phi(r)] = \int_0^\infty F_0(u,r)\,\sqrt{u}\,\mathrm{d}u = \int_{-\Phi(r)}^\infty F_0^{(0)}(\mathcal{W})\sqrt{\mathcal{W} + \Phi(r)}\,\mathrm{d}\mathcal{W}\,. \tag{6.28}$$

It should be stressed that, from the solution for the distribution function in terms of total energy $F_0^{(0)}(\mathcal{W})$, the radially resolved distribution in terms of kinetic energy may be found by retransformation: $F_0^{(0)}\,[\mathcal{W} = u - \Phi(r)] = F_0(u,r)$.

A decisive step towards full self-consistency is the addition of the electrodynamic field relations and their associated boundary value problem. This transforms the important HF heating intensity from the role of a given parameter to a quantity self-consistently determined as necessary for the plasma maintenance.

As has been discussed in Sect. 4.4 in relation to fully two-dimensional solutions, the axial density inhomogeneity is usually much weaker than the radial one. Though the axial inhomogeneity had been accounted for explicitly it had turned out to be relatively unimportant in solving the electrodynamic problem. Therefore, from here on, the field intensities are assumed to vary $\propto \exp[- \int_0^z \bar{k}(z')\,\mathrm{d}z']\exp(\mathrm{i}\omega t)$, with $\bar{k}(z) = \alpha + \mathrm{i}\beta$ giving the local values of the wave number β and the attenuation coeffcient α at a given position z along the plasma column. With the (radially dependent) electron plasma frequency $\omega_\mathrm{p}(r) = [n(r)e^2/m\varepsilon_0]^{1/2}$ (2.8b), the plasma permittivity (2.8a) is in the form

$$\varepsilon(r) = 1 - \frac{\omega_\mathrm{p}^2(r)}{\omega(\omega - \mathrm{i}\nu)}\,, \tag{6.29}$$

or, somewhat more generally [6.65,66],

$$\varepsilon(r) = 1 + \frac{2e^2 n(r)}{3\varepsilon_0 m\omega} \int_0^\infty \frac{u^{3/2}}{\omega - \mathrm{i}\nu(u,r)} \frac{\partial F_0^{(0)}(u,r)}{\partial u}\,\mathrm{d}u\,. \tag{6.30}$$

The latter expression is worth using in cases of an energy-dependent ν (e.g. for argon) when ν/ω is not limited to rather small values.

The field components in the plasma are then determined by

$$\frac{\mathrm{d}^2 E_z}{\mathrm{d}r^2} + \left(\frac{1}{r} + \frac{\bar{k}^2}{\bar{\varkappa}_\mathrm{p}^2(r)} \frac{\mathrm{d}\varepsilon(r)/\mathrm{d}r}{\varepsilon(r)}\right) \frac{\mathrm{d}E_z}{\mathrm{d}r} + \bar{\varkappa}_\mathrm{p}^2(r)E_z = 0\,, \tag{6.31}$$

$$E_r(r) = -\frac{\bar{k}}{\bar{\varkappa}_\mathrm{p}^2(r)} \frac{\mathrm{d}E_z}{\mathrm{d}r}\,, \tag{6.32}$$

where

$$\bar{\varkappa}_\mathrm{p}^2 = k_\mathrm{v}^2\varepsilon(r) + \bar{k}^2 \tag{6.33}$$

and $k_\mathrm{v} = \omega/c$ is the vacuum wave number. In the dielectric and the vacuum, the same equations hold with $\varepsilon(r)$ replaced by ε_d and $\varepsilon_\mathrm{v} = 1$, respectively. $\bar{k}$ – and thus the dispersion relation – has to be determined so that the continuity of both the tangential electric E_z and magnetic H_φ field components is satisfied at the plasma–glass and glass–vacuum boundaries. By this procedure the dispersion relation of the SWs is obtained, i.e. the axial wave number β and the damping coefficient α for various values of $\omega/\overline{\omega_{\rm p}}$. $\overline{\omega_{\rm p}}$ is defined as $\sqrt{\overline{\omega_{\rm p}^2}} \propto \sqrt{\bar{n}}$, with $\bar{n}$, the electron density averaged over the discharge cross-section, being a function of z. The determination of the self-consistent values of the field intensity E required for discharge maintenance amounts to an eigenvalue problem.

Finally, however, the effective damping frequency $\nu_{\rm eff}$ introduced in (6.15), when the collisionless energy transfer is described, requires special attention and modifications of (6.31) and (6.32). The quasi-linear term (6.11) in Boltzmann's equation describes an additional power transfer. On the other hand, the permittivity in Maxwell's equations, i.e. in the field equations (6.31) and (6.32), contains only ν and thus only the damping due to the ohmic energy transfer. The sharpness of the field resonances would be overestimated without introduction of $\nu_{\rm eff}$.

Therefore, in the numerical procedure a larger damping frequency, a multiple $m_{\rm eff}$ of ν, is introduced, which accounts in a heuristic manner for the back reaction of the energy drain from the E_r field via the additional collisionless plasma heating and is actually used for the calculations of the fields:

$$\nu_{\rm eff} = m_{\rm eff}\nu \,. \tag{6.34}$$

With $\nu_{\rm eff}$ an effective permittivity $\varepsilon_{\rm eff}$ is also introduced, using $\nu_{\rm eff}$, whereas ε still indicates the permittivity obtained by use of ν only. As a consequence of the above changes, the following field equations, replacing (6.31) and (6.32), must now be used for the plasma region, accounting for anisotropic damping of E_z and E_r:

$$\frac{{\rm d}^2 E_z}{{\rm d}r^2} + \left(\frac{1}{r} + \frac{\bar{k}^2}{k_{\rm v}^2\varepsilon_{\rm eff}(r) + \bar{k}^2} \frac{{\rm d}\varepsilon_{\rm eff}(r)/{\rm d}r}{\varepsilon_{\rm eff}(r)} \right) \frac{{\rm d}E_z}{{\rm d}r}$$
$$+ \left[k_{\rm v}^2\varepsilon_{\rm eff}(r) + \bar{k}^2 \right] \frac{\varepsilon(r)}{\varepsilon_{\rm eff}(r)} E_z = 0 \,, \tag{6.35}$$

$$E_r(r) = \frac{-\bar{k}}{k_{\rm v}^2\varepsilon_{\rm eff}(r) + \bar{k}^2} \frac{{\rm d}E_z}{{\rm d}r} \,. \tag{6.36}$$

The values of $\nu_{\rm eff}$ are chosen in such a way that consistency between the power flux – based on Maxwell's equations – and the power absorbed by the plasma – based on the power transfer terms obtained from Boltzmann's equation – is obtained. The idea is to define the effective collision frequency in such a way that the power transfer in the "radial" ohmic channel and in the collisionless (quasi-linear) channel, when calculated with the classical frequency ν, is equal to the pure "radial" ohmic power transfer with $\nu_{\rm eff}$. (Here "radial" means due to the radial electric field.) For both cases the field profile calculation is based on (6.35) and (6.36), containing $\nu_{\rm eff}$. By means of this concept the broadening of the field peaks due to the total energy drain from the radial component is accomplished in a consistent way. Concretely, the following scheme has been used, where the power transfer via a certain channel (ohmic via the r or z component, collisionless) is characterized by the power Θ absorbed on average over a discharge cross-section by one electron (for exact definitions see Sect. 6.1.3).

(1) The field profile is calculated from (6.35) and (6.36) using a starting guess for ν_{eff}. The sum of the three transfer channels $\Theta_z + \Theta_r + \Theta_{\mathrm{coll\text{-}less}}$ is calculated, employing ν and the above field profiles.
(2) The same field distributions as in (1) are used. Now the collisionless channel is omitted, Θ_z is calculated with ν, but Θ_r is calculated with ν_{eff}.

In order to equalize the two sums obtained in (1) and (2), ν_{eff} has to be adjusted, which requires an iterative procedure. As a starting value for the iteration, ν_{eff} may be chosen as ν times the square root of the initial ratio of $(\Theta_r + \Theta_{\mathrm{coll\text{-}less}})/\Theta_r$. It should be noted that it is possible to base the determination of ν_{eff} not on considerations using averaging over the discharge cross-section, but rather on radially local considerations as $\nu_{\mathrm{eff}}(r)$. However, since ν_{eff} – if noticeably different from ν – is essentially determined by relatively steep field structures in E_r concentrated in a narrow radial region, such a more detailed radially resolved determination and iteration for ν_{eff} turns out to be of minor importance as a rule, though requiring more computational effort.

6.1.3 Mean Power Absorbed per Electron

Before the final step of the model for the explicit axial structure is described, a useful property, already mentioned, will be defined. The three heating channels resulting from (6.21)–(6.23), *the power shares transferred per electron on average*, are calculated: Θ_z and Θ_r for the two Joule losses connected with $|E_z|^2$ and $|E_r|^2$, respectively, and for the added *collisionless* loss, which results from the *quasi-linear* term (6.11) and is connected with $|E_r|^2$ peaks, here labelled $\Theta_{\mathrm{coll\text{-}less}}$. The terms Θ_z, Θ_r and $\Theta_{\mathrm{coll\text{-}less}}$ are obtained from the macroscopic power balance equation. This equation is obtained by multiplying (6.4) by the total energy and subsequent integration over the whole energy space. (Note that in (6.4) r and $\mathcal{W}$ are independent variables so that the order of integration can be interchanged.) For the three different power transfer channels one obtains

$$\Theta_i = \frac{1}{\bar{n}} \int_0^\infty \left[\frac{\mathrm{d}}{\mathrm{d}\mathcal{W}} \left(\overline{D}_i \frac{\mathrm{d}F_0^{(0)}(\mathcal{W})}{\mathrm{d}\mathcal{W}} \right) \right] \mathcal{W} \, \mathrm{d}\mathcal{W} , \qquad (6.37)$$

with i designating z, r and ql (coll-less), respectively, and D_i are given by (6.21)–(6.23). The resultant three Θ terms constitute the energy gain terms in the macroscopic energy balance of the discharge obtained from (6.4) as stated. They are indeed in balance with the collisional loss terms within 1% accuracy, as can be checked by verifying the numerical method. For the calculation of the Θ terms, of course, the self-consistently calculated $|E_z(r)|^2$ and $|E_r(r)|^2$ field configurations must be employed, as described above.

6.1.4 Axial Structure

The system described appears complete in that it allows simultaneous solution (by numerical iterative procedures) of the unknown quantities. But the axial structure is still hidden in the dependence implied above on the density averaged over the cross-section, $\bar{n} \propto \overline{\omega_{\mathrm{p}}}^2$. The important result for how $\overline{\omega_{\mathrm{p}}}$ actually varies with the axial coordinate z requires additionally invoking the wave–power balance equation (2.27) based on

$$-\mathrm{div}\ \mathrm{Re}\ \left\{\frac{1}{2}\,\boldsymbol{E}(r) \times \boldsymbol{H}^*(r)\right\} = Q(r)\,, \tag{6.38}$$

where Q indicates the power loss per unit volume of plasma.

By the integration of (2.31a) over the cross-section of the whole discharge system in the entire r plane from $r = 0$ to $r \to \infty$ (or $r = R_{\mathrm{m}}$ in the case of a metal shield) a relation between the integrated real part of the power flux P and the power loss $Q = \pi R^2 \bar{n}\Theta_{\mathrm{total}}$ over the entire plane is obtained.

In detail, the desired link is deduced as follows.

$$-\int_0^\infty \left[\mathrm{div}\ \mathrm{Re}\ \left\{\frac{1}{2}\,\boldsymbol{E}(r) \times \boldsymbol{H}^*(r)\right\}\right] 2\pi r\, \mathrm{d}r = \pi R^2 \bar{n}\Theta_{\mathrm{total}}\,. \tag{6.39}$$

The derivative $\partial/\partial z$ of the z component of the Poynting vector can be replaced by $(\partial/\partial\overline{\omega_{\mathrm{p}}}^2)(\mathrm{d}\overline{\omega_{\mathrm{p}}}^2/\mathrm{d}z)$:

$$-\pi\, \mathrm{Re}\Bigg\{\frac{\mathrm{d}\overline{\omega_{\mathrm{p}}}^2}{\mathrm{d}z}\Bigg[\underbrace{\int_0^R \frac{\partial}{\partial\overline{\omega_{\mathrm{p}}}^2}(E_r H_\varphi^*)r\, \mathrm{d}r}_{I_{\mathrm{p}}} + \underbrace{\int_R^\infty \frac{\partial}{\partial\overline{\omega_{\mathrm{p}}}^2}(E_r H_\varphi^*)r\, \mathrm{d}r}_{I_{\mathrm{v}}}\Bigg]\Bigg\}$$

$$= \pi R^2 \bar{n}\Theta_{\mathrm{total}}\,. \tag{6.40}$$

The contribution from the first integral is negative and smaller than that from the second one (as it has been commented on in Sects. 3.2.1 and 3.3.2). With I_{p} and I_{v} as defined in (6.40), one finds

$$\frac{\mathrm{d}\overline{\omega_{\mathrm{p}}}^2}{\mathrm{d}z} = -\frac{R^2\bar{n}\Theta_{\mathrm{total}}}{\mathrm{Re}\{I_{\mathrm{p}}\} + \mathrm{Re}\{I_{\mathrm{v}}\}} \equiv (e^2/\varepsilon_0 m)g(\bar{n})\,. \tag{6.41}$$

The values needed are, to a large extent, already available from calculations of Θ versus $\omega/\overline{\omega_{\mathrm{p}}}$. For H_φ inside the plasma, the expression

$$H_\varphi(r) = \mathrm{i}\,\frac{\omega\varepsilon_0}{r}\int_0^r r\varepsilon E_z\, \mathrm{d}r \tag{6.42}$$

can be used. The axial density profile $\bar{n}(z)$ is obtained from $g(\bar{n})$ via

$$z = \int \frac{\mathrm{d}\bar{n}}{g(\bar{n})} \, . \tag{6.43}$$

Within the comparatively weak axial nonuniformity assumed above, this derivation (using largely the numerically calculated field values) is equivalent to assuming the averaged properties for the cross-section (see (5.77a))

$$-\frac{\mathrm{d}P}{\mathrm{d}z} = 2\alpha P = Q \, , \tag{6.44}$$

and utilizing the previously determined α. As discussed in Sect. 4.3.5, the zero-order solution (in terms of the WKB approximation) for P can be used. Thus the same expression for $\bar{n}(z)$ is obtained as given in Chap. 5 (5.76):

$$\frac{\mathrm{d}\bar{n}(z)}{\mathrm{d}z} = -2\alpha\bar{n}(z) \bigg/ \left(1 - \frac{\mathrm{d}\alpha}{\mathrm{d}\bar{n}} \frac{\bar{n}}{\alpha} + \frac{\mathrm{d}\Theta}{\mathrm{d}\bar{n}} \frac{\bar{n}}{\Theta}\right) \, . \tag{6.45}$$

6.1.5 Results

For the numerical solutions, standard routines were utilized. The field and dispersion calculations were based on Runge–Kutta methods coupled with search routines as employed in Sects. 4.2.3 and 4.2.4. For solving Boltzmann's equation, discretization in the total energy scale ($\mathcal{W}$) was introduced, with subsequent transformation to the kinetic energy u. For the radial profiles of the HF fields, ambipolar potential, $n(r)$, etc., estimated initial values were used, which were thereafter improved by iterative loops until all equations and conditions were simultaneously satisfied within desired accuracies [6.58].

First a typical situation in the microwave range will be addressed: a wave frequency $f \equiv \omega/2\pi = 2.45$ GHz, a tube radius $R = 5$ mm and a wall thickness of $d = 1$ mm ($\varepsilon_\mathrm{d} = 4.7$); for the chosen gas pressure of $p = 133.32$ Pa ($\hat{=} 1$ torr) of argon, ν/ω is about 0.2.

The results for the mean power per electron, Θ, give good access to various phenomena that influence the discharge modelling, aside from the fact that Θ is per se of particular interest in practice.

In Fig. 6.1 the resultant self-consistently calculated $k = \alpha + \mathrm{i}\beta$ is visualized as a function of $\omega/\overline{\omega_\mathrm{p}} \propto 1/\bar{n}^{1/2}(z)$. The small kinks are due to the turn-back of β from the usual monotonic behaviour caused by the combined influence of collisions and radial inhomogeneity, discussed in detail in Chap. 4. With a relatively strong (self-consistently calculated) radial density inhomogeneity the first turn-back, which is due to the effects of collisions and inhomogeneity, see Chaps. 3 and 4, almost coincides with the second turn-back, which indicates the start of resonance absorption (plasmon mode, see Chap. 4): for $\omega/\overline{\omega_\mathrm{p}} \gtrsim 0.23$ the density at the discharge wall has dropped enough that local plasma resonances are possible. Towards higher $\omega/\overline{\omega_\mathrm{p}}$ this fact dominates the phase diagram; the growing α remains less than or of order β up to about

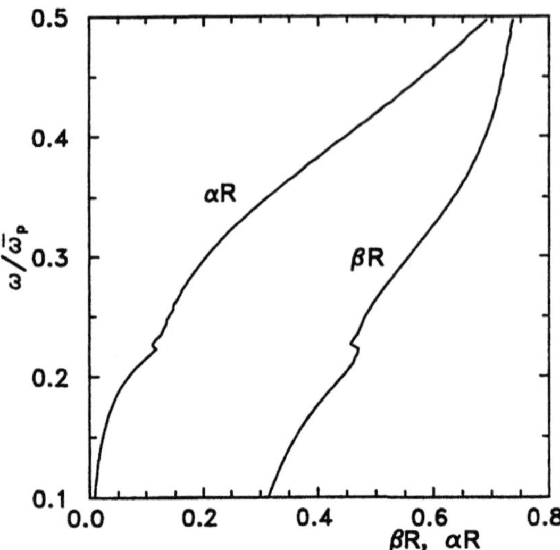

Figure 6.1. Axial wave number β and damping coefficient α of an SW sustaining a microwave ($f \equiv \omega/2\pi = 2.45$ GHz) discharge in argon ($p = 1$ torr) in a glass tube (internal radius $R = 5$ mm, glass thickness $d = 1$ mm, $\varepsilon_\mathrm{d} = 4.7$) ([6.58], Fig. 8)

$\omega/\overline{\omega}_\mathrm{p} = 0.5$. Modifications in the phase diagram as compared with the "standard" one (tending to level off towards $\omega/\overline{\omega}_\mathrm{p} \approx 0.42$, as should be the case in a homogeneous collisionless plasma) are apparent in Fig. 6.1.

Figure 6.2, depicting the three contributions to Θ_total, confirms the increasing importance of the E_r field component for $\omega/\overline{\omega}_\mathrm{p} \geq 0.23$. Not surprisingly, the collisionless heating $\Theta_\mathrm{coll\text{-}less}$, to be expected from the formation of steep resonance peaks in the intensity of the E_r field component, has the highest contribution from then on. The Joule heating by the E_r component, though, crosses that of the E_z component much later in this case, at $\omega/\overline{\omega}_\mathrm{p} \approx 0.4$. The additional collisonless heating via the E_r component consistently shows up in the increased effective damping frequency ν_eff, which indeed has a maximum at $\omega/\overline{\omega}_\mathrm{p} \approx 0.28$.

Therefore, the distribution of the power required per electron on average over the three input channels varies quite significantly along the discharge. Moreover, the distribution is predictably influenced by the external parameters: the radius R and pressure p. The calculations confirm that with increasing radius the domain of dominance by Joule input from the E_z component becomes smaller along the discharge: whereas for $R = 3$ mm this channel stays dominant to $\omega/\overline{\omega}_\mathrm{p} \approx 0.32$, for $R = 9$ mm it loses its dominance at $\omega/\overline{\omega}_\mathrm{p} \approx 0.21$. The pressure p acts in a similar manner. This is understandable from diffusion theory, since the expected value of the electron density near the wall (in relation to the central density) scales with $1/R$ and $\lambda_\mathrm{f.p.(i)}$ (the ion mean free path), and thus with $1/(pR)$. On the other hand, lowered

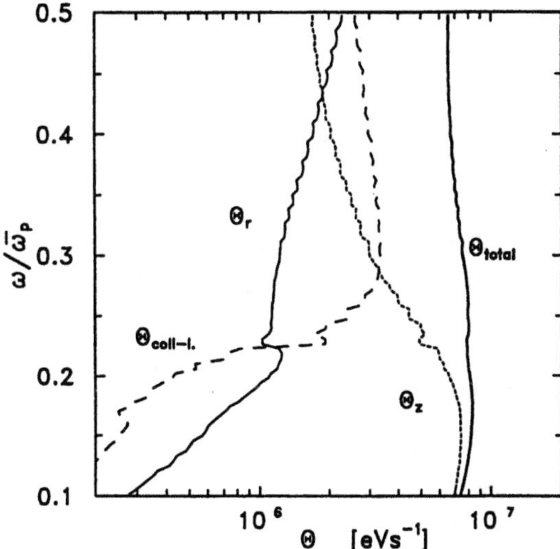

Figure 6.2. Power absorbed per electron on average Θ_{total}. Θ_z and Θ_r represent Joule heating via $|E_z|^2$ and $|E_r|^2$, respectively; $\Theta_{\text{coll-less}}$ represents collisionless contributions via $|E_r|^2$. $\overline{\omega_p}$ is the plasma frequency of the electron density averaged over the discharge cross-section. The conditions are the same as in Fig. 6.1 ([6.58], Fig. 1)

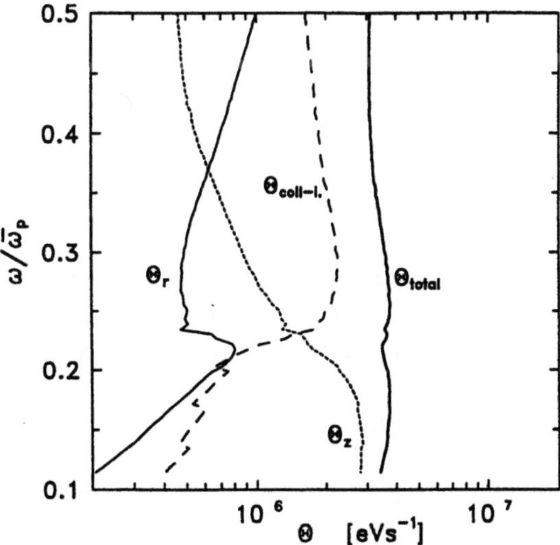

Figure 6.3. Conditions as for Fig. 6.2, exept that p is reduced to 0.5 torr and R increased to 9 mm ([6.58], Fig. 4)

pressure favours the relative strength of collisionless power transfer compared to the Joule input, consistent with Fig. 6.3 for $R = 9$ mm, $p = 66.66$ Pa ($\hat{=}$ 0.5 torr). Figure 6.3 also confirms the above statements on the domain of Θ_z along the discharge: with an increase of the radius R effecting a shortening of the domain, and a decrease of the pressure p a lenghtening. Thus, the changes of R and p in comparison with Fig. 6.2 together yield a shorter domain of Θ_z, from $\omega/\overline{\omega_p} \approx 0.28$ to 0.23, the influence of R obviously winning in this respect. These trends are underlined in Fig. 6.4 for a case of a relatively low value of the product pR (argon, $p = 33.33$ Pa $\hat{=}$ 0.25 torr, $R = 5$ mm).

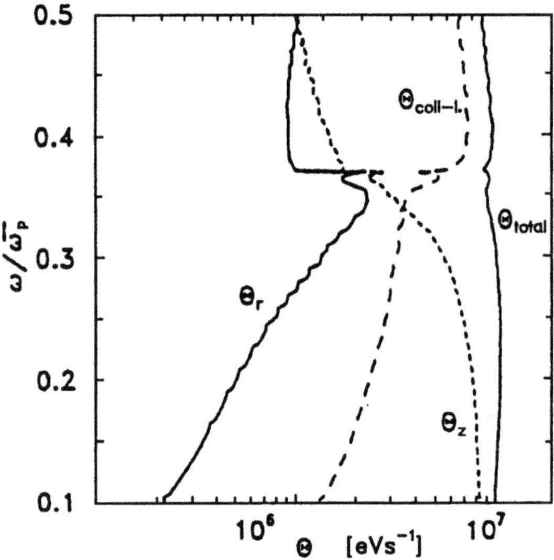

Figure 6.4. As for Fig. 6.2, but pressure decreased by a factor 4 ([6.59], Fig. 8)

Since for helium the nonlocal approximation used here is expected to extend to values of pR higher than for argon, a helium example is depicted in Fig. 6.5. The phenomena are quite similar to those for argon discharges, but owing to different atomic data noncollisional contributions become important at a somewhat larger $\omega/\overline{\omega_p} \approx 0.36$; for $R = 9$ mm and $p = 0.5$ torr in helium this occurs at about 0.32.

However, in spite of the noticeable changes possible in the distribution of the input channels along the discharge length (with increasing $\omega/\overline{\omega_p}$), the changes in the total Θ remain comparatively small, barely noticeable on the logarithmic scale used for Θ in the cases shown. This remains true even if the collisionless heating mechanisms are excluded; both the absolute values of the total Θ and its variation with $\omega/\overline{\omega_p}$ do not change appreciably for the same conditions [6.58]. Qualitatively this is not surprising in diffusion-dominated discharge regimes: since the diffusion losses are not too sensitive to potential

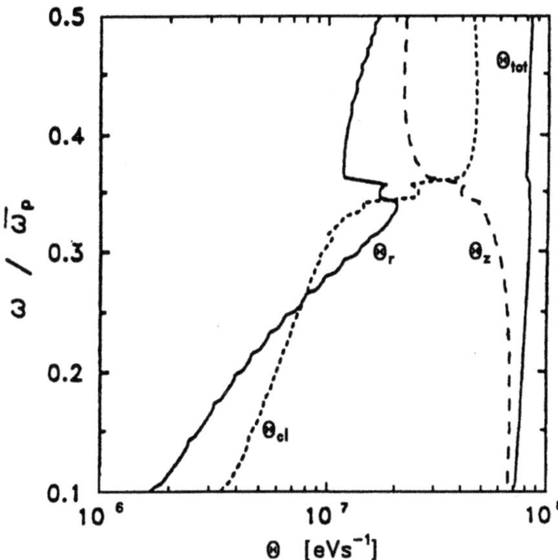

Figure 6.5. As for Fig. 6.2, but for helium ([6.59], Fig. 9)

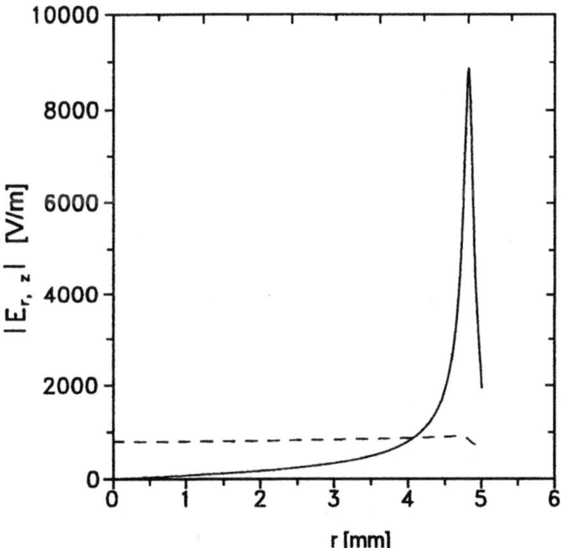

Figure 6.6. Radial field profiles of the E_r (*solid curve*) and E_z (*dashed curve*) field component amplitudes for $\omega/\overline{\omega}_\mathrm{p} = 0.45$ in a discharge under the same conditions as in Fig. 6.3, except $p = 0.5$ torr

changes in the energy distribution function along the discharge column, the discharge will choose self-consistently just enough field strength of any radial field profile to achieve just enough ionization for maintaining the particle balance. Since Θ_{total} largely reflects inelastic losses, it may be expected not to vary very much. In a case where collisionless heating were absent, the distribution between the E_z and E_r field strengths would be different, and somewhat more E field intensity would be required as compared with the situation with $\Theta_{\text{coll-less}} \neq 0$.

There remain small changes of the total Θ along the discharge, the origin of which will be addressed below. Prior to doing this, some phenomena of collisionless heating – other than effects on the total Θ – which can be expected and checked, are now treated.

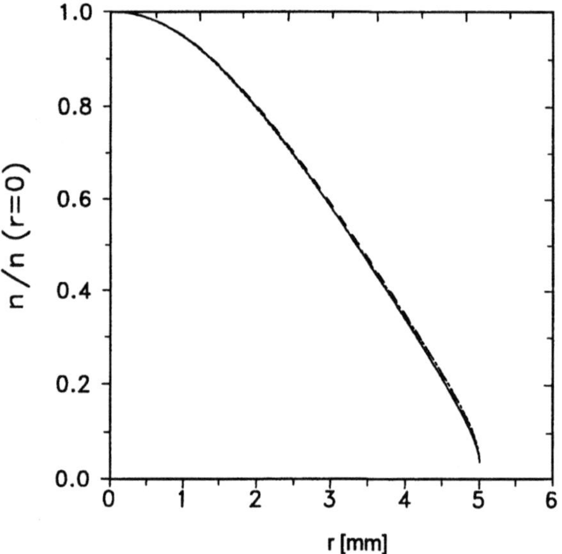

Figure 6.7. Radial electron density profiles along the discharge length, i.e. for different $\omega/\overline{\omega_p}$ values: $\omega/\overline{\omega_p} = 0.51$ (*solid curve*), 0.36 (*dashed curve*), 0.27 (*dotted curve*). Discharge conditions as in Fig. 6.6

The computations reveal the presence of resonance peaks in $|E_r|$ in the appropriate situations, as demonstrated in Fig. 6.6, although the self-consistent radial density profile $n(r)$ (Fig. 6.7) does not indicate their presence, as expected in the nonlocal scenario here; virtually no changes with different axial positions (and $\omega/\overline{\omega_p}$) are to be discerned. Therefore, these peaks are to be considered as typical of the influence of a strong radial inhomogeneity, and may be addressed in experimental checks. This is pointed out in the following chapter on experimental aspects, and also for the case of SW-like "inverted" arrangements (with a vacuum inside and a plasma outside).

Another consequence of the radial inhomogeneity leading to a stronger role of the E_r field towards the discharge end is the change (discussed in Sects. 5.5.2 and 5.6.3) from a linear axial density profile to a steeper one,

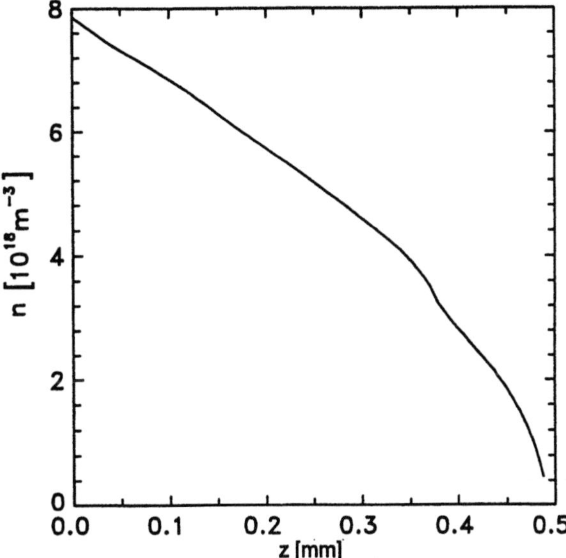

Figure 6.8. Axial field profile $\bar{n}(z)$ starting with $z = 0$ at $\omega/\overline{\omega_{\mathrm{p}}} = 0.15$. Conditions of Fig. 6.2 ([6.58], Fig. 7)

demonstrated in Fig. 6.8 for the conditions of Fig. 6.2. Possible experimental evidence for this will also be addressed in Chap. 7.

At this point it should be remarked that the presentations in the previous figures using the scaling with $\omega/\overline{\omega_{\mathrm{p}}}$ compress the main and mostly longer part of the discharge into about the range from $\omega/\overline{\omega_{\mathrm{p}}} = 0.1$ to 0.2, whereas the entire remainder of the higher $\omega/\overline{\omega_{\mathrm{p}}}$ values addresses and emphasizes a shorter part towards the discharge end. This can also be recognized in Fig. 6.8, which starts at $\omega/\overline{\omega_{\mathrm{p}}} = 0.15$. The transition from a linear density profile to a steeper one in the range of resonance absorption – dominantly noncollisional – sets in after about the first 30 cm of the total 50 cm shown (see Fig. 6.2).

There is a specific effect of noncollisional heating via the E_r field. As pointed out in the previous comments on the quasi-linear term, it favours fast electrons. The expected tendency to an increase in the distribution function towards higher energies is, of course, tempered in the self-consistent situation by the limitation on the ionization rate, which should not be too large for the diffusion losses in the particle balance. This puts a limit on the required electric-field strength, and thus too drastic increases in the distribution function towards high energies should not be expected. Nevertheless, a change in the shape of the distribution function because of the noncollisional heating term can show up, as demonstrated by Fig. 6.9. For larger radii this effect can be even more important and typical, as again included in the following chapter.

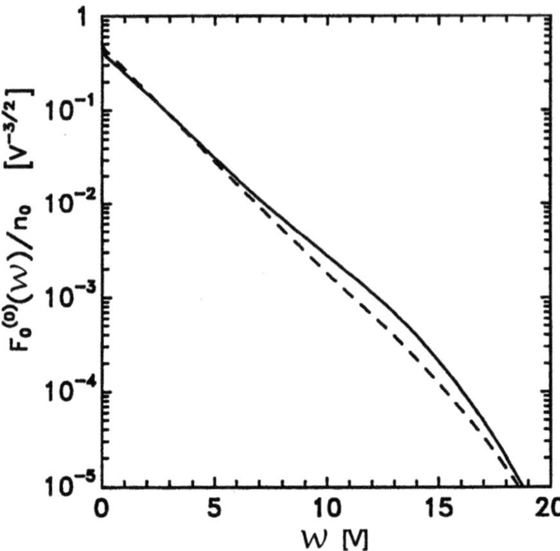

Figure 6.9. Electron energy distribution function $F_0^{(0)}$ normalized to the electron density n_0 at the centre of the discharge, for the conditions of Fig. 6.3, with $\omega/\overline{\omega_{\mathrm{p}}} = 0.25$. The two situations with collisionless absorption included (*solid curve*) and excluded (*dashed curve*) are compared ([6.58], Fig. 12)

6.1.6 Axial Changes of Electric-Field Intensity

The self-consistent electric-field intensities resulting from the above numerical modelling are closely connected to the average power loss per electron Θ. This connection is obviously given in a nonlocal way by integration, which is appropriate in the nonlocal modelling scenario considered here in any case. This loss, averaged over the discharge cross-section, required to maintain the discharge (to cover, e.g., the main losses by inelastic atomic processes) can be obtained via energy conservation from the balance to the power input by the HF field, as given by (6.38). The sum is for the total Θ over the input channels of the E_z and E_r field components. Thus, the electric-field intensity for discharge maintenance, averaged over the discharge cross-section, is contained in the diffusion coefficients in energy space. From the Schottky approximation (2.67) for diffusion-dominated regimes, equating ground-state ionization and diffusion particle losses, simple – usually reasonable – estimates of the field intensity can be obtained. However, as has been discussed in Sects. 5.1 and 5.3.4, such estimates are independent of the electron density, which is in contradiction to daily experience, since the discharge, when turned on, chooses its self-consistent electric-field strength, electron density and power values and their spatial distributions. The formation of the discharge is obviously a strongly nonlinear phenomenon. For full self-consistency, processes nonlinear in the electron density have to be taken into account, even though their contributions may appear, for instance in particle balance considerations, to be small compared with the linear ones and may not lead to drastic changes of Θ along the discharge.

As a relevant nonlinear atomic process step ionization has already been pointed out in Chap. 5, as well as its tendency to saturation towards higher electron densities, and the influence of recombination finally arises at very high densities [6.67]. This will be taken up in connection with confirming observations in Chap. 7. Another (weak) nonlinear effect, accessible in a kinetic treatment, is the influence of electron–electron collisions. There can be axial changes in the radial electron density profiles; they are usually small in a nonlocal scenario. In addition to these general phenomena, there can be effects more specific to the particular discharge type, in this case to plasmas maintained by travelling SWs. The resultant variations of Θ, even small ones, contribute to the aspect of self-consistency. It might be mentioned that in discharges with an additional strong static magnetic field the (nonlocal) axial diffusion transport can be a governing nonlinear effect [6.68].

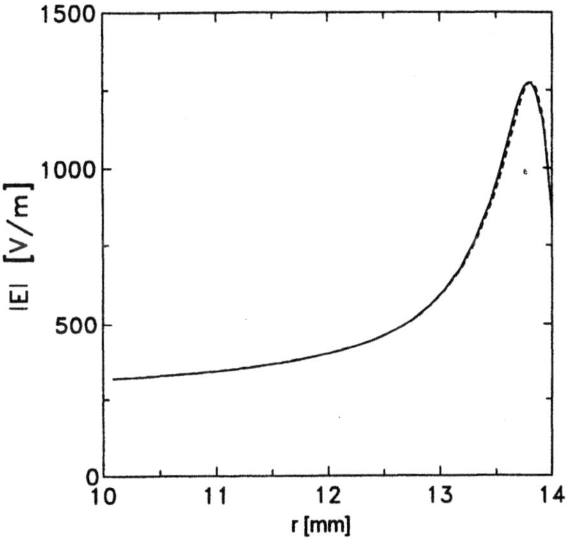

Figure 6.10. Radial field profiles of the total field $|E|$, self-consistently calculated for $\omega/\overline{\omega_p} = 0.4$ with (*solid curve*) and without (*dashed curve*) taking into account the ponderomotive force potential. Discharge in argon gas ($p = 0.2$ torr) at frequency $f \equiv \omega/2\pi = 136$ MHz in a glass tube (internal radius $R = 1.4$ cm, glass thickness $d = 0.2$ cm, $\varepsilon_d = 4.7$) surrounded by a metal cylinder of radius $R_m = 4.5$ cm ([6.70], Fig. 14)

Though effects of step ionization are only taken into account here in an approximation indicating the trend, the slight increase from the smallest $\omega/\overline{\omega_p}$ values considered to higher ones should be noticed in the figures above showing Θ versus $\omega/\overline{\omega_{\mathrm{p}}}$ as being in line with expectations, for helium essentially throughout the range and for argon up to about $\omega/\overline{\omega_{\mathrm{p}}} = 0.2$, i.e. in the main part of the discharge. The role of step ionization is taken up again in the next chapter together with confirming observations. On the specific behaviour in argon towards the lowest electron densities, the associated aspects of decreased diffusion have been discussed in [6.30].

6.1.7 Influence of Ponderomotive-Force Effects

The occurence of relatively strong radial field changes brings to mind the potential influence of ponderomotive effects. Estimates are possible, according to previous studies [6.69], by using the ponderomotive potential $e|E|^2/4m(\omega^2 + \nu^2)$ and comparing it with the ambipolar potential, for example, usually predicting a minor influence. Indeed, calculations where the total energy $\mathcal{W}$ is extended in its definition to include this ponderomotive potential confirm the expected small difference within the framework of a nonlocal treatment, as demonstrated in Fig. 6.10 for a self-consistently calculated radial $|E|$ profile even for the low-frequency case.

6.2 Local Approach

Above, the nonlocal approximation was used for the calculation of the EEDF in helium and argon. The applicability range of this approach will be commented on in more detail in Sect. 6.3. But it should already be pointed out that the examples considered above for helium discharges are well in the estimated applicability range, with values of the product of radius and gas pressure below the critical value $pR \approx 1.4$ torr cm. The argon cases, though, are close to or even slightly beyond the corresponding critical value of $pR \approx 0.2$ torr cm; these were chosen with a view to many argon experiments for such conditions and in order to show trends in the influence of the inhomogeneity. Though the domain appropiate to the opposite limiting approach using a local scenario has been avoided, in some cases corrections due to the transition from one regime to the other one may still have to be expected.

The other limiting approach will now be presented and commented on. This scenario can be expected to be an accurate one – not requiring corrections – in cases when the above values are much surpassed. This calls not only for small mean free paths $\lambda_{\mathrm{f.p.}} \ll R$, but also for the energy relaxation length to comply at least with $\lambda_\varepsilon < R$. Nevertheless, it is of some interest to study features of the local approach when only the former, but not the more stringent latter requirement is met, and to include comparisons with experimental findings for such situations.

6.2.1 Set of Equations

To a large extent the local approach calls for the same equations and the same boundary conditions for the EEDF and the EM fields as the nonlocal scenario described before. The same types of relations are used (Sect. 6.1.2): the fluid equations, incorporating the presence of the plasma ions, and the field equations are employed. However, the ambipolar potential is neglected. Moreover, noncollisional effects – of less importance at higher collisionality – are not taken into account below. The essential difference from the nonlocal approach is in the treatment of Boltzmann's equation.

In situations when the electron motion is strongly collisional, energy diffusion and collisional effects may be much more important than the spatial motion of the electrons and the associated thermal conduction. This implies that effects which originate explicitly from the spatial plasma inhomogeneity become negligible. Two approximations may hold, as follows.

- The EEDF at every (radial) position may be determined from Boltzmann's equation for a homogeneous plasma. This equation is often referred to as the homogeneous Boltzmann's equation.
- The EEDF at every position is determined by the local electric-field strength (and the local values of collisional terms).

It should be mentioned, though, that in some cases the simplification of a cross-section-averaged electric field is employed, retaining an effective density profile.

A local scenario specifically based on the two assumptions above is considered from here on: it can also be referred to as the "local-field approximation".

In this approximation the space-charge field E_A connected to a spatial plasma inhomogeneity becomes unimportant and the EEDF is essentially only locally determined by the effective HF electric-field intensity (2.41b). The first four terms in (2.41b) are neglected. Thus the simplified Boltzmann's equation reads as (6.3) with the first term missing:

$$\frac{2e}{3m}\frac{\partial}{\partial u}\left\{\frac{u^{3/2}}{\nu}\frac{1}{2}\left[E_z^2(r)+E_r^2(r)\right]\frac{\nu^2}{\omega^2+\nu^2}\right\}\frac{\partial F_0}{\partial u}=S_0\,. \tag{6.46}$$

Since no space-charge potential is taken into account, the kinetic energy u is used in the following rather than the total energy $\mathcal{W}$. The fact that the local values of the electric-field strength (and also, conceivably, radially changing values for ν and S_0) are present in (6.46) now leads directly to a spatially (radially) varying EEDF. This is obviously a radial dependence of the EEDF different from the one found in the case of the nonlocal approach, which resulted from the reverse transformation from the total to the kinetic energy. Equation (6.46) is formally equivalent to (6.4) – again with $\mathcal{W}$ replaced by u – except that the coefficients $\nu_{\mathcal{W}}$, $g(\mathcal{W})$, $h(\mathcal{W})$ and ν_{0_k} ((6.6)–(6.9)) are now used without averaging. The heating coefficient D_ε (6.20) is simply given by the sum of D_z and D_r ((6.21) and (6.22)), the quasi-linear term (6.23) not being added. Equation (6.46) is again an ordinary differential equation in a single variable (u), to be solved via iteration for each value of r with the rest of the set of equations.

6.2.2 Essential Features

This type of approach has been the basis for many modelling and related studies, applied to large ranges of conditions, even for cases of relatively low pR values, where the nonlocal approximation may become relevant, as commented on in Sect. 6.3. In most of these studies great attention has been paid to the influence of various atomic and molecular processes, greatly contributing to progress in discharge modelling.

Two interesting and important features of the local approach will be treated with numerical examples.

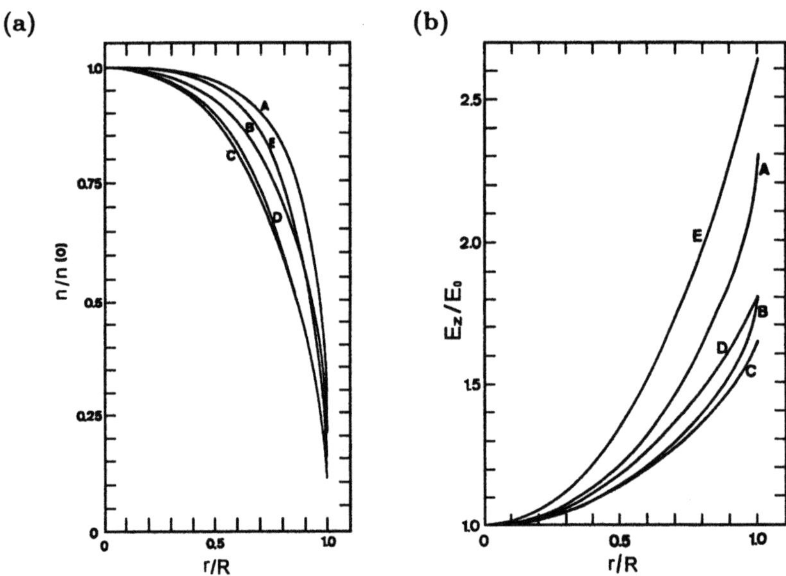

Figure 6.11. Radial electron density distribution (a) and radial distribution of the axial SW electric-field component (b) normalized to the value of the corresponding quantity in the discharge centre. Discharge conditions: SW frequency $f \equiv \omega/2\pi = 433$ MHz, argon gas with $p = 30$ mtorr, discharge tube with internal radius $R = 3.8$ cm, quartz thickness $d = 0.2$ cm surrounded by a metal cylinder of radius $R_\mathrm{m} = 9.5$ cm. Values of $\bar{n}$ (and of $n(r = 0)$, the plasma density in the discharge centre, in parentheses) in 10^{10} cm^{-3}: 1.0 (1.2) (A); 1.25 (1.70) (B); 2.2 (3.3) (C); 3.4 (5.0) (D); 7.6 (10) (E). $\omega/\overline{\omega_\mathrm{p}} = 0.17$–$0.487$ ([6.23], Figs. 2, 1)

In the local scenario the EEDF is obtained as a local function of radius essentially owing to the variation of the electric field strength, which is, in turn, influenced by the nonuniformity of the radial density. As a consequence the

EEDF contains more electrons in the inelastic range at radial positions outside the centre (closer to the wall), where the electric-field intensity is higher, increasing the ionization frequency there. This is demonstrated by Fig. 6.11, from [6.23], for argon discharges. Different axial positions with different radial field intensity changes are addressed. Resulting from the competition between ionization and radial diffusion, the density profiles (Fig. 6.11a) are somewhat higher away from the centre, most noticeably so near the wall, where the field intensity rise is strongest. There is some tendency to form a flat part of the profile before the final drop towards the wall. This is substantiated by inspection of Fig. 6.11b, depicting the associated $|E_z|$ profile, self-consistently calculated. The density $\bar{n}$ increases from case A to case E. The slightly modified behaviour of the $|E_r|$ profiles does not essentially change the situation. It should be noted that for the situations under consideration the density profiles do not drop enough towards the wall to encounter the resonance condition $\omega_\mathrm{p}(r) = \omega$. The results in Fig. 6.11 for $f \equiv \omega/2\pi = 433$ MHz ($R = 38$ mm) are complemented in [6.23] by similar findings for $f \equiv \omega/2\pi = 2.45$ GHz ($R = 15$ mm). The variation of the calculated radial profiles for different $n(r = 0)$ is displayed in [6.23], though in that situation there is a monotonic sequence of the steepening of the field profiles and of the shifting of the density profiles with growing $\bar{n}(z)$; only rather high values of $\bar{n}$ are considered ($\omega/\overline{\omega_\mathrm{p}} = 0.03$–$0.11$). In view of the comparatively high degrees of ionization in this situation, the calculations are simplified by assuming Maxwellian distributions; see the remark on this aspect related to (6.4)–(6.9).

The above feature of the density profiles – and also the noticeable changes axially – is different from that of the nonlocal scenario, where the EEDF is dependent on radius only through the transformation from kinetic to total energy. The nonlocal approach tends to concentrate the ionization and also the electron density in the discharge centre, with only a rather weak dependence of the radial density profile on the axial position. In Fig. 6.12 the curves calculated in the nonlocal scenario for different axial positions become virtually indistinguishable, the different symbols just increasing the thickness of the curves. Moreover, in this scenario axial variations in the field intensity profiles enter relatively weakly into $F_0^{(0)}(\mathcal{W})$, though there may be a stronger influence when the correction $F_0^{(1)}(\mathcal{W}, r)$, (6.2), becomes more important, pointing towards a transition to a local scenario.

Thus, generally, in the radial density profile obtained by the local approach there is a trend to flattening in the central part of the discharge that is absent in the nonlocal scenario. This favours a more gradual decay, not so much limited to the region close to the wall. This behaviour is related to the even more pronounced and sensitive behaviour in the radial profiles of line emission (Chap. 7). The underlying phenomenon is that in the local scenario the ionization (and excitation) frequency clearly follows the increasing field intensity close to the wall, whereas in the nonlocal scenario there is a virtu-

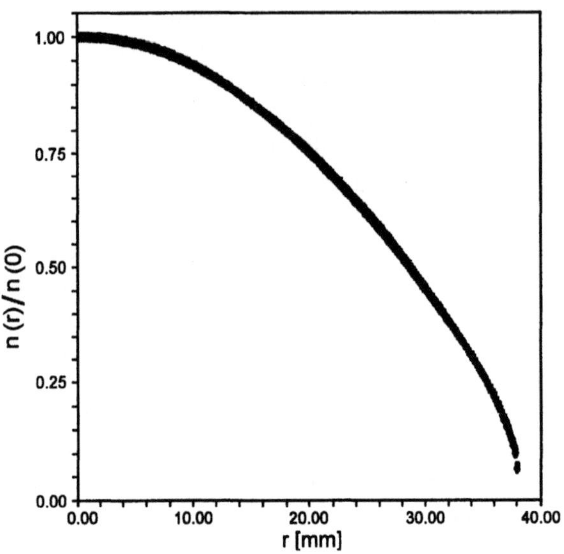

Figure 6.12. Radial electron density profiles as in Fig. 6.11a, but calculated in the nonlocal approximation (including electron–electron collisions) ([6.59], Fig. 3)

ally monotonic decrease with growing radius, ionization and excitation being largely concentrated in the central region of the discharge.

A second feature is apparent from calculations performed within the local approach: for values of the average power transferred per electron, Θ, which can reasonably be compared with experiment, inclusion of a suffciently detailed and accurate atomic database is important; this is true of many theoretical studies. Examples demonstrating the trends for argon discharges were given in [6.28] for $f \equiv \omega/2\pi = 900$ MHz and $R = 1.3$ cm, and for $f \equiv \omega/2\pi = 2.45$ GHz and $R = 0.15$ cm. Figure 6.13, for the 900 MHz case, is taken from [6.28], demonstrating that the inclusion of both step ionization and electron–electron collisions is generally important for arriving at absolute values of Θ close to the observed values. It should be pointed out that the right-hand side of Fig. 6.13 reaches the estimated domain of the local approach (use of the homogeneous Boltzmann's equation) and that the middle part of the figure constitutes an expected transition region, whereas the left-hand side might be in the domain of a nonlocal situation, at least according to the estimates reported in the next section.

Finally, considerable related work on helium can be found in [6.24,29].

6.3 Transition Regime

The importance of a radial density inhomogeneity tends to increase with the product pR, as is apparent from the results in Sect. 6.1 and also from basic

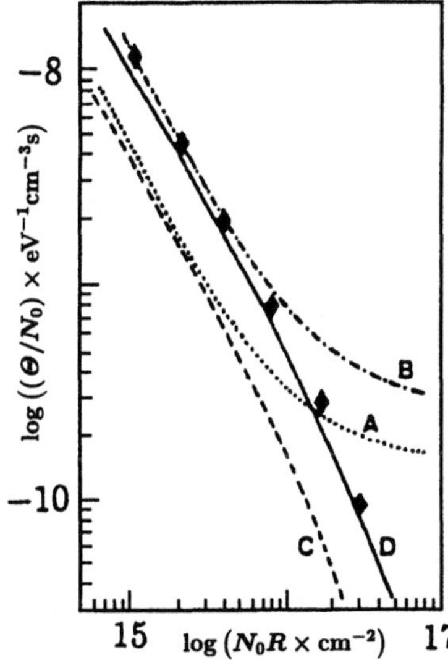

Figure 6.13. Characteristics of Θ/N_0 versus N_0R for a microwave discharge operated at 900 MHz in a cylindrical tube with $R = 1.3$ cm in which $n = 2 \times 10^{11}\,\mathrm{cm}^{-3}$. The *curves* represent calculations: A, without influence of electron–electron collisions and multistep ionization; B, including the former influence; C, including the latter influence; D, including both influences. The data points are from experiments on SW discharges (Montreal and Orsay groups) ([6.28], Fig. 7)

considerations on diffusion-controlled density profiles. On the other hand, for the applicability of a nonlocal approximation this product should not be too large. Vice versa, to keep the effects of radial density nonuniformity down – for example, to exclude radial electron density profiles allowing local plasma resonances – pR should be limited, whereas for the applicability of a local approximation a larger value of this product is desirable. Therefore it is not surprising that some of the experiments considered above encompass conditions in transition to a scenario opposite to the one assumed in the calculations. Moreover, many interesting basic experiments just happen to be performed in a transition regime or nearby.

The estimates below largely follow the considerations given in [6.7] for the applicability regime of the nonlocal approximation. The inequalities presented below have, of course, to be turned around when considering the local regime. The requirement that the energy relaxation length L_ε (6.1) should be large compared with the typical (radial) inhomogeneity scale L_n can obviously be reformulated as

$$\frac{1}{\sigma_\mathrm{c}\sigma_*} \gg (N_0L_n)^2 , \tag{6.47}$$

where σ_c is the elastic cross-section and σ_* is the total inelastic cross-section for a typical energy, for example a cross-section-averaged energy, and N_0 is the neutral-gas density.

This relation is recovered if the terms of (6.2) are compared when checking the requirement $|F_0^{(1)}| \ll |F_0^{(0)}|$, taking the leading terms in the central region of the discharge and in a region close to the wall. In the spatially averaged form of (6.3), i.e. in (6.4), the leading terms are the heating term, containing $\bar{D}_{\mathcal{W}}$, and the inelastic term, containing an (effective) inelastic collision frequency ν_*, roughly balancing each other. In the central discharge region, where the variation of the ambipolar potential is small and $W \approx u \approx$ const., the inelastic term dominates. With this approximation from (6.3), the requirement for $F_0^{(1)}$ can be formulated as

$$|F_0^{(1)}| \Big/ |F_0^{(0)}| \approx L_n^2 \nu_*(W)/D_e \ll 1 , \tag{6.48a}$$

with

$$D_e = \frac{2e}{3m\nu} u \tag{6.48b}$$

being the spatial diffusion coefficient. This leads to

$$\frac{1}{3\,\sigma_c\sigma_*} \gg (N_0 L_n)^2 . \tag{6.49}$$

Therefore, inequality (6.47) is recovered, strengthed by a factor $1/3$. For the outer discharge regions, where the ambipolar potential is relatively high, inelastic collisions may be considered to be of less importance than the HF heating term. Estimating the heating term by (6.4) for energies at the beginning of the EEDF tail, the inequality (6.49) is again recovered.

Using the somewhat more stringent relation (6.49) – instead of (6.47) – and taking the tube radius R for L_n, the following "cautious" estimates are obtained as requirements that the nonlocal approximation might hold without essential errors:

- argon, $pR < 0.2$ torr cm
- neon, $pR < 1.0$ torr cm
- helium, $pR < 1.4$ torr cm.

Thus, the requirements for using the nonlocal approximation appear most severe for argon, whereas for helium the use of the nonlocal approximation may be extended to significantly higher pressure values.

The estimates derived from (6.49) are supported by model calculations using numerical solutions of the inhomogeneous Boltzmann equation in the form of a partial differential equation in both W and r, as given by (6.3) for the cylindrical case. The resulting estimates, as reported in [6.7], will be commented on. In doing this a loss cone of the thermal electron flux to the wall is taken into account in the boundary conditions: the kinetic energy at the perpendicular velocity has to be equal to at least the sheath potential $\Phi_{wall}-\Phi_{sh}$, with Φ_{sh} being the potential at the plasma–sheath boundary. More details, in particular about the boundary conditions, were given in [6.7,42].

In a model calculation for a positive column with a spatially constant, fixed heating field $E_\parallel$ given by $E_\parallel/N_0 = 2.5 \times 10^{-20}$ V m^2, with $R = 1$ cm [6.7], it was shown that for $N_0 = 3 \times 10^{21}$ m^{-3} an EEDF results which to high accuracy is solely a function of the total energy for all radii, as expected for a nonlocal scenario. However, for $N_0 = 3 \times 10^{23}$ m^{-3} a trend towards a depletion in the discharge centre by inelastic processes can be seen, as well as towards a drain of energy near the wall ([6.7], Figs. 4a,b). This is reflected in the radial dependence of the mean kinetic energy $\bar{u}$ ([6.7], Fig. 5). Whereas for $N_0 = 5 \times 10^{21}$ m^{-3} $\bar{u}$ reveals a distinct variation, as to be expected in a nonlocal situation, the variation has become weak for $N_0 = 1 \times 10^{23}$ m^{-3} and $\bar{u}$ is virtually constant for $N_0 = 1 \times 10^{24}$ m^{-3}. This also shows up in the development of $F_0(\mathcal{W}, r)$ with growing N_0 as compared with the nonlocal solution $F_0^0(\mathcal{W})$, where one and the same function of *total energy* holds for all radial positions ([6.7], Figs. 6a,b). Obviously, for $N_0 = 5 \times 10^{21}$ m^{-3} and even $N_0 = 1 \times 10^{22}$ m^{-3} ($\hat{\approx} 0.3$ torr cm for pR), the deviation from the nonlocal approximation remains tolerable, but the correction becomes quite noticeable for $N_0 = 3 \times 10^{22}$ m^{-3} ($\hat{=} 1$ torr cm), particularly near the wall.

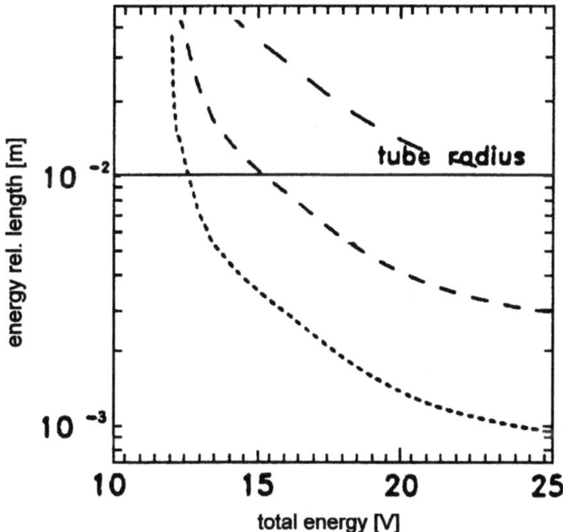

Figure 6.14. Energy relaxation lengths as a function of total energy for different neutral densities $N_0 = 3 \times 10^{21}$ m^{-3} (*long-dashed curve*), 1×10^{22} m^{-3} (*short-dashed curve*) and 3×10^{22} m^{-3} (*dotted curve*) ([6.7], Fig. 7)

This behaviour is consistent with the calculated values of the cross-section-averaged energy relaxation lengths for these conditions, shown in Fig. 6.14 as a function of total energy. From these test calculations it appears, therefore, that the above estimate of pR seems reasonable so that, for instance, in the case of argon for $pR \lesssim 0.2$ torr cm the nonlocal scenario should be tolerable as a qualitative description without the quantitative errors being too drastic, whereas for larger values corrections become necessary, finally leading to a transition to the local scenario. A slight modification to lower critical values of pR may be caused by strong radial field inhomogeneity.

This aspect of radially inhomogeneous heating was illustrated in Fig. 8a–c of [6.7] by model calculations as discussed above, except that a radial increase

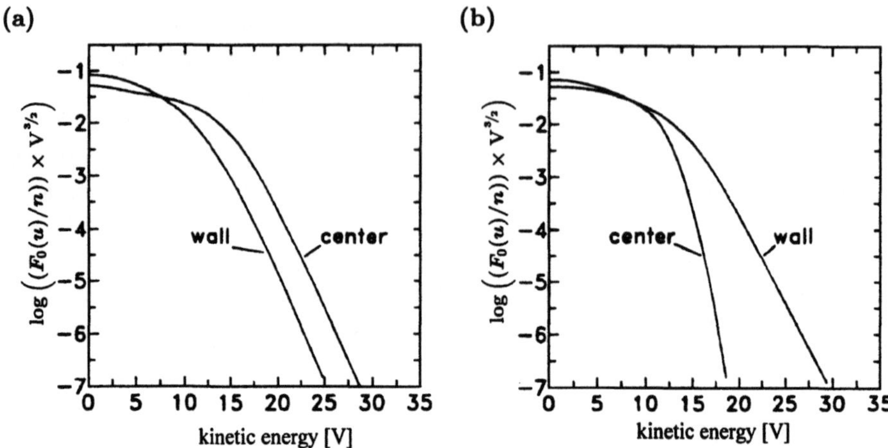

Figure 6.15. Normalized isotropic part of EEDF versus kinetic energy u in the presence of radial field inhomogeneity, at the discharge centre ($r = 0$) and near the wall ($r = 0.9R$), calculated in the nonlocal **(a)** and the local **(b)** scenario ([6.7], Figs. 8b,c)

of the heating-field strength was allowed, proportional to $[1 + 4(r/R)^2]$. As compared with the situation above with $E_\parallel$ at the same level everywhere, the tail of $F_0(u)$ is lifted; it is lifted somewhat more, but not dramatically so, towards the central region. But the typical feature of the nonlocal scenario is retained, namely that $F_0(u)$ is reduced near the wall ($r = 0.9R$) compared with $F_0(u)$ at the centre ($r = 0$), as shown in Fig. 6.15a. The situation could be quite different – with an opposite trend – in a local scenario, as the results of a local-model calculation in [6.7] demonstrate. The tail near the wall is *not* lowered compared with the centre, but rather clearly raised owing to the local effect of the higher field near the wall, as shown in Fig. 6.15b. The solution obtained from the complete kinetic equation without approximation – nonlocal or local – is given in Fig. 6.16 for the conditions of Fig. 6.15. F_0 is plotted versus the total energy $\mathcal{W}$. It is now weakly dependent on r even on the $\mathcal{W}$ scale. But at the discharge centre $F_0(\mathcal{W})$ is rather close to the value of $F_0(u)$ in Fig. 6.15a, i.e. the nonlocal approximation seems still to be good under these conditions.

It should be mentioned that an improvement to the nonlocal approximation helpful in the transition region can be the evaluation of the first-order correction $F_0^{(1)}(\mathcal{W}, r)$ obtained from integration of (6.4) in the one-dimensional case. In particular, in this way the influence of a radially inhomogeneous electric-field configuration would be slightly larger. For a truly higher-dimensional case, with, for example, comparatively strong r and z dependences, though, the situation may be expected to be more complicated and sensitive. A relevant reference for observations on the transition region

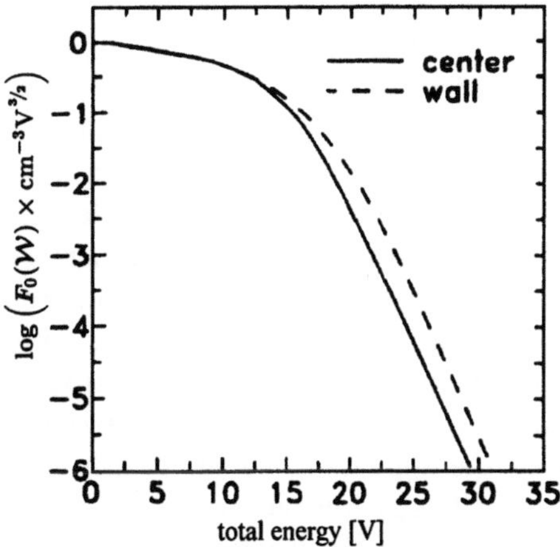

Figure 6.16. Isotropic part of EEDF $F_0(W)$ versus total energy W for the conditions of Fig. 6.15, but obtained from the solution of the complete kinetic equation without an approximating scenario. Normalized by setting $F_0(0) = 1$. $F_0(W)$ is weakly dependent on r ([6.7], Fig. 8a)

in the case of a capacitively coupled HF discharge [6.71] should be pointed out in this context.

In order to emphasize the usefulness and validity of the local approximation towards higher pressures, numerical solutions of (6.3) for the same conditions as above were included in [6.7], though without the radial field increase just considered. They demonstrate a transition from $N_0 = 10^{22}\,\text{m}^{-3}$ to $N_0 = 10^{23}\,\text{m}^{-3}$. At the higher pressures virtually no differences in the kinetic-energy scale show up between the centre and wall regions.

Of course, the estimates presented above contain inaccuracies due to the assumptions and model conditions used and depend also on the atomic data used. For instance, in the numerical solution of (6.3) above a simplified ambipolar potential has been used; in [6.72,73] a more detailed approach is presented. These references add to the hope of complete self-consistent models for different types of discharges in the near future. Such generalizations, still essentially based on the two-term approximation (2.34a), are expected to give good physical insight into the interesting transition region between the local and nonlocal scenarios.

The energy relaxation length L_ε (6.1) is obviously most critical with respect to the range of inelastic processes in the tail of the EEDF, and the cautious estimates above for the critical pR values are essentially based on this. There is, however, an interesting aspect concerning the degree of symmetry of the electron density contours and the role of step ionization in that context. With an increase of pR (at a given R) these should be expected to be more affected by an asymmetry in the electric-field configuration, away from the symmetric contours in the nonlocal regime and approximation. This seems, indeed, to be confirmed by calculations (and observations as well) for, for example, an inductive discharge [6.73]. The full solutions of Boltzmann's equation, avoiding a nonlocal or local approximation, and even in two spatial dimensions, with improved boundary conditions (using a loss-cone concept) have been obtained within self-consistent modelling. In the transition from the lowest (argon) pressure to one a factor 3 higher, the symmetry is largely lost in the calculated (and measured) density contours. However, an increase by another factor 3 in the pressure brings in metastable densities and step ionization, which were not important before at low p (and n), hand in hand with an again enhanced symmetry in the electron density contours, in spite of the metastable contours being asymmetric. Obviously, at lower energy in the EEDF the central region of the discharge appears to be sufficiently pronounced to yield, from the competition of (step) ionization and diffusion, almost symmetric electron density contours. In such situations the density contours are apparently less sensitive to the localization of the field asymmetry again. It should be pointed out that the profiles of population densities of excited, in particular, fast radiative states are more sensitive to a transition from a nonlocal to a local regime than density profiles (Chap. 7).

7. Experimental Aspects

Although some preliminary indications of the maintenance of gas discharges by SW propagation were given in the late 1960s [7.1], substantial experimental activity in this field started in 1974 with the pioneer studies by Moisan et al. [7.2]. At this time the "surfatron" was introduced as a launching structure for SW-sustained discharges. Many data concerning the experiments on wave-produced discharges (launchers, experimental techniques and results) have been summarized in the review papers by Marec et al. [7.3], Moisan et al. [7.4], Moisan and Zakrzewski [7.5], Chaker et al. [7.6] and Moisan and Zakrzewski [7.7]. Excellent surveys of the experimental arrangements used and experimental findings are also outlined in [7.8–10], together with rather comprehensive collections of literature citations.

This chapter provides only a brief summary of experiments on SW-produced discharges and deals with

(1) methods which have been applied or developed for the diagnostics of these discharges
(2) experimental results which show basic features of the discharge or give indications of experimental verification of the recent developments in discharge modelling discussed in the preceding chapters.

The discussion follows generally the presentation of the experimental activity on SW-sustained discharges in [7.11,12]. At the end of the chapter problems of current activity in the field of applications of this type of discharge are listed.

7.1 Experimental Conditions

The experimental conditions in which SW-sustained discharges can be produced cover wide ranges of variation in

- wave frequency: $f \equiv \omega/2\pi = 500$ kHz to 10 GHz
- gas pressure: $p = 1$ mPa to 1 atm
- size of the discharge vessel, e.g. radius of the gas discharge tube $R = 0.5$ mm to 15 cm, and even larger dimensions in recent applications.

The density of the plasmas produced cover the complete range of variation of plasma densities in gas discharges: $n = (10^8 \text{ to } 5 \times 10^{15})\,\text{cm}^{-3}$.

The power applied for producing the discharge is not too high: $P_0 \leq$ 200 W, or up to 700 W at $p = 1$ atm.

Many different gases have been studied, the variety still increasing with the widening of the applications. The basic studies, though, have been performed largely in inert gases (argon, helium, neon). In many cases a fixed gas filling of the apparatus or a very weak gas flow maintaining a constant pressure has been utilized. In some cases – in particular, for applications – stronger gas flows may be applied (e.g. to prevent impurity accumulation), which may lead to axial gradients in gas density and gas temperature, which have to be taken into account in the analysis of such discharges.

The flexibility with respect to the experimental conditions makes SW-sustained discharges very attractive for use in gas-discharge applications. Some of their advantages are:

- all types of gas-discharge plasma (nonisothermal, nonequilibrium plasmas at low gas pressure, quasi-equilibrium plasmas at atmospheric pressure and the transition regimes between them) can be produced by the same type of discharge
- the plasma extends far away from the region of the launcher, which provides flexibility in constructing plasma reactors
- SW-sustained discharges are sources of a dense plasma (see the discussion in Sect. 5.1), which could also be exploited in some applications.

The extended research [7.2,7,13–20] on designing wave launchers which ensure single-mode operation and high efficiency of the power transfer from the generator to the launched wave (and very good impedance-matching conditions for this transfer) is the basis of the interest in studying SW-sustained discharges. This important aspect of the work on SW-produced discharges makes them attractive for use as plasma sources.

The "family" [7.7] of SW launchers for excitation of azimuthally symmetric SWs is shown in Fig. 7.1. These are as follows.

(1) The surfatron device (Fig. 7.1a), an integrated wave launcher of coaxial type with a circular gap for forming the field configuration and a capacitive coupler for transferring the HF power to the plasma; operation frequency from 100 MHz up to 2.4 GHz.

(2) The surfaguide (Fig. 7.1b), a device consisting of standard waveguide elements, by which the operation frequency of SW-sustained discharges is extended up to 10 GHz and the applied power up to several kW.

(3) The waveguide surfatron (Fig. 7.1c), an integrated wave launcher which, combining waveguide and coaxial components and being a waveguide equivalent of the surfatron, keeps the good tuning capability of the surfatron and the high power capability of the surfaguide.

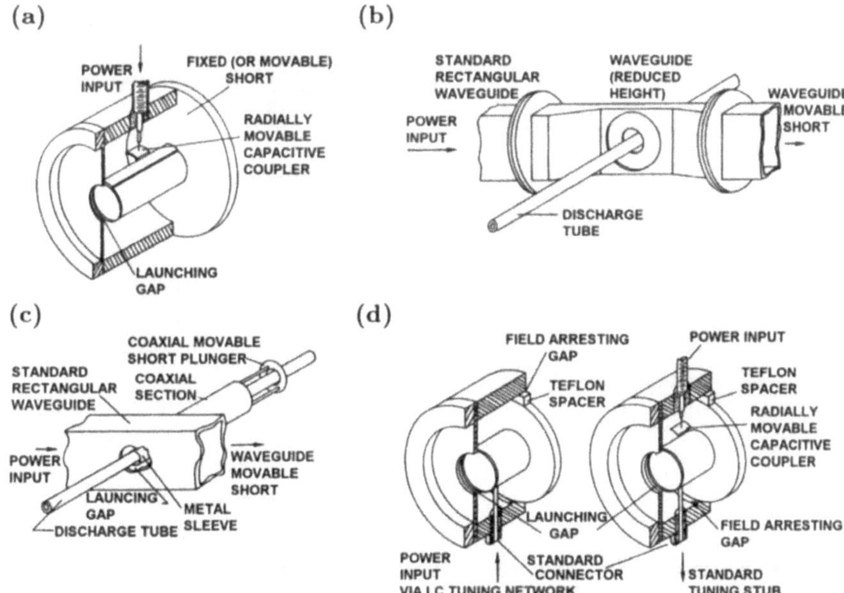

Figure 7.1. The family of SW launchers for excitation of azimuthally symmetric SWs: **(a)** surfatron, **(b)** surfaguide, **(c)** waveguide surfatron, **(d)** Ro-boxes with *LC* impedance-matching network (on the *left*) and with capacitive coupling and stub tuner (on the *right*) ([7.7], Fig. 12)

(4) The Ro-box (Fig. 7.1d), with its two designs (*LC* Ro-box for an operation frequency from 1 MHz up to 100 MHz and stub Ro-box for operation frequencies above 100 MHz), by which the frequency of existence of SW-discharges is extended towards lower frequencies, still keeping the launcher structure compact (by using an isolated endplate, reducing the launcher length below about a quarter of the vacuum wavelength).

The basic principle of all these launchers is the creation of concentrated HF electric-field intensities in a circular gap close to the cylindrical discharge vessel which are essentially axially oriented and well suited to the SW mode, and thus effectively starting the SWs along the vessel outside the launcher.

The recent activity [7.21–23] in constructing launchers more closely directed towards applications combines the previous experience in designing SW launchers with experience in designing plasma reactors.

Besides flexibility with respect to the operation frequency there is also flexibility as to the shape of the vessel, since SWs have the ability to follow rather well changes in the diameter and form of the dielectric surface bounding the discharge.

Examples of discharge vessels are depicted in Fig. 7.2. The case of a discharge in a straight, cylindrical glass tube (Fig. 7.2a) is complemented (Fig. 7.2b,c,d) by modifications of the discharge vessel introduced with regard

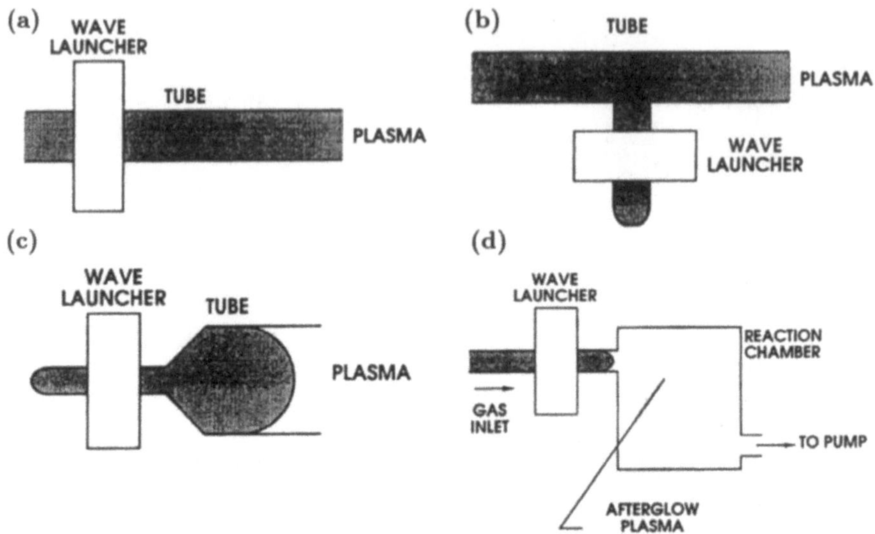

Figure 7.2. Examples of plasma vessels used in SW plasma sources: cylindrical tube **(a)**, symmetric T-tube **(b)**, tapered tube **(c)** and reactor chamber for afterglow plasma **(d)** ([7.7], Fig. 30)

to the applications of the discharge (for spectral lamps and gas lasers, plasma chemistry and plasma processing).

It should also be mentioned that systems may be used with launchers at both ends of the discharge or arrangements allowing standing-wave operation, as one way of reducing the axial nonuniformity [7.24]. Also, in "inverse" arrangements, where the vacuum and the main energy flux are inside an inner glass tube and the plasma is in a large volume surrounding the tube, power can be supplied from both ends [7.25].

7.2 Diagnostic Methods

Over the years quite a variety of diagostic methods has been used in studies of SW-discharges, basically the whole spectrum of methods in the diagnostics of low-temperature plasmas [7.26–33]. However, it should be emphasized that there are some specific aspects of diagnostic methods brought into play by the very nature of the discharges sustained by SWs: the dispersion properties of the SW modes lend themselves to diagnostic use.

Independently of the large amount of activity in the diagnostics of SW-sustained discharges, it may be mentioned that compared with the output from the research on the theory and applications of SW-sustained discharges, the amount of information accumulated from the diagnostic is not so large. On the one hand, this can be attributed to caution in using some of the standard methods of plasma diagnostics. This is understandable since in diagnosing SW-sustained discharges care must be taken not only not to disturb the plasma too much but also not to disturb the wave propagation by introducing changes in the waveguide configuration. In SW-sustained discharges, the wave sustains the discharge and disturbance of the wave properties causes changes of the plasma characteristics. On the other hand, the fact that there are not too many data for the discharge properties can be associated with the diagnostic purposes of the experiments. The axial density inhomogeneity specifies these discharges and it is understandable that its determination has largely been aimed at in experiments. As has been mentioned above, methods applicable particularly to these discharges have been developed. As a result, a lot of data have been accumulated about $\bar{n}(z)$. The other type of data, accumulated from many experiments, is about the spatial distribution of spectral-line intensities $I_L(r)$ and excited-state population densities $N_k(r)$. Such studies are motivated both by the development of the kinetic modelling of the discharges and by the research on their applications. However, there is still a lack of results about other parameters such as the electron temperature T_e (or mean electron energy) and, correspondingly, Θ, which are also very important because of the information they give about the mechanisms of discharge maintenance.

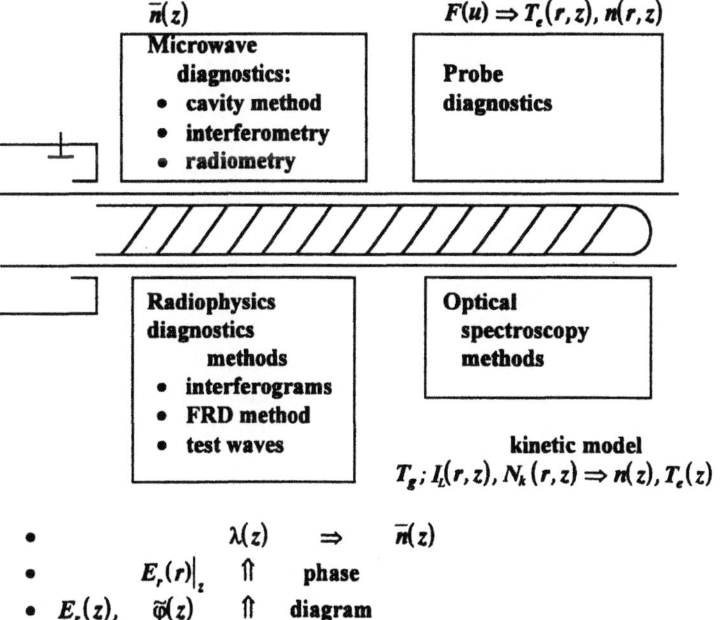

Figure 7.3. Diagram representing methods used in the diagnostics of SW-sustained discharges ([7.11], Fig. 8)

The methods applied to the diagnostics of SW-sustained discharges, with indications of their capabilities for determination of plasma and discharge parameters, are schematically shown in Fig. 7.3.

Microwave, probe and optical-spectroscopy diagnostics are the standard methods which have been used. The cavity method was used in the first experiment [7.34] on the diagnostics of SW-sustained discharges, for calibration of data from measurements of the total light emission to obtain the $\bar{n}(z)$ dependence. The results were combined with measurements of the SW wavelength λ, and it was shown that the experimentally obtained phase diagram was quite close to that of SWs in homogeneous plasma columns. The latter were widely used in the development of radiophysics diagnostic methods. Microwave interferometry, although attractive as a method that does not disturb the discharge (the complete structure of plasma and wave field), is surprisingly applied seldom in the diagnostics of SW-sustained discharges. Even less often used is microwave radiometry. Probe diagnostics, despite their capability for getting quite complete information – the EDF $F(u)$, spatial distribution of electron temperature T_e (i.e., the mean electron energy) and plasma density n – create difficulties for HF gas discharges in general and questions arise about their applicability to SW-sustained discharges. The optical-spectroscopy methods seem to be the most proper tool for the diagnostics of SW-sustained discharges although they are of indirect nature with respect to the plasma parameters. Data for the gas temperature T_g, $I_L(r)$ and $N_k(r)$ are the direct results from the experiments. Further extension of the method to obtain results for the plasma parameters (n, T_e) involves kinetic modelling of the discharge.

The methods denoted in Fig. 7.3 as "radiophysics methods" widely used over the years in the diagnostics of SW-sustained discharges, include the interferogram method and the field radial decay (FRD) method as well as the test wave method applied recently. These methods, developed especially for diagnostics of SW-sustained discharges, are based on measurements of the propagation characteristics of the SWs and on the relation of the plasma parameters to the wave behaviour. Data for $\bar{n}(z)$ are the final results. The radiophysics diagnostic methods are, in general, indirect. The direct data from the measurements concern the space distribution of the SW field in free space. Obtaining results for $\bar{n}(z)$ involves use of theoretical SW phase diagrams.

A well-advanced experimental arrangement of different techniques – the interferogram method combined with measurements of the axial distribution of the SW power (related to the E_r SW field component) and of the total light emission from the discharge, as well as the cavity method – developed in the first experiment on diagnostics of SW-sustained discharges by Zakrzewski et al. [7.34] is shown in Fig. 7.4.

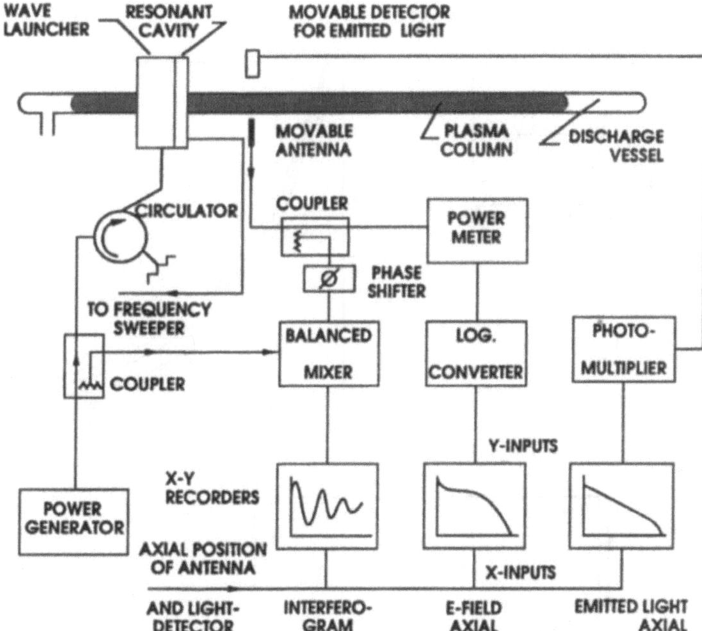

Figure 7.4. The experimental set-up in the first experiment on diagnostics of SW-sustained discharges ([7.34], Fig. 1)

7.2.1 Radiophysics Diagnostic Methods

The first two methods [7.4,34–43] – the interferogram method and the FRD method – are based on measurements of the field distribution of the SW which sustains the discharge. The third method [7.44,45] – the test wave method – involves measurements of the propagation characteristics (amplitude and phase) of small-amplitude (test) SWs with frequencies which are different from the frequency of the powerful signal applied for creating the discharge.

In all these methods, the signal is taken from a radially oriented antenna, i.e. the E_r field component in free space is picked up. Experience from many applications reported in the literature shows that the perturbations of the discharge from antennae placed outside the plasma can be kept to a tolerable level.

In the interferogram method the SW wavelength is measured from the recorded interferogram (Fig. 7.5a) obtained by mixing the signal from the antenna (which is movable in the z direction, i.e. along the discharge) with a reference signal. The different ways of using the interferogram method (as a single-interferogram method, a modified single interferogram method and a shifted-interferogram method), as well as the accuracy and the spatial resolution of the method, are discussed in [7.40]. In the FRD method, the antenna is movable in both the z and the radial r direction. In this case the radial field distribution at a given z position is measured (Fig. 7.5b). Fitting with

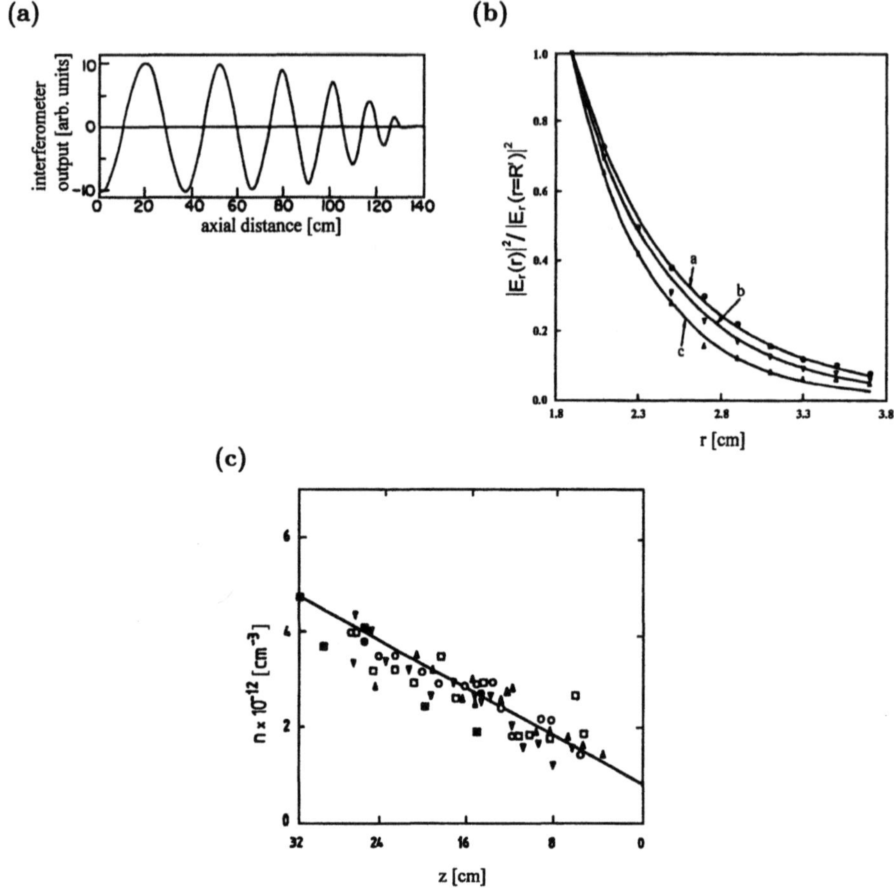

Figure 7.5. (a) Recorded interferogram ([7.34], Fig. 2c); (b) experimental results for the radial variation of $|E_r|^2$ normalized to the corresponding value at R' for different z positions together with the theoretical field profiles (*solid curves*), obtained with β as the fitting parameter ([7.41], Fig. 4); (c) axial density profile obtained by using interferogram technique (*open symbols*) and FRD method (*filled symbols*) ([7.41], Fig. 5)

a theoretical field distribution of the SW field in the transverse direction leads to a result for the wave number β. In both cases the experimental data are for the wave number, although in the case of the FRD method these data are indirect. The result for the plasma density $\bar{n}$ comes out after using theoretical phase diagrams. In a way, the FRD method is doubly indirect. However, the results it provides are in good agreement (Fig. 7.5c) with those from the interferogram method. Up to now, in the application of these two methods to the diagnostics of SW-produced discharges the phase diagrams of SWs in homogeneous collisionless plasmas have been used. This is based on results (Fig. 7.6) for the dispersion behaviour of SWs in SW-produced discharges, obtained using β taken from interferograms and $\bar{n}$ obtained from

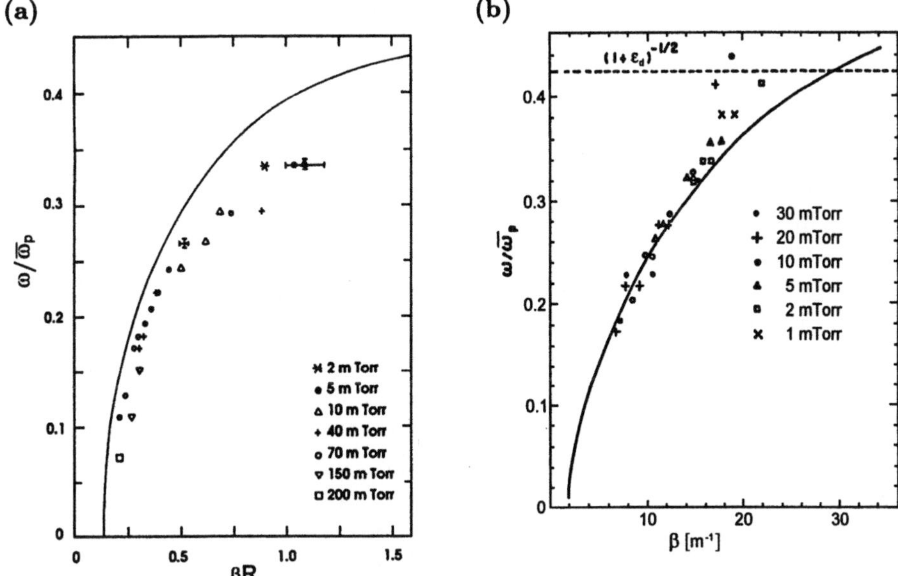

Figure 7.6. Phase diagrams obtained in SW-sustained discharges in argon with different values of the gas pressure. Comparison of the experimental data (given by the *symbols*) with theoretical phase diagrams (*solid curves*) for a homogeneous, collisionless plasma waveguide. **(a)** $f \equiv \omega/2\pi = 500$ MHz, glass tube with internal diameter $2R = 2.5$ cm, glass thickness $d = 0.25$ cm and permittivity $\varepsilon_d = 4.52$ ([7.4], Fig. 9); **(b)** $f \equiv \omega/2\pi = 100$ MHz, glass tube with internal diameter $2R = 6.4$ cm, glass thickness $d = 0.35$ cm and permittivity $\varepsilon_d = 4.52$ ([7.36], Fig. 4)

cavity measurements (combined with measurements of the axial distribution of the total light emission from the discharge). Although the deviation of the experimental results from the theoretical phase diagrams of SWs in homogeneous collisionless plasma waveguides is not large, the influence of transverse plasma inhomogeneity (as an effect of weak inhomogeneity, Chap. 4) and of the combined effects of collisions and transverse plasma inhomogeneity (Chap. 4) are evident in Figs. 7.6a,b, respectively. A determination of the elastic-collision frequency ν by comparing the experimental results for the axial density profile with corresponding results taken from discharge models is the further extrapolation of these methods. However, the accuracy of such data is strongly influenced by the discharge model used and the simplifications involved in it. Simultaneous measurements of β and α could provide more reliable data for the collision frequency.

Although also indirect, the test wave method provides more possibilities for the fitting procedure, since the test wave frequency can be varied. By using the results for the SW dispersion behaviour discussed in Chap. 4, the sensitivity of the wave propagation characteristics to different plasma parameters (the ν/ω ratio and the parameters related to the radial and axial inhomogeneity) could be exploited.

No matter that the interferogram method has been used in numerous experiments and has been established as a basic tool – at least up to now – in the diagnostics of SW-sustained discharges, the manner of its application and of the application of the radiophysics diagnostic methods, in general, require caution and further checks. This is said because of developments in the theory of SW propagation in collisional inhomogeneous plasmas (Chap. 4). First, the applicability of the methods require the validity of the geometrical-optics approach (WKB approximation): $|dk^{-1}(z)/dz| < 1$. Secondly, the experimental evidence on the influence of plasma inhomogeneity and collisions on SW dispersion behaviour (Sect. 7.3) shows that these effects should be taken into account. Particularly, the further extrapolation of these methods to obtain indirect data for other plasma parameters (e.g. determination of the collision frequency ν from comparison of experimental and theoretical results for the axial density gradient) needs reconsideration [7.41]. Therefore, it looks as if the basis of the radiophysics methods needs to be checked. For example, more experimental results for the SW dispersion behaviour are necessary. Although the test wave method can be very helpful in this direction, the main efforts should probably aim at direct measurement of the plasma parameters: measurement of $\bar{n}$ by, for example, microwave interferometry [7.46,47], and of T_e by other methods of plasma diagnostics.

7.2.2 Probe Diagnostics

The lack of experimental data on T_e (as an effective measure of the mean electron energy for non-Maxwellian energy distributions) was the original reason for including probe measurements [7.24,48] in the diagnostics of SW-sustained discharges. The start towards kinetic modelling [7.4,49–51] and the need for experimental results [7.42,47,52–61] on $F(n)$ for verification of the models were also a motivation for efforts to apply probe diagnostics to SW-sustained discharges. The probe methods can also provide valuable information on the spatial variation of the electron density and temperature.

Concerning the interpretational side there is, of course, the difficulty of choosing the appropiate theory for the probe [7.26–29] for the parameter range under consideration, sometimes requiring the use of other diagnostic methods for corroboration and calibration. On the other hand, there is much literature and experience available [7.62–65]. This is also true concerning details of (active and passive) compensation measures in the presence of HF fields, which are required mainly because of the sheath impedance of the probes. But the problems noted here are well known in the literature and in spite of the difficulties, probes are often resorted to in order to gain data that is otherwise accessible only with more effort and investment. Moreover, probe diagnostics are virtually unique in providing rather direct information on the EEDF: various methods [7.66–73] have been described and used to obtain from the current–voltage characteristics of probes the second derivative and, thus, the EEDF itself. This distribution function has gained growing

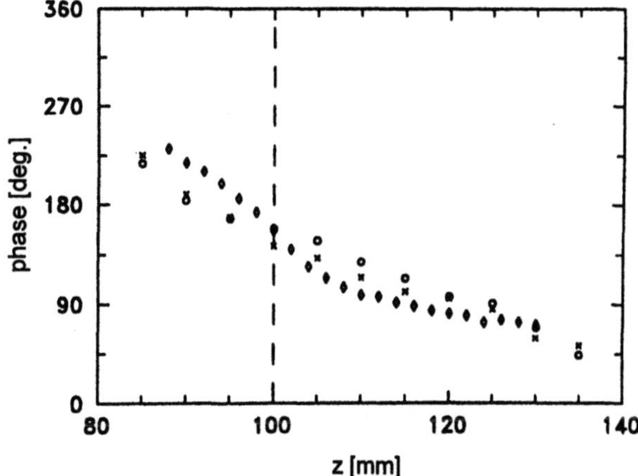

Figure 7.7. Phase measurements in the presence of a probe in a radial arrangement (*crosses*) and an axial arrangement (*diamonds*), compared with measurements without a probe present (*circles*), at 2.45 GHz and 1 torr argon ([7.55], Fig. 6)

interest over the past few years. For the situations considered here it is not measurable by other methods, since Thomson scattering methods – which are in theory possible – have not yet developed to a point where they can easily be applied with sufficient energy resolution. However, one final cautioning remark is obviously called for: in SW-sustained discharges the perturbation caused by probes inside the plasma and the resultant potential inaccuracies are aggravated by the fact that the wave pattern, the basis of plasma maintenance, can be quite noticeably perturbed even outside the plasma, e.g. by reflection from the probe [7.55], thus enlarging possible deviations of the measured plasma parameters from their unperturbed values. Figure 7.7 demonstrates measured phases with and without a radial and with an axial probe present. The radial probe arrangement obviously delivers better results in this case ($R = 0.7$ cm), with a deviation in the resultant absolute value of the electron density of about a factor of 1.2, as compared with a factor of 0.5 with the axial arrangement. Great care has to be taken in the choice of relative dimensions, geometry, position of auxiliary electrodes and compensation measures, in order to minimize the unavoidable deviations. In spite of the difficulties noted, probe diagnostics appear to be indispensable, particularly when information on a non-Maxwellian EEDF is sought. Probes may well be incorporated for determination of parameters in plasma reactors, provided no reactive constituents are present that would degrade the quality of the probes.

It should be mentioned that probes/antennae have also been used to detect HF electric fields and to pick up magnetic-field changes inside the plasma [7.47,74]. The problem of perturbations discussed above recurs, though it can be reduced by the proper choice of dimensions.

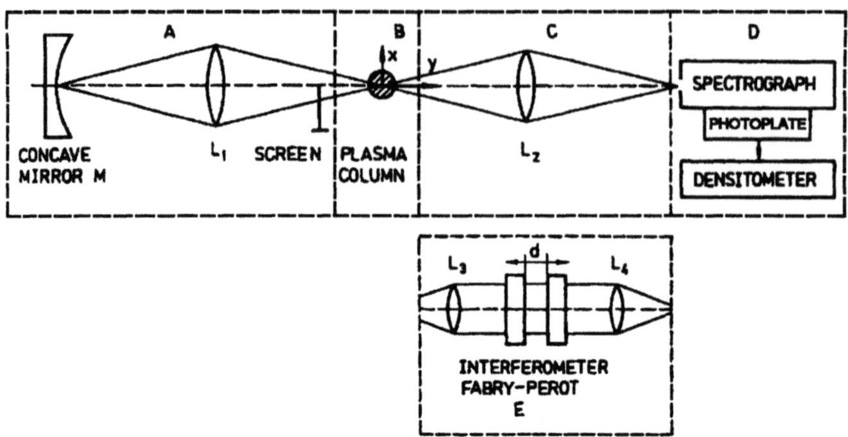

Figure 7.8. Example of experimental set-up for emission and absorption (self-absorption) measurements via a photographic plate (or optionally, a detector array) and determination of T_g by a Fabry–Pérot interferometer ([7.43], Fig. 1)

7.2.3 Microwave Diagnostics

There are microwave methods which have occasionally been used to obtain the important parameter $\bar{n}$ [7.2,4,36,46,47]. *Cavity methods* make use of the detuning of cylindrical resonance structures outside the plasma caused by the plasma coaxially inserted into the resonance structure. The values obtained for $\bar{n}_e$ are slightly dependent on the assumed radial electron density profile [7.75,76]. This is also the case with methods based on the *transmission of microwaves*, employing interferometer arrangements [7.76,77]. The evaluation requires carefully taking into account the transmission bypassing the plasma cross-section, and refractive effects. The improved availability of signals with lower wavelengths, down from 8 mm to the millimetre range, allows simplified analysis and even gives access to radial profiles (via Abel inversion). The difficulty of the very small plasma shifts that have to be measured can be overcome by means of double interferometric techniques. In this context it is worth pointing out that microwave transmission techniques can be extended and complemented by *microwave radiometry*, as recent studies indicate [7.78].

7.2.4 Optical Spectroscopy Methods

Optical and spectroscopic methods are certainly tempting to use and are important. The techniques are rather well known, and the equipment required is usually quite affordable, if modern, intricate laser techniques are left aside [7.26–33]. These methods have been widely employed with SW-sustained discharges [7.4,5,43,79–89]. For an example of an experimental arrangement, see Fig. 7.8. For radial resolution Abel inversion techniques or two-channel tomography can be used [7.43,87].

The demands of the optical/spectroscopic methods are more on the side of the necessary theoretical/modelling basis needed for determining the essential plasma parameters. This is not the case for the determination of gas temperatures on the basis of measurements of the Doppler broadening or of population densities on the basis of absorption and emission techniques (in the latter case, with proper handling of self-absorption). However, relating the measured intensities to parameters such as the electron density or mean energy requires knowledge of the type of equilibrium (thermodynamic equilibrium relations not being usable in most cases), the potentially non-Maxwellian behaviour of the EEDF and the scenario appropriate to the way it is formed (local or nonlocal). Even if complete numerical modelling is not necessary, proper estimates of the situations are at least required. This also includes judgements on the importance of the various atomic processes, for example step processes. On the other hand, spectroscopy – especially measurements of emission intensities of lines or of ratios thereof – may offer great possibilities to check and study open problems and to validate the theory and modelling, in some cases combined with other methods. Generally, spatial resolution can be provided, and systems with various and even many constituents can be investigated. Therefore optical and spectroscopic methods appear to be quite appropriate tools to be used with SW-sustained discharges, except for direct determination of EEDFs. But for the scenario of the formation of the energy distribution – nonlocal or local – valuable results are obtainable, in particular from line emission spectroscopy.

7.3 Summary of Observations

7.3.1 Observations on Basic Features

The large body of experimental work on SW-sustained and related discharges has grown out of concentrated and systematic series of studies by several groups. It also benefited from valuable contributions of less extensive or differently oriented programmes from almost all over the world. The work has profited from a lot of international cooperation and joint activity. This is also reflected in the references encountered if an attempt is made to find just a few examples of the main lines of approach, for example measurement of the plasma parameters. To give full credit to all valuable contributions, many more references could be cited.

Quite a large number of observations have been concerned with *electron densities*, the results having been obtained mainly by the radiophysics diagnostic methods [7.4,5] mentioned above. Stark broadening has been used, e.g. in [7.90]. In this way, plenty of results have been accumulated about the axial profile of the plasma density averaged over the discharge cross-section. As has been mentioned in Sect. 7.2, local (at a given axial position) measurements of the plasma density by other methods (the cavity method [7.2,36], microwave interferometry [7.46,47], probe diagnostics [7.42] and optical-spectroscopy methods [7.43,82,91]) have been used for calibration or in developing diagnostic procedures.

As has been stressed in Chaps. 5 and 6, the axial density profile is a characteristic consequence of the self-consistent interdependence between the guided wave and the generated plasma. For most conditions the self-consistent axial density profile $\bar{n}(z)$ is – as expected theoretically – to good accuracy a linear function of the axial coordinate z. Figure 7.9 exhibits results for argon

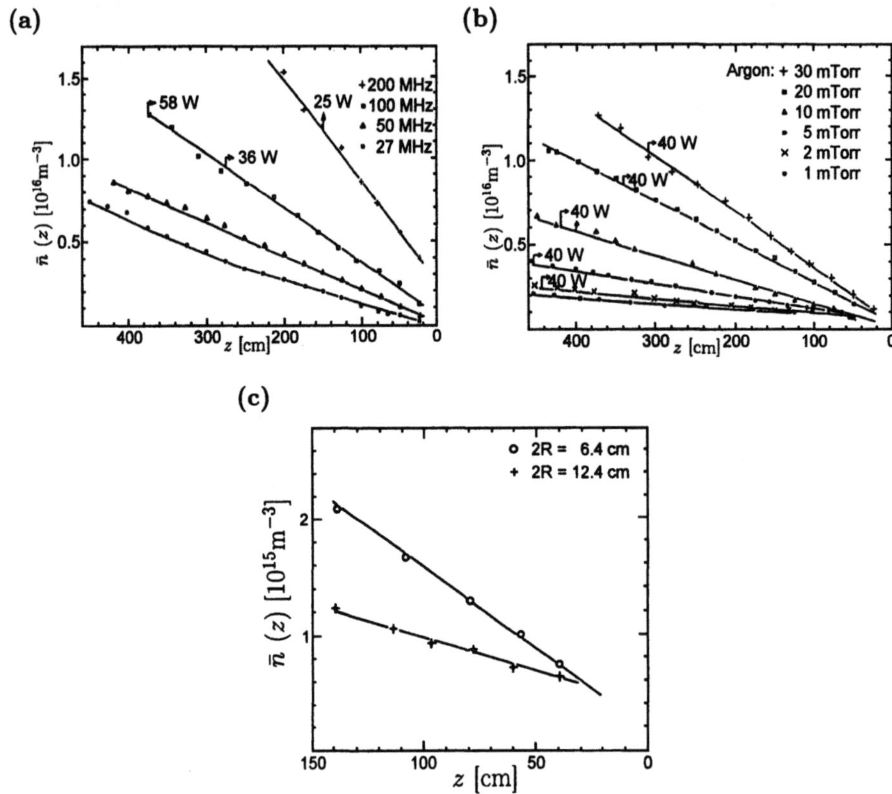

Figure 7.9. Axial distribution of the electron density $\bar{n}(z)$ with $R = 32$ mm (wall thickness 3.5 mm): **(a)** for various frequencies at 30 mtorr argon; **(b)** for various pressures at 100 MHz; **(c)** $\bar{n}(z)$ for various radii at 100 MHz and 10 mtorr argon ([7.5], Fig. 3–5)

discharges for various wave frequencies ω, pressures p and tube radii R. The predicted linear axial decrease of the plasma density is basically proved, as well as a dependence of the type $\mathrm{d}\bar{n}/\mathrm{d}z \propto \omega p/R$ for its gradient, which agrees with the result $\mathrm{d}n/\mathrm{d}z \propto \nu\omega/R$ (5.68) obtained under conditions of weak collisions and Joule heating in the plasma volume.

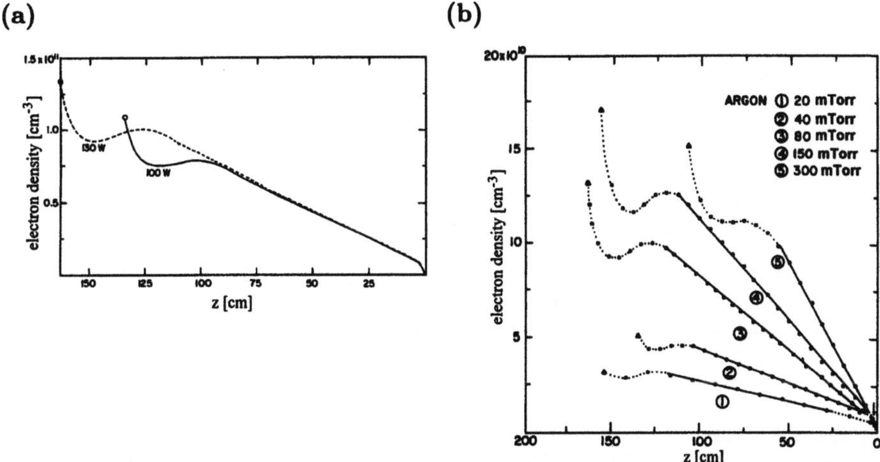

Figure 7.10. Axial density profiles in argon discharges: **(a)** $f \equiv \omega/2\pi = 360$ MHz, $2R = 2.5$ cm at $p = 80$ mtorr argon ([7.4], Fig. 2); **(b)** same conditions, but with different gas pressures; axial position z counting from the end of the column ([7.4], Fig. 20)

Figures 7.9a and 7.10a also indicate a typical behaviour of SW discharges: turning up the settings of the power source and launcher to allow higher power results in adding a higher-density part at the beginning of the discharge, yielding, in a simple manner, an increase of the discharge length with the same axial density gradient.

However, Fig. 7.10 shows that complications and deviations from the linear density profile occur close to the launcher and towards the end of the discharge. At the onset of the discharge, i.e. in the region near the wave launcher, the light emission intensity (which is assumed to be proportional to $\bar{n}(z)$) drops fast, forming a minimum, and then slightly increases in the transition to the long length of the discharge where the density profile is linear. This behaviour has usually been related to the radiation field near the launcher [7.4]. However, it can also be associated with a charged-particle flux from the launcher, which causes axial diffusion to be the mechanism ensuring self-consistency in this part of the discharge (see [7.92] and the comments in Sect. 5.1). The fast drop of the density at the end of the discharge is commented on in the next section.

Figure 7.11 shows experimental results [7.93] for the axial density profile in a discharge sustained under conditions of strong collisions ($\nu > \omega$). As has been commented on in Sect. 5.6.3, Joule heating in the plasma volume under conditions of strong collisions leads to a quadratic dependence of $\bar{n}$ on the axial coordinate.

As mentioned at the beginning of this chapter, a stronger gas flow leads to axial density gradients which can differ strongly from those in discharges at constant pressure. The experimental results presented in Fig. 7.12 show

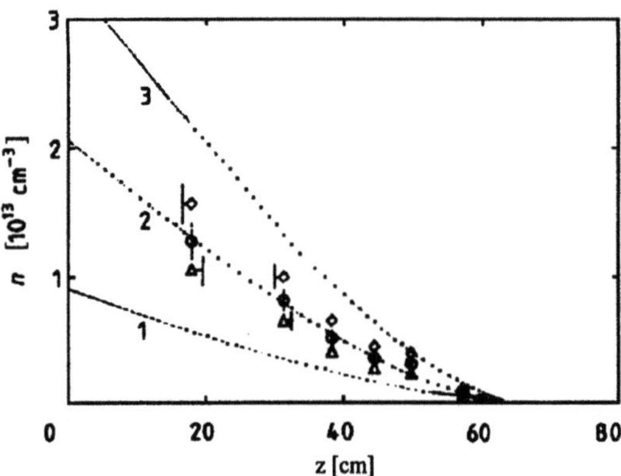

Figure 7.11. Axial density profiles in an argon discharge sustained by an SW of frequency $f \equiv \omega/2\pi = 210$ MHz; values of $\nu/\omega = 1$ ($\triangle$), 2 ($\circ$) and 3 ($\diamond$) ([7.93], Fig. 4)

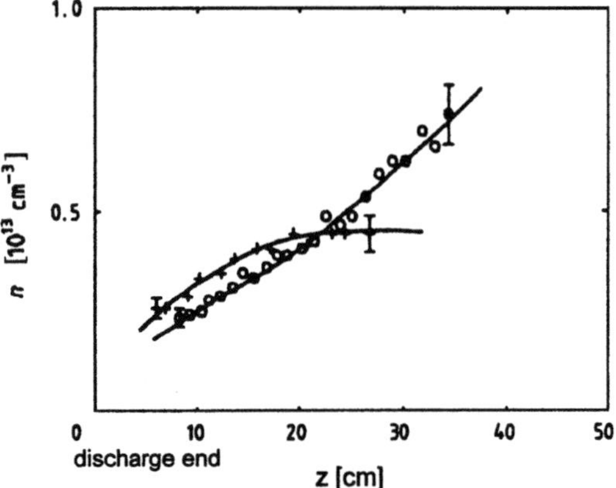

Figure 7.12. Experimental electron densities in a helium discharge at $f \equiv \omega/2\pi = 2.45$ GHz, gas flow of 33 sccm; gas flow and wave propagation in the same direction ($\circ$) and in the opposite direction ($+$) ([7.85], Fig. 4)

the influence of a finite gas flow on the axial density profiles in SW-sustained discharges.

As has been discussed in Sect. 7.2.4, the optical-spectroscopy methods are very valuable, since quantities such as spectral line-intensities I_L and excited-state population densities N_k, which are of great importance for applications of the discharges, are directly measured. Experimental data for these quantities are also important for verification of theoretical predictions and of results from kinetic modelling of the discharges. For example, on the basis of the spatial shapes of the line intensity profiles measured, conclusions can be drawn about the locality/nonlocality of the discharge regime. Depending on the discharge conditions, different shapes of the radial distribution of the spectral-line intensities I_L and of the excited-state population densities

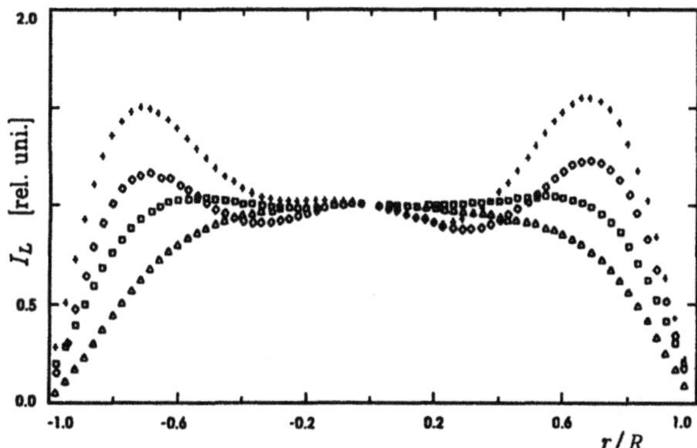

Figure 7.13. Radial distribution of the emission intensity of the ArI 549.6 nm spectral line at constant cross-section-averaged density ($\bar{n} = 1.1 \times 10^{11}\,\mathrm{cm}^{-3}$) in a SW-sustained discharge at $f \equiv \omega/2\pi = 900$ MHz and various gas pressures: $p = 50$ mtorr ($\triangle$), 150 mtorr ($\square$), 500 mtorr ($+$) and 1000 mtorr ($\diamond$) ([7.87], Fig. 11b)

N_k have been observed. In Fig. 7.13 results for the radial distribution of the ArI 549.6 nm line are depicted showing the influence of the gas pressure on the radial distribution of the emission intensity. The three-dimensional plots

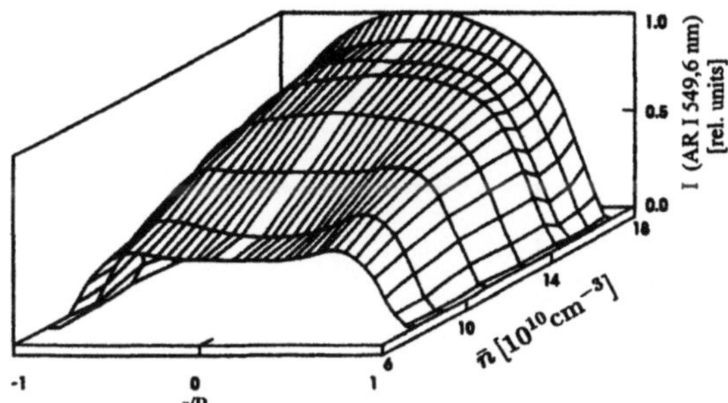

Figure 7.14. 3D plots showing the spectral-line intensity as a function of the radial position and of the cross-section-averaged plasma density; conditions as in Fig. 7.13 ([7.87], Fig. 7d)

in Fig. 7.14 show the changes in the radial distribution of the intensity of the same line with the varying plasma density along the discharge column. Figure 7.15 shows the radial density distributions of excited helium atoms for different frequencies of the wave sustaining the discharge.

Measurements of the gas temperature T_g have largely been performed with a view to a sizeable increase above room temperature with increasing power and frequency, and also with a view to its axial structure [7.94] and the influence of the gas flow (if this is not drastically reduced); see Fig. 7.16.

The difficulties in the use of the optical-spectroscopy diagnostics appear when experimental results on plasma parameters – n, T_e – are aimed at.

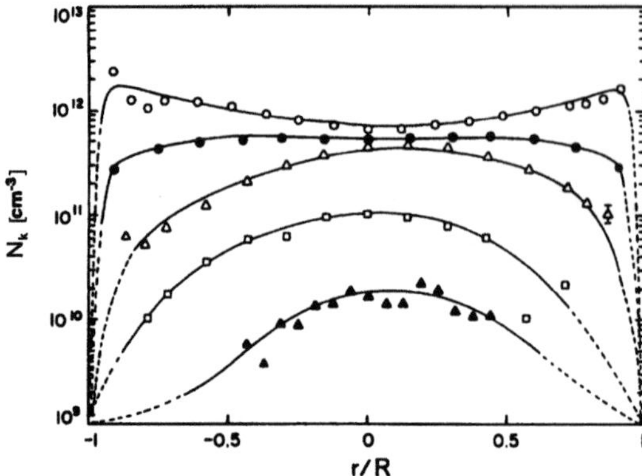

Figure 7.15. Radial distribution of the population density of the He (2^3S) excited state in a SW-sustained discharge at $p = 1$ torr and $2R = 2.6$ cm and different frequencies: $f \equiv \omega/2\pi = 2.45$ GHz (o), 900 MHz (•), 50 MHz ($\triangle$), 4 MHz ($\square$), 600 kHz ($\blacktriangle$) ([7.83], Fig. 9a)

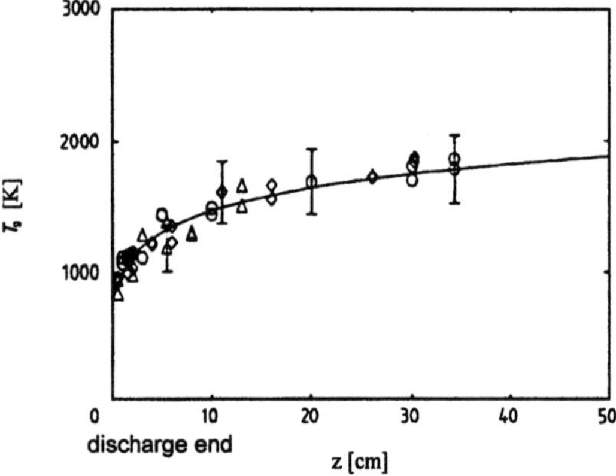

Figure 7.16. Gas temperature measured along the discharge by a Fabry–Pérot interferometer at different values of the applied power P_0 in the case of opposite directions of the gas flow and the wave propagation; gas flow 33 sccm, 2.45 GHz discharge in helium ([7.85], Fig. 3)

In this case the method becomes indirect and involves kinetic modelling of the discharge. For example, from comparison of theoretical and experimental results for the population density of excited states, the pair (n, T_e) can be obtained as fitting parameters (Fig. 7.17).

As to electron energies, EEDFs have been of interest, using probe techniques and studying the types of deviation from a Maxwellian distribution, including variations with radius, at the low degrees of ionization encountered in most HF discharges [7.42,48,52–60,95]. Attention has been paid also to axial variations [7.56]. The use of Thomson scattering, to avoid perturbation by probes, is still in the initial stages of use for low-density discharges

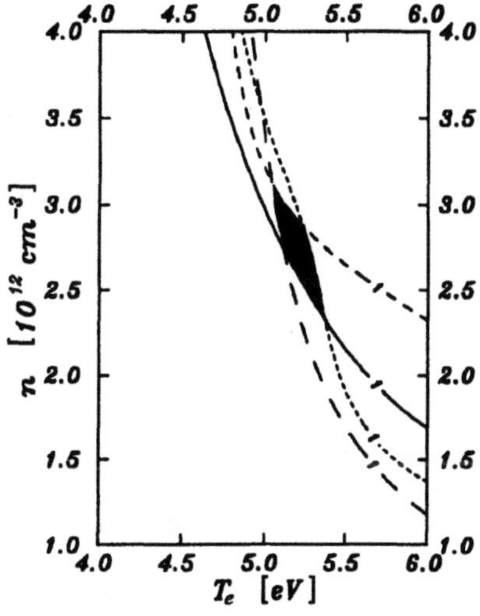

Figure 7.17. Topographic presentation of theoretical results (from kinetic modelling of a helium discharge at $f \equiv \omega/2\pi = 2.45$ GHz, $p = 600$ mtorr) for the population densities of the 2^3P (*solid curve*), 2^1P (*long-dashed curve*), 4^3S (*short-dashed curve*) and 4^1S (*dotted curve*) levels, normalized to the corresponding experimental data. The *black region* gives the intervals determined for the possible values of the plasma density n and electron temperature T_e (as a measure of the averaged energy $\bar{u}$, with $T_e = 2/3\,\bar{u}$) ([7.91], Fig. 9)

[7.96]. Studies such as [7.24,42,60] were also concerned with the mean electron energies and their dependences on discharge parameters.

An interesting feature of SW discharges is the power absorbed (and lost) per electron on average Θ, closely connected to $\langle |E|^2 \rangle$, as discussed in detail in Chap. 5. In recombination-controlled regimes of high electron density Θ is expected to vary $\propto \bar{n}(z)$ and thus to decrease along the discharge. In a diffusion-controlled regime, the axial variations of Θ are slight. The behaviour of Θ is taken up in the following subsection, together with some recent observations used for comparison with the expectations from modelling.

7.3.2 Observations Relevant to Features of Self-Consistency

Over the last few years observations have been accumulated which reflect more specific features of current theory and modelling. Such findings and some recent evidence relevant to some more detailed aspects of theory and modelling described in the previous chapters will be summarized in this subsection.

The expected non-Maxwellian character of the EEDFs shows up in probe measurements, which have been made feasible for SW-sustained discharges in spite of the difficulties mentioned in Sect. 7.2. Figure 7.18 gives results clearly depicting a lowering in the tail of the EEDF – compared with a Maxwellian distribution – which is caused by heavy inelastic energy losses that are not sufficiently counterbalanced by electron–electron collisions in the case of a low degree of ionization ($\ll 10^{-2}$). Results for different radial positions are plotted. With increasing radius r and increasing ambipolar potential $|\Phi|$ (determined by means of probes) the curves are shifted. The normalization follows the radially decreasing value of the electron density. Thus, it becomes evident that the case $r = 0$ constitutes an envelope for all curves with satisfactory

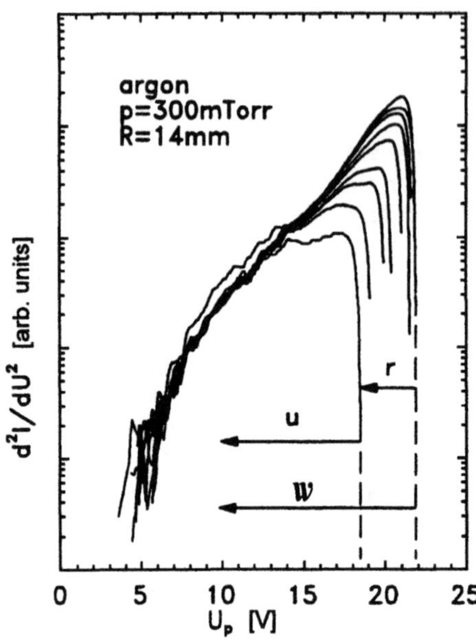

Figure 7.18. Measured EEDFs for various radii; the envelope corresponds to expectation for the nonlocal scenario; 200 MHz, $R = 14$ mm, $p = 300$ mtorr argon. $\mathcal{W} = 0$ at probe voltage $U_p \approx 22$ V. The curves and the points $u = 0$ shift with r to the left ([7.58], Fig. 2)

accuracy, as expected from the nonlocal approach to the EEDF discussed in Chap. 6. This indicates that this scenario is at least a useful approximation.

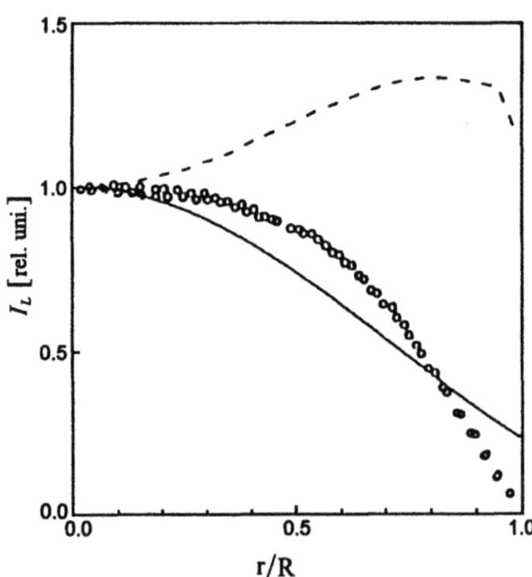

Figure 7.19. Tomographically radially resolved line intensity I_L (of the argon 594.6 nm spectral line) from [7.87], and values expected from local (*solid curve*) and nonlocal (*dashed curve*) scenario; 600 MHz, $R = 26$ mm, $p = 50$ mtorr, $n = 1.1 \times 10^{17}\,\mathrm{m}^{-3}$ ([7.97], Fig. 5)

Figure 7.19 demonstrates a radial profile of line emission, obtained by tomography [7.87]. The measured values are much closer to the behaviour (purely monotonic) which results from a strictly nonlocal approximation than to the rather nonmonotonic behaviour predicted by a strictly local approximation. The point made with regard to Fig. 7.18 is emphasized here: the process that determines the EEDF clearly does not locally reflect the radial inhomogeneity of the HF electric-field intensity, which, in turn, is accentuated by the radial density nonuniformity. Figure 7.19 also clearly demonstrates that emission measurements are sensitive to deviations from a strictly non-local approximation, which appears quite reasonable for the situation under consideration, though leaving room for improvement. The growing deviations from observation with increasing pressure are taken up below.

The measurements presented next also reflect influences connected with the radial density nonuniformity, in situations where the drop from the central value of $n(r)$ at $r = 0$ to the value at the wall (actually at the sheath) is pronounced enough to allow the occurrence of plasma resonance effects at $\omega_p(r_r) = \omega$ in the outer regions of the plasma. As pointed out in the previous chapters, such situations are associated with peaks in the radial component of the electric field $|E_r|$, which move slightly away from the wall towards the discharge end. Direct evidence of this can be obtained with a special probe/antenna [7.12,47,61]. Figure 7.20 plots contours of $|E_r|$ measured in a

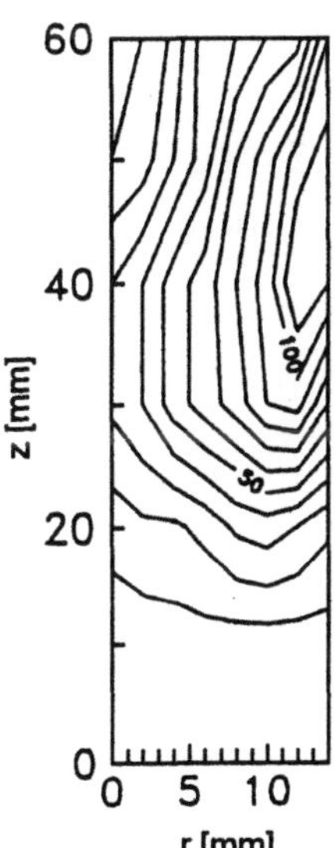

Figure 7.20. Contour plots of $|E_r|$ (relative units) at distance z from the visible end of an argon discharge; $R = 15$ mm, with a reduced value in the launcher ([7.47], Fig. 6.3.2)

discharge of rather large diameter, yielding the expected trend towards the discharge end, in spite of the finite experimental spatial resolution due to the inherent perturbation by the antenna inside the plasma. Recent Japanese experiments with a planar SW device confirm the $|E_r|$ peaks and their expected behaviour [7.98].

In earlier work [7.99] a peak in the space-to-floating potential difference of a radially movable probe in a cylindrical SW plasma column was observed, suggesting a local $\omega_p(r_r) = \omega$ resonance at the corresponding radial position.

The occurrence of the above-mentioned resonance in the radial structure of $|E_r|$ cannot be expected to be reflected in the radial density profile as a back reaction due to locally enhanced ionization as long as the process determining the EEDF is a nonlocal one to a good approximation. For larger products pR, when a transition to a local scenario occurs, a structuring in $n(r)$ may become possible, though increased collisionality can be expected to smear out the field peaks. Moreover, the diffusion process participating in the determination of the density profile also tends to smear out the structuring, whereas in the excitation and line emission radial profiles the latter effect of smearing out is absent. Nevertheless, a local scenario can be reached at modest pressures if the radial dimensions become large enough. This can be accomplished in the arrangements, previously mentioned, which may be termed "inverse" SW devices, for which the EM power flux is essentially provided within an inner glass tube and the plasma created is situated outside the tube, filling a large volume. For such situations $|E_r|$ peaks, accompanied by peaks in the electron temperature (measured by probes) and even with indications of structuring in the observed electron density, have been reported for the positions given by $\omega_p(r_r) = \omega$ [7.25]. Figure 7.21 (see also [7.25]) depicts measured E_r field

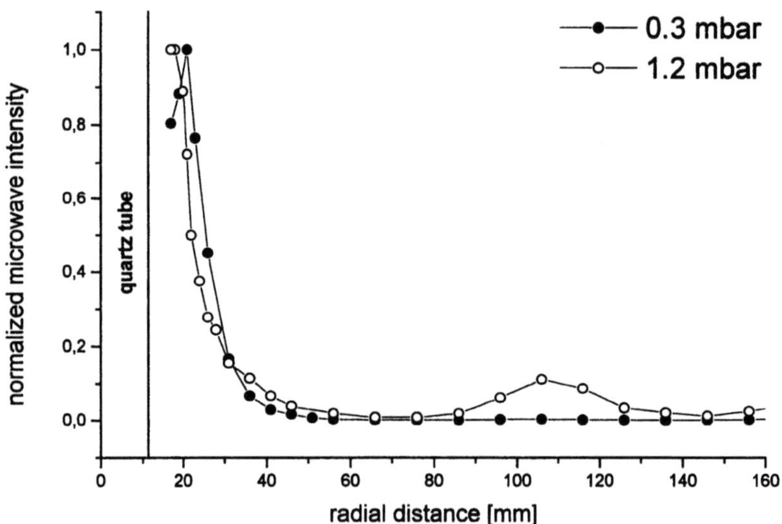

Figure 7.21. Measured values of E_r field intensity versus radius for an H_2 plasma in an "inverse" SW arrangement (duo-plasma line) at 2.45 GHz: for 0.3 mbar ($\bullet$) and for 1.2 mbar ($\circ$), (see [7.25])

intensities with two maxima. These occur when the electron density decreases from a maximum value in the external region towards both the quartz wall and larger radial distances, passing through the situation $\omega_p(r_r) \approx \omega$ in both cases.

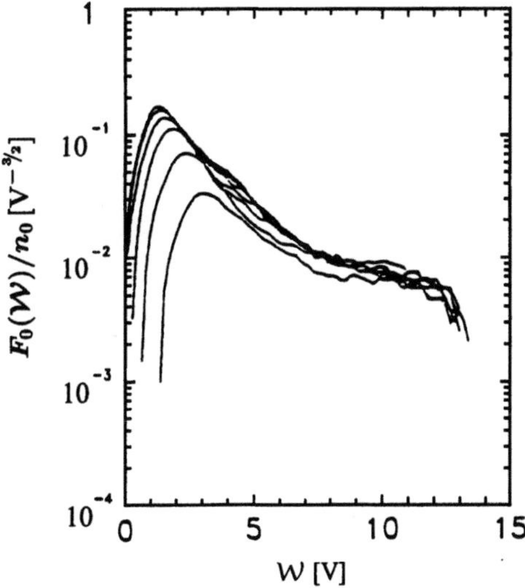

Figure 7.22. Measured EEDFs, normalized versus the central density n_0, at different radial positions, at about 10 cm from the discharge end versus total energy $\mathcal{W}$; $f \equiv \omega/2\pi = 2.45$ GHz, $R = 15$ mm, $p = 0.3$ torr argon. ([7.47], Fig. 7.3.6)

Peaks of the $|E_r|$ field component are expected to result in appreciable noncollisional SW damping and electron heating [7.100], direct evidence of which is not easy to observe. Since, however, this type of heating tends to lead to a bump in the EEDF at energies near the inelastic threshold, accentuated for large pR (Chap. 6) when $\nu/\omega < 1$, the results in Fig. 7.22 [7.12,47,61] may be considered as at least consistent with expectations. It should be stressed that the bumps discernible in the EEDFs of Fig. 7.22 show up only towards the discharge end and not in sections of the discharge where no peaks of $|E_r|$ due to $\omega_p(r_r) = \omega$ can be present. Noncollisional damping was discussed in [7.12,101], together with experimental evidence from other types of HF discharge. In particular, in [7.102] clear evidence is reported of collisionless power transfer in inductively coupled discharges.

Indirect evidence for the occurrence of radial resonance effects in SW discharges can be gained by observing consequences of the phenomenon in the wave dispersion behaviour and in the axial density profile. Steep radial density gradients may lead to appreciable deviations towards larger $\omega/\overline{\omega_p}$ from the phase curves in homogeneous, collisionless plasma waveguides (such as considered in Fig. 7.6). In [7.47] such deviations were indicated, as shown in Fig. 7.23 for relatively large ν/ω. When $\omega_{p_{\text{wall}}} < \omega$ can be realized, the dispersion behaviour is even more complicated associated with the "type 3" branch shown in Fig. 4.11. Such a trend can be seen from measurements [7.60] in nitrogen SW discharges at lower $\nu/\omega = 0.5$ (Fig. 7.24): the measurements for $\omega/\overline{\omega_p} \geq 0.5$ are in the range dominated by the radial plasma resonance effect. They seem to agree with the behaviour of βR which has been calculated

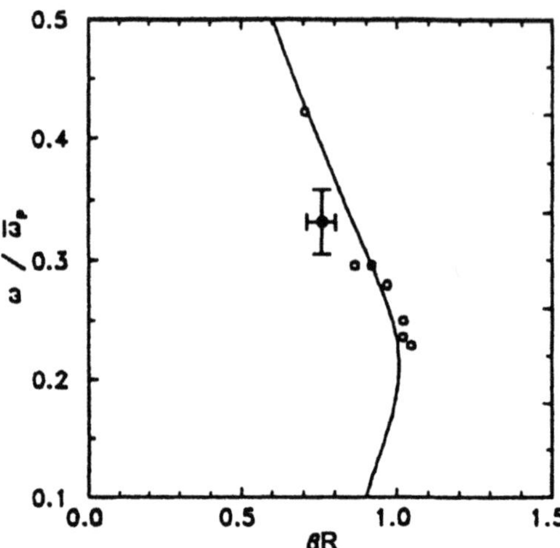

Figure 7.23. Measured phase diagrams (○). *Solid curve*: WKB calculations for 2.45 GHz, $a = 15$ mm, $\nu/\omega = 1$ (argon), $\mu = 2.3$ ([7.47], Fig. 6.2.4)

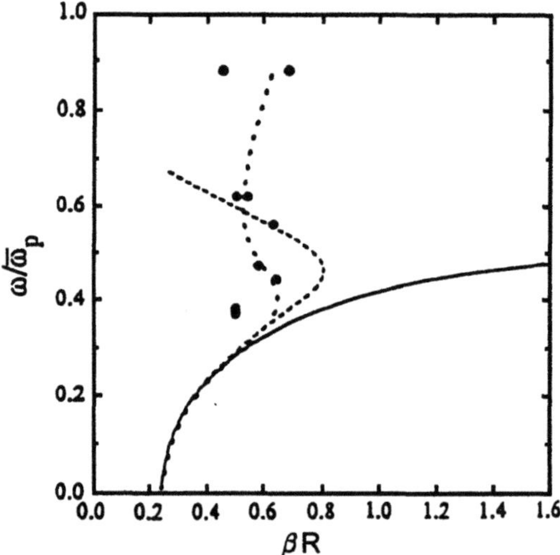

Figure 7.24. Measured phase diagrams (●) for 500 MHz, $R = 23$ mm, $\nu/\omega = 0.5$ (nitrogen). WKB calculation for $\mu = 0$ (*dashed line*) and $\mu = 2.1$ (*dotted line*). *Solid curve* for $\mu = 0$, $\nu/\omega = 0$. See also [7.60]

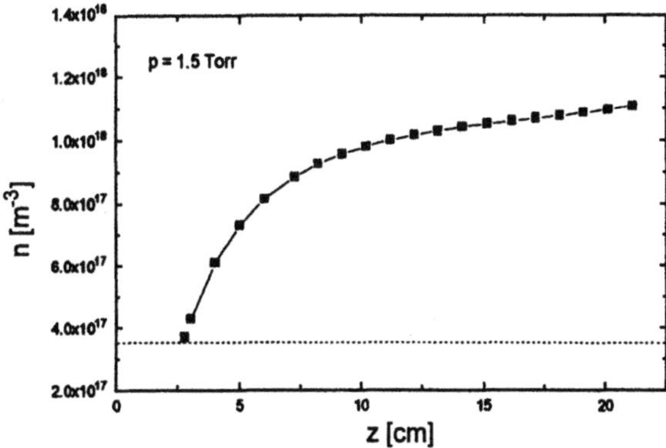

Figure 7.25. Axial profile of plasma density $\bar{n}$ versus z determined from 9.46 MHz radiometry measurements; $f \equiv \omega/2\pi = 2.45$ GHz, $p = 200$ Pa argon, $R = 14$ mm ([7.78], Fig. 1)

on the basis of a radial density profile obtained from probe measurements. This behaviour implies that for about the last 6 cm of the discharge (total length $\approx$ 14 cm) radial plasma resonances are present.

The presence of such local resonances can be expected to steepen the axial density profile to a faster descent towards the discharge end. This is, of course, detectable only in suitable situations and with sufficient resolution of the density determination. In [7.12,59] 2 mm microwave measurements revealed a descent steeper than at the higher densities of the preceding discharge section, which had a linear axial density variation. The evidence for this seems to be clearer in Fig. 7.25, exhibiting recent results from the application of microwave radiometry. More details were reported in [7.78]. It should be pointed out that in this situation the interesting region towards the discharge end extends over a length of many times the radius, in contrast to the situation of Fig. 7.24, for example, where the length of this region amounts only to a few times the radius.

The effects just considered are largely connected to radial density nonuniformity. There is also an effect connected to the axial density gradient, considered in Chap. 5: the variation of the relative roles of stepwise contributions to ionization with varying axial density $\bar{n}(z)$. It can be expected that for the maintenance of the discharge at higher $\bar{n}(z)$ less electric-field strength $|E|^2$ (and a lower Θ) is needed, though the resultant field decrease and the associated decrease in mean electron energy may be relatively small. The measurements of Θ at relatively low pressures and ν/ω depicted in Fig. 7.26 have indicated, for most of the discharge length, a slight increase of Θ which becomes stronger towards the discharge end [7.103].

However, in the excitation of line emission the effect of the variation of Θ due to stepwise ionization may become clearly visible. Recent observations [7.104,105] of the line emission from argon discharges as a function of $\bar{n}(z)$

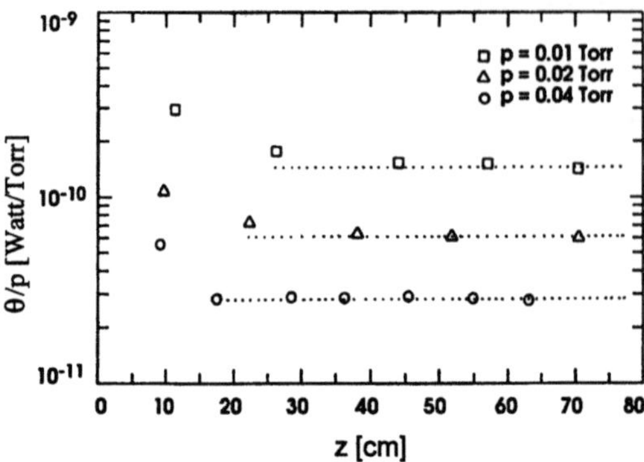

Figure 7.26. Measured Θ/p values along an SW-produced argon plasma column, as a function of axial distance from the end of the column, at three different gas pressures; 200 MHz, $R = 13$ mm, $d = 1$ mm. ([7.103], Fig. 11)

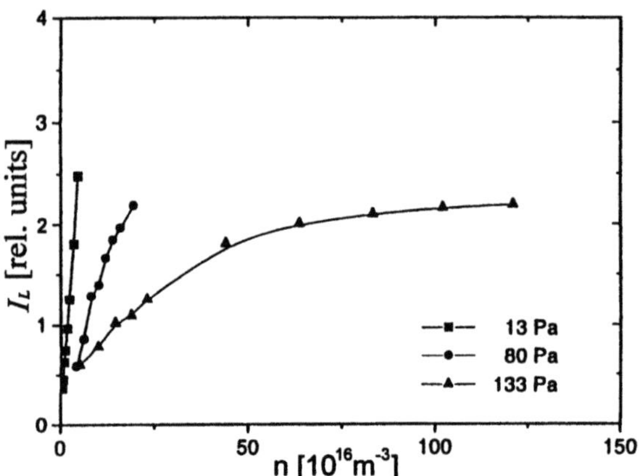

Figure 7.27. Measured H_α line intensity versus electron density n along a discharge ([7.105], Fig. 10)

reveal a linear behaviour at lower $\bar{n}(z)$, but with a variation weaker than a linear one at larger $\bar{n}(z)$, which can be explained by a decreasing mean electron energy $\bar{u}$ (Chap. 5). Since the population density of the emitting excited state is governed by a balance between collisonal excitation and spontaneous decay, the line intensity can be expected to be essentially proportional to $\bar{n}(z) \exp(-U_*/\bar{u})$, with U_* being the excitation energy and $\bar{u}$ some averaged electron energy. An increased contribution from step ionization allows a lower $\bar{u}$ to be sufficient for discharge maintenance. Here $\bar{n}(z)$ was determined by radiophysics methods, taking into account effects of the energy-dependent collision frequency. In Fig. 7.27 H_α line emission, due to very weak admixtures of hydrogen in the argon discharge, is depicted. For these admixtures the influence of self-absorption can be disregarded. Collisional depopulation processes are unimportant. The influence of a decreased mean electron energy $\bar{u}$ becomes quite apparent with increasing $\bar{n}$. It can be estimated that a decrease of $\bar{u}$ by 10% – indicative of roughly a 30% lowering in $|\boldsymbol{E}|^2$ or Θ – leads to a lowering of the line intensity by about 30%. In view of the large values of ν/ω at 1 torr, the observations are unlikely to be connected to an increase in $\bar{u}$ towards the discharge end influenced by a system resonance phenomenon.

7.4 Open Questions and Related Areas of Research

Without doubt the experimental work reported above calls for broadening and extension. On the other hand, theoretical progress on various points is desirable to allow better understanding of the observations and checks between experiment and theory.

One of these points concerns the very end of the discharge where inclusion of additional losses, in particular from axial diffusion, and a full two-dimensional treatment appear necessary. Another aspect is the need for more exact theoretical investigation of the transition regime between the nonlocal and local scenarios in the determination of the EEDF. The available observations clearly show a transition from virtually monotonic radial emission profiles to nonmonotonic ones with growing gas pressure.

There is current activity on topics which are beyond the scope of this book, but which are attracting increasing attention. Such areas include

- stability problems
- pulsed discharge operation
- self-organization of discharges at higher gas pressures and filamentation phenomena
- discharges at atmospheric pressure
- discharges with production of a magnetized plasma
- sheath effects, particularly in metal-bounded systems.

7.5 Applicational Aspects

Waveguided plasma systems have already been applied for many purposes, as pointed out in the Preface. Recent reviews on applications are contained in [7.9,106] and, in particular, [7.10]. In the latter, more details are given of the aspects mentioned below, as well as specific references. Owing largely to the flexibility of SW-sustained plasmas concerning the operating conditions – including the choice of almost any shape of plasma vessel and the ease of impedance matching – the range of applications is rather wide.

Various uses of the *light emission* of these discharges are of interest. They permit applications such as

- detectors for elemental analysis
- lasers
- lighting sources.

In the first case the discharges are applied in connection with optical emission and absorption spectroscopy. The laser applications are concerned with hydrogen fluoride chemical lasers, high-frequency pumped He–Ne lasers and argon ion lasers. As to lighting sources, a recent extensive review on electrodeless high-frequency sources may be referred to [7.107].

The situation is similar concerning the realization of *particle sources*. Interesting examples are

- nitrogen and hydrogen atom sources
- high-velocity beam sources
- ion sources
- metastable-beam sources.

A high-velocity oxygen neutral beam was employed for spacecraft materials testing, as were plasma jet systems. Krypton ion sources designed for remote space applications use a surfatron as a coupler, whereas argon ion sources employ a holey-plate system. A metastable argon beam source developed in Princeton [7.108] is mainly utilized in Penning ionization spectroscopy.

A currently expanding range of applications is the area of surface treatment, for tasks such as *deposition of films*. These include

- diamond films
- polymer and similar films
- amorphous silicon films.

The peculiarities and advantages of using SW discharges for diamond films were discussed in [7.109]. Details of depositing hydrocarbon and fluorocarbon films by means of SW plasmas were described in [7.110]. A circularly bent so-called "SLAN" system has been employed, mainly for the production of scratch-resistant optical-grade films [7.111]. Whether the SLAN system operates in an SW-like mode is not yet quite clear. Silicon oxycarbide films have been produced, as well as silicon nitride films. Two different systems for depositing amorphous silicon (e.g. for low-cost solar cells) were described in [7.112].

Numerous applications for *etching and ashing* include

- etching of metals
- etching of polymers
- stripping and surface modification.

Methods of ion-assisted plasma etching of metals have been realized, as well as etching of polymers with mixtures of O_2 and CF_4. Ashing, cleaning and modification of surfaces have been performed in conjunction with oxygen, for instance using a reactor with a surfaguide launcher or dielectric slabs illuminated by antennae. The SLAN systems mentioned above were applied in various cases.

There is an increasing interest in deploying SW discharges for *detoxification* of environmentally detrimental agents, particularly in the microelectronics industry. Moreover, *sterilization*, in particular for medical purposes, seems a promising application [7.113].

The developments in the last three classes of applications in particular increasingly add phenomena of *reactive plasmas*, which can both complicate and interestingly modify the behaviour of SW discharges. Of course, this behaviour remains governed by the interplay of electrodynamics with "standard" discharge plasma physics. However, in many applicational situations aspects of *dusty plasmas* and of formation of *particles* of various sizes will enter, e.g. [7.114]. Information on reactive plasmas was given in [7.115]. For such discharges additional methods of diagnostics become important, such as actinometry, chemical luminescence, filtration, optogalvanic methods and, increasingly, mass spectrometry [7.115]. Various modern laser methods may be used to obtain information on molecular processes, for example laser-induced fluorescence, coherent anti-Stokes Raman scattering (CARS) and similar laser procedures [7.116].

The development of applications indicated above encompasses a trend to SW discharges with larger dimensions of vessels and plasmas. As pointed out in [7.117], such developments may be classed into

- tubular discharges
- planar discharges.

The tubular discharges are relatively short, but the wave is excited over a circumferential interface closing on itself with a diameter large compared with the axial dimension. The SLAN (slot antennae) system touched on already belongs to this class of systems, consisting of a bend in a rectangular waveguide with inwardly positioned radiating slots, coupled to a rectangular waveguide fed from a capacitive coupler. But rectangular waveguides with a diameter above the free-space wavelength at 2.45 GHz and an array of slots in the inward wall can be fed from a standard waveguide with a plunger. Standard-type launchers with the usual circular coupling gap (waveguide surfatron, Ro-box, surfaguide) can be designed with oversized diameters (partly operating in an azimuthally nonsymmetric mode). Gap-type launchers have been applied in variants of "Okamoto cavities" to achieve large diameters [7.118]. Systems with an abrupt transition from a tube of small diameter to one of large diameter are suited for deposition work [7.109].

To the planar configurations belong the systems already mentioned with dielectric slabs irradiated by antennae, working with standing surface waves as does a planar SLAN system. Standing waves are again employed in developments using dielectric discs, centrally fed from a tapered waveguide or circumferentially fed from a radial line. There are systems at Nagoya University, Japan, with dielectric discs top-fed from inclined slots [7.119]; another variant was described in [7.120]. Similar discs top-fed from circular slots were described in [7.121]. Fully metallic structures were studied in [7.122] and [7.123]. Finally, the "inverse SW" arrangement mentioned in [7.25] should be pointed out as also being suitable and used for large-area applications.

References

Chapter 2

2.1   V.L. Ginzburg: *The Propagation of Electromagnetic Waves in Plasmas* (Pergamon, Oxford 1964)

2.2   A.F. Alexandrov, L.S. Bogdankevich, A.A. Rukhadze: *Principles of Plasma Electrodynamics*, Springer Ser. Electrophys., Vol. 9 (Springer, Berlin, Heidelberg 1984)

2.3   Th.H. Stix: *Waves in Plasmas* (American Institute of Physics, New York 1992)

2.4   L. Brillouin: *Wave Propagation and Group Velocity* (Academic Press, New York 1960)

2.5   I.P. Shkarofsky, T.W. Johnston, M.P. Bachynski: *The Particle Kinetics of Plasmas* (Addison-Wesley, London 1966)

2.6   V.E. Golant, A.P. Zhilinsky, I.E. Sakharov: *Fundamentals of Plasma Physics* (Wiley, New York 1980)

2.7   B.M. Smirnov: *Physics of Weakly Ionized Gas* (Nauka, Moscow 1978) [in Russian]

2.8   Yu.M. Aliev, I.D. Kaganovich, H. Schlüter: Phys. Plasmas **4**, 2413 (1997)

2.9   I.B. Bernstein, T. Holstein: Phys. Rev. **94**, 1475 (1954)

2.10 C.M. Ferreira, J. Loureiro, A. Ricard: J. Appl. Phys. **57**, 82 (1985)

2.11 U. Kortshagen, A. Shivarova, E. Tatarova, D. Zamfirov: J. Phys. D **27**, 301 (1994)

2.12 C.M. Ferreira, J. Loureiro: J. Phys. D **16**, 2471 (1983)

2.13 A.D. MacDonald: *Microwave Breakdown in Gases* (Wiley, New York 1966)

2.14 L.M. Biberman, V.S. Vorob'ev, I.T. Iakubov: *Kinetics of Nonequilibrium Low–Temperature Plasmas* (Consultants Bureau, New York 1987)

2.15 Yu.P. Raizer: *Gas Discharge Physics* (Springer, Berlin, Heidelberg 1991)

2.16 S.M. Levitskii: *Book of Problems and Calculations in Physical Electronics* (Naukova Dumka, Kiev 1964) [in Russian]

2.17 J.H. Ingold: Z. Physik **233**, 89 (1970)

2.18   Yu.M. Aliev, M. Georgieva, S. Grosse, H. Schlüter, A. Shivarova: Contrib. Plasma Phys. **36**, 573 (1996)
2.19   E.W. McDaniel: *Collision Phenomena in Ionized Gases* (Wiley, New York 1964)
2.20   W. Schottky: Phys. Z. **25**, 635 (1924)
2.21   E. Spenke: Z. Phys. **127**, 221 (1950)
2.22   M. Pahl: Z. Naturforschg. **12a**, 632 (1957)
2.23   F.G. Bass, Yu.G. Gurevich: Sov. Phys.-Usp. **14**, 113 (1971)
2.24   F.G. Bass, Yu.G. Gurevich: *Hot Electrons and Strong Electromagnetic Waves in Plasmas of Semiconductors and Gas Discharges* (Nauka, Moscow 1975) [in Russian]
2.25   A.V. Gurevich, A.B. Shvartsburg: *Nonlinear Theory of Radio Wave Propagation in the Ionosphere* (Nauka, Moscow 1973) [in Russian]

Chapter 3

3.1   L.D. Landau, E.M. Lifshitz: *Electrodynamics of Continuous Media* (Pergamon, New York 1960)
3.2   A.W. Trivelpiece, P.W. Gould: J. Appl. Phys. **30**, 1784 (1959)
3.3   Yu.B. Fainberg, M.F. Gorbatenko: ZhTF **29**, 549 (1959)
3.4   S.M. Levitskii, N.S. Branchuk: Izv. VUZ-Radiofizika **4**, 1078 (1961)
3.5   A.A. Oliner, T. Tamir: J. Appl. Phys. **33**, 231 (1962)
3.6   T. Tamir, A.A. Oliner: Proc. IEEE **51**, 317 (1963)
3.7   R.H. Ritchie: Progr. Theor. Phys. **29**, 607 (1963)
3.8   I. Akao, Y. Ida: J. Appl. Phys. **34**, 2119 (1963)
3.9   V.L. Granatstein, S.P. Schlesinger, A. Vigants: IEEE Trans. Antenna Prop. **AP-11**, 489 (1963)
3.10   L.S. Napoli, G.A. Swartz: Phys. Fluids **16**, 918 (1963)
3.11   F.W. Crawford, G.S. Kino, S.A. Self, J. Spalter: J. Appl. Phys. **34**, 2186 (1963)
3.12   Yu.A. Romanov: Izv. VUZ-Radiofizika **7**, 242, 828 (1964)
3.13   I. Akao, Y. Ida: J. Appl. Phys. **35**, 2565 (1964)
3.14   R.N. Carlile: J. Appl. Phys. **53**, 1384 (1964)
3.15   M.F. Gorbatenko, V.I. Kurilko: ZhETF **34**, 1136 (1964)
3.16   V.L. Granatstein, S.P. Schlesinger: J. Appl. Phys. **35**, 2846 (1964)
3.17   L.S. Napoli, G.A. Swartz, H.T. Wexler: Phys. Fluids **8**, 1142 (1965)
3.18   B. Kerzar, P. Weissglas: Electron. Lett. **1**, 43 (1965)
3.19   B. Kerzar, P. Weissglas: J. Appl. Phys. **36**, 2479 (1965)
3.20   B. Kerzar, K. Abrahamsen, P. Weissglas: Appl. Phys. Lett. **7**, 155 (1965)
3.21   V.L. Granatstein, S.P. Schlesinger: J. Appl. Phys. **36**, 3503 (1965)
3.22   B. O'Brien, R.W. Gould, J. Parker: Phys. Rev. Lett. **14**, 630 (1965)

3.23   P. Diament, V.L. Granatstein, S.P. Schlesinger: J. Appl. Phys. **37**, 1771 (1966)

3.24   B. Aničin: Brit. J. Appl. Phys. **17**, 117 (1966)

3.25   V.I. Miroshnichenko: ZhTF **36**, 1008 (1966)

3.26   S.F. Paik, R.J. Briggs, J.M. Apsechuk: J. Appl. Phys. **37**, 2475 (1966)

3.27   B.C. Gregory: Int. J. Electron. **20**, 495 (1966)

3.28   P. Leprince, J. Pommier: Proc. IEEE **113**, 588 (1966)

3.29   B.B. O'Brien Jr.: Plasma Phys. **9**, 369 (1967)

3.30   Y. Akao, I. Ida: Jpn. J. Appl. Phys. **6**, 525 (1967)

3.31   V.L. Granatstein, P. Korn, A. Ojo, S.P. Schlesinger: J. Appl. Phys. **38**, 1969 (1967)

3.32   B.C. Gregory, G. Mourier: Ann. Radioélectricité **22**, 121 (1967)

3.33   I. Zhelyazkov, I. Stoychev, N. Nikolov: Electron. Lett. **3**, 253 (1967)

3.34   R.N. Carlile, H.W. Swinford: J. Appl. Phys. **39**, 3268 (1968)

3.35   Yu.A. Romanov: Izv. VUZ-Radiofizika **11**, 1533 (1968)

3.36   B.A. Aničin: Phys. Lett. **27A**, 56 (1968)

3.37   B.A. Aničin: Fizika **1**, 69 (1968)

3.38   R.C. Ajmera, K.E. Lonngren: J. Appl. Phys. **39**, 2265 (1968)

3.39   R.L. Guernsey: Phys. Fluids **12**, 1852 (1969)

3.40   V.M. Ristic, S.A. Self, F.W. Crawford: J. Appl. Phys. **40**, 5244 (1969)

3.41   J.B. McBride, P.K. Kaw: Phys. Lett. **33A**, 72 (1970)

3.42   A.V. Rezvov: Izv. VUZ-Radiofizika **8**, 688 (1970)

3.43   D.L. Landt, K.E. Lonngren: Phys. Lett. **33A**, 170 (1970)

3.44   E.H. Klevans, J.D. Mitchell: Phys. Fluids **13**, 2721 (1970)

3.45   D.B. Ilić, B.A. Aničin: Int. J. Electron. **28**, 41 (1970)

3.46   B.A. Aničin, D.B. Ilić: J. Appl. Phys. **41**, 2737 (1970)

3.47   A.V. Rezvov: ZhTF **41**, 465 (1971)

3.48   O. Demokan, H.C.S. Hsuan, K.E. Lonngren, B.A. Aničin: Plasma Phys. **13**, 29 (1971)

3.49   B.A. Aničin, D.B. Ilić: Fizika **3**, 109 (1971)

3.50   S.M. Levitskii, V.I. Burykin: Radiotekhnika i elektronika **19**, 1894 (1971)

3.51   H.C. Barr, T.J.M. Boyd: J. Phys. A **5**, 1108 (1972)

3.52   A.N. Kondratenko: ZhTF **42**, 743 (1972)

3.53   B.A. Aničin, V.M. Babović, K.E. Lonngren: J. Plasma Phys. **7**, 403 (1972)

3.54   S. Nonaka, Y. Akao: J. Appl. Phys. **44**, 644 (1973)

3.55   P.C. Clemmou, J. Elgin: J. Plasma Phys. **12**, 91 (1974)

3.56   A. Shivarova: J. Phys. A **7**, 881 (1974)

3.57   D.L. Landt, H.C.S. Hsuan, K.E. Lonngren: Plasma Phys. **16**, 407 (1974)

3.58   T. Stoychev, A. Shivarova, V. Raychev: Phys. Lett. **94A**, 249 (1974)

3.59   K.E. Lonngren, H.C.S. Hsuan, D.L. Landt, C.M. Burde, G. Joyce, I. Alexoff, W.D. Jones, H.J. Doucet, A. Hirose, H. Ikezi, S. Aksornkitti, M. Widner, K. Estabrook: IEEE Trans. **PS-2**, 93 (1974)

3.60   S.M. Levitskii, Yu.I. Burykin: Radiotech. Electron. **19**, 1894 (1974)

3.61   M. Fuse, S. Ichimaru: J. Phys. Soc. Jpn. **38**, 559 (1975)

3.62   L.J. Gagliardi: Phys. Fluids **18**, 73 (1975)

3.63   A. Shivarova, T. Stoychev, S. Russeva: J. Phys. D **8**, 383 (1975)

3.64   P. Heymann: Beitr. Plasma Phys. **15**, 137 (1975)

3.65   B.A. Aničin, V.M. Babović: Fizika **7**, 71 (1975)

3.66   Yu.I. Burykin, S.M. Levitskii, V.G. Martinenko: Radio Eng. Electron. Phys. **20**, 86 (1975)

3.67   V. Atanassov, I. Zhelyazkov, A. Shivarova, Zh. Ghenchev: J. Plasma Phys. **16**, 47 (1976)

3.68   A. Shivarova, I. Zhelyazkov: Bulg. J. Phys. **5**, 554 (1977)

3.69   J.H. Hopps, W.L. Waldron: Phys. Rev. A **15**, 1721 (1977)

3.70   A. Shivarova, I. Zhelyazkov: Z. Naturforsch. **33a**, 261 (1978)

3.71   H. Prinzler, P. Heymann: Beitr. Plasma Phys. **18**, 171 (1978)

3.72   B. Kampmann: Z. Naturforsch. **34a**, 414 (1979)

3.73   B. Kampmann: Z. Naturforsch. **34a**, 423 (1979)

3.74   J.M. Larsen, F.W. Crawford: Int. J. Electron. **46**, 577 (1979)

3.75   V.I. Baibakov, Yu. Kistovich: Fiz. Plazmy **7**, 192 (1981)

3.76   N.A. Kroll, A.W. Trivelpiece, *Principles of Plasma Physics* (McGraw-Hill, New York 1973)

3.77   A.F. Alexandrov, L.S. Bogdankevich, A.A. Rukhadze: *Principles of Plasma Electrodynamics*, Springer Ser. Electrophys., Vol. 9 (Springer, Berlin, Heidelberg 1984)

3.78   A.W. Trivelpiece, *Slow-Wave Propagation in Plasma Waveguides* (San Francisco Press, San Francisco 1967)

3.79   A.N. Kondratenko: *Plasma Waveguides* (Atomizdat, Moscow 1976) [in Russian]

3.80   A. Shivarova, I. Zhelyazkov: Plasma Phys. **20**, 1049 (1978)

3.81   M. Moisan, A. Shivarova, A.W. Trivelpiece: Plasma Phys. **24**, 1331 (1982)

3.82   A. Shivarova, I. Zhelyazkov: in *Electromagnetic Surface Modes*, ed. by A.D. Boardman (Wiley, Chichester 1982), p. 465

3.83   E. Burstein, F. de Martini (eds.): *Polaritons* (Pergamon, New York 1974)

3.84   A.D. Boardman (ed.): *Electromagnetic Surface Modes* (Wiley, Chichester 1982)

3.85   P. Halevi (ed.): *Spatial Dispersion in Solids and Plasmas* (Elsevier, Amsterdam 1992)

3.86   J. Marec, E. Bloyet, M. Chaker, P. Leprince, P. Nghiem: in *Electrical Breakdown and Discharges in Gases*, ed. by E.E. Kuhnhardt, L.H. Luessen (Plenum, New York 1981) p. 347

3.87   M. Moisan, C.M. Ferreira, Y. Haylaoui, D. Henry, J. Hubert, R. Pantel, A. Ricard, Z. Zakrzewski: Rev. Phys. Appl. **17**, 707 (1982)

3.88   M. Moisan, Z. Zakrzewski: in *Radiative Processes in Discharge Plasmas*, ed. by J.M. Proud, L.H. Luessen (Plenum, New York 1986) p. 381

3.89   A. Shivarova: in *Spatial Dispersion in Solids and Plasmas*, ed. by P. Halevi (Elsevier, Amsterdam 1992) p. 557

3.90   S.V. Vladimirov, M.Y. Yu, V.M. Tsytovich: Phys. Rep. **241**, 1 (1994)

3.91   L. Stenflo: Phys. Scripta **T50**, 15 (1994)

3.92   L. Stenflo: Phys. Scripta **T63**, 59 (1996)

3.93   P. Halevi: in *Electromagnetic Surface Modes*, ed. by A.D. Boardman (Wiley, Chichester 1982) p. 249

3.94   A. Sommerfeld: *Elektrodynamik* (Geest und Portig, Leipzig 1967)

3.95   I. Ghanashev, A. Shivarova, H. Schlüter: Radio Sci. **29**, 1265 (1994)

3.96   Z. Zakrzewski, M. Moisan, V.M.M. Glaude, C. Beaudry, P. Leprince: Plasma Phys. **19**, 77 (1977)

3.97   P.K. Cibin: Plasma Phys. **22**, 609 (1980)

3.98   M. Zethoff, U. Kortshagen: J. Phys. D **25**, 1574 (1992)

3.99   J. Margot, M. Moisan: J. Plasma Phys. **49**, 357 (1993)

3.100 Yu.M. Aliev, A.V. Maximov, U. Kortshagen, H. Schlüter, A. Shivarova: Phys. Rev. E **53**, 6091 (1995)

3.101 S. Grosse, M. Georgieva, I. Ghanashev, M. Schlüter: J. Electromagn. Waves Appl. **11**, 609 (1996)

3.102 E.T. Arakawa, M.W. Williams, R.N. Hamm, R.H. Ritchie: Phys. Rev. Lett. **31**, 1127 (1973)

3.103 D.L. Mills, A.A. Maradudin: Phys. Rev. Lett. **31**, 372 (1973)

3.104 Yu.M. Gerbshtein, D.N. Mirlin: Fizika Tverdovo Tela **16**, 2584 (1974) [in Russian]

3.105 D.N. Mirlin, I.I. Reshina: Fizika Tverdovo Tela **16**, 2241 (1974) [in Russian]

3.106 V.N. Agranovich: Sov. Phys. – Usp. **18**, 99 (1975)

3.107 T. Lopez-Rios: Opt. Commun. **17**, 372 (1976)

3.108 Yu.M. Aliev, O.M. Gradov, Yu.M. Kirii: ZhETF **63**, 112 (1972)

3.109 Yu.M. Aliev, A.V. Maximov, H. Schlüter, A. Shivarova: Phys. Scripta **51**, 257 (1995)

3.110 Yu.M. Aliev, J. Berndt, H. Schlüter, A. Shivarova: J. Electromagn. Waves Appl. **9**, 697 (1995)

3.111 T.H. Stix: *Waves in Plasmas* (American Institute of Physics, New York 1966) p. 244

3.112 J.A. Tataronis, F.W. Crawford: *Proc. XIIth ICPIG, Gradevinska Knjaga* (Belgrade 1966) Vol. 2, p. 244

3.113 M. Krämer, K. Lucks, H. Schlüter, F. Wiesemann: Phys. Lett. **96A**, 195 (1983)

3.114 Yu.M. Aliev, I. Ghanashev, H. Schlüter, A. Shivarova, M. Zethoff: Plasma Sources Sci. Technol. **3**, 216 (1994)

3.115 Z. Zakrzewski, M. Moisan: in *Proc. XVIIth ICPIG, Budapest 1985*, ed. by Z.S. Bakos, Z. Sörlei (Roland Eötvös Physical Society, Budapest 1985) p. 762

3.116 Z. Zakrzewski: J. Phys. D **16**, 171 (1983)

3.117 C.M. Ferreira: J. Phys. D **16**, 1673 (1983)

3.118 Yu.M. Aliev, A.G. Boev, A.P. Shivarova: Phys. Lett. A **92**, 235 (1982)

3.119 Yu.M. Aliev, K.M. Ivanova, M. Moisan, A.P. Shivarova: Plasma Sources Sci. Technol. **2**, 145 (1993)

3.120 Yu.M. Aliev, A.V. Maximov, I. Ghanashev, A. Shivarova, H. Schlüter: IEEE Trans. **PS-23**, 409 (1995)

3.121 Yu.M. Aliev, A.G. Boev, A.P. Shivarova: J. Phys. D **17**, 2233 (1984)

3.122 K. Ivanova, I. Koleva, A. Shivarova, E. Tatarova: Phys. Scripta **47**, 224 (1993)

3.123 P. Nghiem, M. Chaker, E. Bloyet, P. Leprince, J. Marec: J. Appl. Phys. **53**, 2920 (1982)

3.124 C. Boisse-Laporte, A. Granier, E. Dervisevic, P. Leprince, J. Marec: J. Phys. D **20**, 197 (1987)

3.125 S. Daviaud, C. Boisse-Laporte, P. Leprince, J. Marec: J. Phys. D **22**, 770 (1989)

3.126 C. Boisse-Laporte: in *Microwave Discharges: Fundamentals and Applications*, ed. by C.M. Ferreira, M. Moisan (Plenum, New York 1993) p. 25

Chapter 4

4.1   S.S. Iha, G.S. Kino: J. Electron. Control **14**, 167 (1963)

4.2   Yu.A. Romanov: Sov. Phys. – JETP **20**, 1424 (1965)

4.3   K.N. Stepanov: Sov. Phys. – Techn. Phys. **10**, 773 (1965)

4.4   D.K. Sarkar, K.E. Lonngren: IEEE Trans. **AP–15**, 716 (1967)

4.5   D. Ilić: Int. J. Electron. **24**, 439 (1968)

4.6   D.E. Baldwin, D.W. Ignat: Phys. Fluids **12**, 697 (1969)

4.7   V.V. Demchenko, V.V. Dolgopolov, K.N. Stepanov: Izv. VUZ–Radiofizica **12**, 1371 (1969)

4.8   V.V. Dolgopolov, A.Yu. Omel'chenko: Sov. Phys. – JETP **31**, 741 (1970)

4.9   P.K. Kaw, J.B. McBride: Phys. Fluids **13**, 1784 (1970)

4.10   V.V. Demchenko, K.E. Zayer: Physica **59**, 385 (1972)

4.11   Yu.A. Romanov: Sov. Phys. – Techn. Phys. **17**, 1447 (1973)

4.12   V.V. Demchenko, N.M. El-Siragy, A.M. Hussein, K.E. Zayed: Nucl. Fusion **13**, 665 (1973)

4.13   A.N. Kondratenko: *Plasma Waveguides* (Atomizdat, Moscow 1976) [in Russian]

4.14 R. Darchicourt, S. Pasquiers, C. Boisse-Laporte, P. Leprince, J. Marec: J. Phys. D **21**, 293 (1988)

4.15 M. Zethoff, U. Kortshagen: J. Phys. D **25**, 1574 (1992)

4.16 Yu.M. Aliev, A.V. Maximov, U. Kortshagen, A. Shivarova: Phys. Rev. E **51**, 6091 (1995)

4.17 S. Grosse, M. Schlüter, M. Georgieva, I. Ghanashev: J. Electromagn. Waves Appl. **11**, 609 (1997)

4.18 Yu.M. Aliev, J. Berndt, H. Schlüter, A. Shivarova: J. Electromagn. Waves Appl. **9**, 67 (1995)

4.19 Yu.M. Aliev, V.Yu. Bychenkov, A.V. Maximov, H. Schlüter: Plasma Sources: Sci. Technol. **1**, 126 (1992)

4.20 Yu.M. Aliev: Some aspects of nonlinear theory of ionizing surface plasma waves, in *Microwave Discharges: Fundamentals and Applications*, ed. by C.M. Ferreira, M. Moisan, NATO ASI Series B 302 (Plenum, New York 1993), p. 105

4.21 Yu.M. Aliev, A.V. Maximov, H. Schlüter: Phys. Scripta **48**, 464 (1993)

4.22 Yu.M. Aliev, A.V. Maximov, H. Schlüter, A. Shivarova: J. Plasma Phys. **52**, 321 (1994)

4.23 Yu.M. Aliev, A.V. Maximov, H. Schlüter, A. Shivarova: Phys. Scripta **51**, 257 (1995)

4.24 Yu.M. Aliev, S. Grosse, H. Schlüter, A. Shivarova: Phys. Plasma **3**, 3162 (1996)

4.25 M. Moisan, C. Beaudry, P. Leprince: Phys. Lett. **50A**, 125 (1974)

4.26 V.M.M. Glaude, M. Moisan, R. Pantel, P. Leprince, J. Marec: J. Appl. Phys. **51**, 5693 (1980)

4.27 Yu.M. Aliev, A.G. Boev, A.P. Shivarova: Phys. Lett. **92A**, 235 (1982)

4.28 M. Moisan, C.M. Ferreira, Y. Hajlaoui, D. Henry, J. Hubert, R. Pantel, A. Ricard, Z. Zakrzewski: Rev. Phys. Appl. **17**, 707 (1982)

4.29 M. Moisan, Z. Zakrzewski: in *Radiative Processes in Discharge Plasmas*, ed. by J.M. Proud, L.H. Luessen (Plenum, New York 1986) p. 381

4.30 C.M. Ferreira, M. Moisan (eds.): *Microwave Discharges: Fundamentals and Applications*, NATO ASI Series B 302 (Plenum, New York 1993)

4.31 V.L. Ginzburg: *The Propagation of Electromagnetic Waves in Plasmas* (Pergamon, London 1964)

4.32 G.G. Lister, T.R. Robinson: Surface waves in highly collisional plasmas, in *Europhysics Conf. Abstracts, Xth ESCAMPIG, Orleans, 1990*, ed. by B. Dubreuil (European Physical Society, Geneva 1990) Vol. 14E, p. 322

4.33 G.G. Lister, T.R. Robinson: J. Phys. D **24**, 1993 (1991)

4.34 G.G. Lister: J. Phys. D **25**, 1649 (1992)

4.35 G.G. Lister: Strongly damped surface waves in plasmas, in *Microwave Discharges: Fundamentals and Applications*, ed. by C.M. Ferreira, M. Moisan, NATO ASI Series B 302 (Plenum, New York 1993) p. 85

4.36   Yu.M. Aliev, J. Berndt, H. Schlüter, A. Shivarova: Surface waves in nonuniform plasmas and their absorption at the localization of surface plasmons, in *Proc. Xth Topical Conf. on Radio Frequency Power in Plasmas, Boston, 1990*, ed. by M. Porkolab, J. Hosea, AIP Proc. **289**, p. 309 (American Institute of Physics, New York 1990)

4.37   Yu.M. Aliev, J. Berndt, H. Schlüter, A. Shivarova: Plasma Phys. Contr. Fusion **36**, 937 (1994)

4.38   J. Berndt: *Untersuchungen zum Ausbreitungsverhalten von Oberflächenwellen in axial inhomogenen Plasmen*, Dissertation, Ruhr-Universität Bochum (1996)

4.39   M. Schlüter: *Ausbreitung von Oberflächenwellen in zylindrischen Plasmen mit axialer und radialer Elektronendichtinhomogenität*, Dissertation, Ruhr-Universität Bochum (1997)

4.40   M. Schlüter: J. Phys. D **30**, L11 (1997)

4.41   S. Grosse, H. Schlüter, M. Schlüter: Surface wave sustained discharges, in *Electron Kinetics and Applications of Glow Discharges*, ed. by U. Kortshagen, L.D. Tsendin, NATO ASI Series B 367 (Plenum, New York 1998) p. 423

4.42   L. Brillouin: *Wave Propagation and Group Velocity* (Academic Press, New York 1960)

4.43   Yu.M. Aliev, H. Schlüter: J. Phys. D **29**, 2519 (1996)

4.44   L.D. Landau, E.M. Lifshitz: *The Classical Theory of Fields* (Pergamon, London 1962)

4.45   Yu.M. Aliev, K.M. Ivanova, M. Moisan, A.P. Shivarova: Plasma Sources Sci. Technol. **2**, 145 (1993)

4.46   M. Moisan, A. Shivarova, A.W. Trivelpiece: Plasma Phys. **24**, 1331 (1982)

4.47   Yu.M. Aliev, I. Ghanashev, H. Schlüter, A. Shivarova, M. Zethoff: Plasma Sources Sci. Technol. **3**, 216 (1994)

4.48   J. Berndt, D. Grozev, H. Schlüter: Studies on surface wave dispersion in discharges produced by surface waves in view of diagnostics, J. Phys. D, submitted; report 76-B8-99 Sonderforschungsbereich 191, Ruhr-Universität Bochum (1999)

4.49   A. Shivarova: Nonlinear Surface Modes, in *Spatial Dispersion in Solids and Plasmas*, ed. P. Halevi (Elsevier, Amsterdam 1992) p. 557

Chapter 5

5.1   V.A. Bailey: Nature **139**, 838 (1937)

5.2   A.V. Gurevich: *Nonlinear Phenomena in the Ionosphere* (Springer, Berlin Heidelberg 1978)

5.3   V.B. Gil'denburg, S.V. Golubev: Sov. Phys. – JETP **40**, 46 (1974)

5.4   V.B. Gil'denburg, A.V. Kochetov, A.G. Litvak, A.M. Feigin: Sov. Phys. – JETP **57**, 28 (1983)

5.5   A.G. Litvak, V.A. Mironov, G.M. Fraiman, A.D. Yunakovskii: Sov. J. Plasma Phys. **1**, 31 (1975)

5.6   A.G. Litvak (ed.): *High-Frequency Discharges in Wave Fields* (Institute of Applied Physics, USSR Academy of Sciences, Gor'kii 1988) [in Russian]

5.7   A.L. Vikharev, O.A. Ivanov, A.N. Stepanov: Sov. J. Plasma Phys. **14**, 32 (1988)

5.8   V.E. Semenov: Fizika Plazmy **8**, 613 (1982)

5.9   Yu.L. Bogomolov, S.F. Lirin, V.E. Semenov, A.M. Sergeev: JETP Lett. **45**, 680 (1987)

5.10   R.S. Smith, G.E. Thomas, D.C. Coleman, T.J. Pappalardo: IEEE Trans. **PS–14**, 63 (1986)

5.11   A.G. Boev, A.V. Prokopov: Sov. Phys. JETP **42**, 617 (1976)

5.12   A.G. Boev: Sov. Phys. JETP **50**, 47 (1979)

5.13   A.G. Boev: Sov. J. Plasma Phys. **8**, 410 (1982)

5.14   M. Moisan, C. Beaudry, P. Leprince: Phys. Lett. **50A**, 125 (1974).

5.15   M. Moisan, C.M. Ferreira, Y. Hajlaoui, D. Henry, J. Hubert, R. Pantel, A. Ricard, Z. Zakrzewski: Rev. Phys. Appl. **17**, 707 (1982)

5.16   V.M.M. Glaude, M. Moisan, R. Pantel, D. Leprince, J. Marec: J. Appl. Phys. **51**, 5693 (1980)

5.17   Yu.M. Aliev, A.G. Boev, A.P. Shivarova: Phys. Lett. **92A**, 235 (1982)

5.18   C.M. Ferreira: J. Phys. D **16**, 1673 (1983)

5.19   Z. Zakrzewski: J. Phys. D **16**, 171 (1983).

5.20   Yu.M. Aliev, A.G. Boev, A.P. Shivarova: J. Phys. D **17**, 2233 (1984)

5.21   W. Schottky: Phys. Z. **25**, 635 (1924)

5.22   V.L. Granovsky: *Electrical Current in Gases: Stationary Current* (Nauka, Moscow 1971) [in Russian]

5.23   J.H. Ingold: Z. Physik **233**, 89 (1970)

5.24   M. Mitchner, C.H. Kruger: *Partially Ionized Gases* (Wiley, New York 1973)

5.25   A.V. Gurevich, A.B. Scharzburg: *Nonlinear Theory of Radio Wave Propagation in Ionosphere* (Nauka, Moscow 1973) [in Russian].

5.26   G.B. Withman: J. Fluid Mech. **22**, 273 (1965)

5.27   A. Shivarova: in *Spatial Dispersion in Solids and Plasmas*, ed. by P. Halevi (Elsevier, Amsterdam 1992) p. 557

5.28   Yu.M. Aliev, K.M. Ivanova, M. Moisan, A.P. Shivarova: Plasma Sources Sci. Technol. **2**, 145 (1993)

5.29   V.E. Golant, A.P. Zhilinski, I.E. Sakharov: *Fundamentals of Plasma Physics* (Wiley, New York 1980).

5.30   Yu.M. Aliev, A.V. Maximov, I. Ghanashev, A. Shivarova, H. Schlüter: IEEE Trans. **PS–23**, 409 (1995)

5.31   Yu.M. Aliev, H. Schlüter, A. Shivarova: Plasma Sources Sci. Technol. **5**, 514 (1996)

5.32 Yu.M. Aliev, M. Georgieva, A. Shivarova, H. Schlüter: J. Phys. D **28**, 1997 (1995)

5.33 Yu.M. Aliev, M. Georgieva, S. Grosse, H. Schlüter; A. Shivarova: Contr. Plasma Phys. **36**, 573 (1995)

5.34 Yu.M. Aliev, S. Grosse, H. Schlüter, A. Shivarova: Phys. Plasmas **3**, 3162 (1996)

5.35 Yu.M. Aliev, A.V. Maximov, H. Schlüter, A. Shivarova: J. Plasma Phys. **52**, 321 (1994)

5.36 E. Spenke: Z. Phys. **127**, 221 (1950)

5.37 M. Pahl: Z. Naturforsch. **12a**, 682 (1957)

5.38 I. Zhelyazkov, V. Atanassov: Phys. Rev. **55**, 79 (1995)

5.39 U. Kortshagen, H. Schlüter: *Proc. Int. Conf. Plasma Physics, Innsbruck 1992*, ed. by V. Bethge (European Physical Society, Geneva 1992) Part III, p. 2145

5.40 J. Berndt, U. Kortshagen, H. Schlüter: J. Phys. D **27**, 1470 (1994)

5.41 Yu.M. Aliev, I. Ghanashev, H. Schlüter, A. Shivarova, M. Zethoff: Plasma Sources Sci. Technol. **3**, 216 (1994)

5.42 P. Leprince, J. Marec: J. Physique **42**, 1421 (1981).

5.43 P. Nghiem, M. Chaker, E. Bloyet, P. Leprince, J. Marec: J. Appl. Phys. **53**, 2920 (1982)

5.44 M. Chaker, P. Nghiem, E. Bloyet, P. Leprince, J. Marec: J. Physique Lett. **43**, L71 (1982)

5.45 M. Chaker, M. Moisan: J. Appl. Phys. **57**, 91 (1985)

5.46 A. Granier, C. Boisse-Laporte, P. Leprince, J. Marec, P. Nghiem: J. Phys. D **20**, 204 (1987)

5.47 C. Boisse-Laporte, A. Granier, E. Dervisevic, P. Leprince, J. Marec: J. Phys. D **20**, 197 (1987)

5.48 C. Boisse-Laporte, A. Granier, E. Bloyet, P. Leprince, J. Marec: J. Appl. Phys. **61**, 1740 (1987)

5.49 R. Darchicourt, S. Pasquiers, C. Boisse-Laporte, P. Leprince, J. Marec: J. Phys. D **21**, 293 (1988)

5.50 S. Daviaud, C. Boisse-Laporte, P. Leprince, J. Marec: J. Phys. D **22**, 770 (1989)

5.51 C.M. Ferreira: J. Phys. D **22**, 705 (1989)

5.52 A. Granier, S. Pasquiers, C. Boisse-Laporte, R. Darchicourt, P. Leprince, J. Marec: J. Phys. D **22**, 1487 (1989)

5.53 S. Daviaud, G. Gousset, J. Marec, E. Bloyet: J. Phys. D **23**, 856 (1990)

5.54 A.B. Sá, C.M. Ferreira, S. Pasquiers, C. Boisse-Laporte, P. Leprince, J. Marec: J. Appl. Phys. **70**, 4147 (1991)

5.55 Z. Rakem, P. Leprince, J. Marec: J. Phys. D **25**, 943 (1992)

5.56 C.M. Ferreira, M. Moisan (eds.): *Microwave Discharges: Fundamentals and Applications* (Plenum, New York 1993)

5.57 Z. Zakrzewski, M. Moisan: Plasma Sources Sci. Technol. **4**, 379 (1995)

5.58   M. Moisan, Z. Zakrzewski: in *Radiative Processes in Discharge Plasmas*, ed. by Y.M. Proud, L.H. Luessen (Plenum, New York 1986) p. 381
5.59   Yu.M. Aliev, A.V. Maximov, H. Schlüter: Phys. Scripta **48**, 464 (1993)
5.60   Yu.M. Aliev, A.V. Maximov, H. Schlüter, A. Shivarova: Phys. Scripta **51**, 257 (1995)
5.61   S. Nonaka: Jpn. J. Appl. Phys. **33**, 4226 (1994)

Chapter 6

6.1   I.B. Bernstein, T. Holstein: Phys. Rev. **94**, 1475 (1954)
6.2   D.J. Rose, S.C. Brown: Phys. Rev. **98**, 310 (1955)
6.3   C.M. Ferreira, M. Moisan: Phys. Scripta **38**, 382 (1988)
6.4   U. Kortshagen, H. Schlüter, A. Shivarova: J. Phys. D **24**, 1571 (1991)
6.5   L.D. Tsendin: Plasma Sources Sci. Technol. **4**, 200 (1995)
6.6   V.I. Kolobov, V.A. Godyak: IEEE Trans. Plasma Sci. **23**, 503 (1995)
6.7   U. Kortshagen, C. Busch, L.D. Tsendin: Plasma Sources Sci. Technol. **5**, 1 (1996)
6.8   Yu.B. Golubovskii, Yu.M. Kagan, R.I. Lyagushchenko: ZhTF **38**, 1934 (1969).
6.9   Yu.M. Kagan, R.I. Lyagushchenko, N. Khristov: ZhTF **41**, 2053 (1972)
6.10   S.D. Rockwood: Phys. Rev. A **8**, 2348 (1973)
6.11   L.A. Schlie: J. Appl. Phys. **47**, 1397 (1976)
6.12   F. Dothan, Yu.M. Kagan: J. Phys. D **12**, 2155 (1979).
6.13   C.M. Ferreira: J. Phys. D **14**, 1811 (1981)
6.14   C.M. Ferreira, J. Loureiro: J. Phys. D **16**, 2471 (1983)
6.15   C.M. Ferreira, J. Loureiro: J. Phys. D **17**, 1175 (1984)
6.16   M. Capitelli, M. Dilonardo, R. Winkler, J. Wilhelm: Contr. Plasma Phys. **26**, 443 (1986)
6.17   R. Winkler, M. Dilonardo, M. Capitelli, J. Wilhelm: Plasma Chem. Plasma Proc. **7**, 125 (1987)
6.18   R. Winkler, M. Dilonardo, M. Capitelli, J. Wilhelm: Plasma Chem. Plasma Proc. **7**, 245 (1987)
6.19   M. Capitelli, C. Gorse, R. Winkler, J. Wilhelm: J. Appl. Phys. **62**, 4398 (1987)
6.20   M. Capitelli, R. Celiberto, C. Gorse, R. Winkler, J. Wilhelm: J. Phys. D **21**, 691 (1988)
6.21   M. Capitelli, R. Celiberto, C. Gorse, R. Winkler, J. Wilhelm: Plasma Chem. Plasma Proc. **8**, 175 (1988)
6.22   E.V. Karoulina, Y.A. Lebedev: J. Phys. D **21**, 411 (1988)
6.23   A.B. Sá, C.M. Ferreira, S. Pasquiers, C. Boisse-Laporte, P. Leprince, J. Marec: J. Appl. Phys. **70**, 4147 (1991)

6.24   L.L. Alves, C.M. Ferreira: J. Phys. D **24**, 581 (1991)

6.25   C.M. Ferreira, L.L. Alves, M. Pinheiro, A.B. Sá: IEEE Trans. Plasma Sci. **19**, 229 (1991)

6.26   U. Kortshagen, H. Schlüter, A.V. Maximov: Phys. Scripta **46**, 450 (1992)

6.27   E.V. Karoulina, Y.A. Lebedev: J. Phys. D **25**, 401 (1992)

6.28   P.A. Sá, J. Loureiro, C.M. Ferreira: J. Phys. D **25**, 960 (1992)

6.29   L.L. Alves, G. Gousset, C.M. Ferreira: J. Phys. D **25**, 1713 (1992)

6.30   A. Böhle, U. Kortshagen: Plasma Sources Sci. Technol. **3**, 80 (1994)

6.31   L. Dountchev, I. Koleva, A. Shivarova: Plasma Sources Sci. Technol. **5**, 531 (1996)

6.32   L.D. Tsendin: Sov. Phys. – JETP **39**, 805 (1974)

6.33   I.D. Kaganovich, L.D. Tsendin: IEEE Trans. Plasma Sci. **20**, 86 (1992)

6.34   J. Behnke, Y.B. Golubovskii: Opt. Spectrosc. **73**, 73 (1992)

6.35   V.I. Kolobov, L.D. Tsendin: Phys. Rev. A **46**, 7837 (1992)

6.36   U. Kortshagen: J. Phys. D **26**, 1691 (1993)

6.37   U. Kortshagen, L.D. Tsendin: Appl. Phys. Lett. **65**, 1355 (1994)

6.38   V.I. Kolobov, D.F. Beale, L.J. Mohoney, A.E. Wendt: Appl. Phys. Lett. **65**, 537 (1994)

6.39   J. Behnke, Y.B. Golubovskii: Sov. Phys. Techn. Phys. **39**, 38 (1994)

6.40   U. Kortshagen, I. Pukropski, L.D. Tsendin: Phys. Rev. E **51**, 6063 (1995)

6.41   V. Kolobov, W.N.G. Hitchton: Phys. Rev. E **52**, 972 (1995)

6.42   C. Busch, U. Kortshagen: Phys. Rev. E **51**, 280 (1995)

6.43   M.J. Kushner: J. Appl. Phys. **54**, 4958 (1983)

6.44   M.J. Kushner: J. Appl. Phys. **61**, 2784 (1987)

6.45   R.W. Boswell, I.J. Morey: Appl. Phys. Lett. **52**, 21 (1988)

6.46   W.N.G. Hitchton, D.J. Koch, J.B. Adams: J. Comput. Phys. **83**, 79 (1989)

6.47   T.J. Sommerer, W.N.G. Hitchton, J.E. Lawler: Phys. Rev. A **39**, 6356 (1989)

6.48   M. Surenda, D.B. Graves: IEEE Trans. **PS-19**, 144 (1991)

6.49   T.J. Sommerer, M.J. Kushner: J. Appl. Phys. **71**, 1654 (1992)

6.50   V. Vahedi, C.K. Birdsall, M.A. Lieberman, T.D. Rognlien: Plasma Sources Sci. Technol. **2**, 261 (1993)

6.51   V. Vahedi, C.K. Birdsall, M.A. Lieberman, G. diPeso, T.D. Rognlien: Plasma Sources Sci. Technol. **2**, 273 (1993)

6.52   V. Vahedi, C.K. Birdsall, M.A. Lieberman, G. diPeso, T.D. Rognlien: Phys. Fluids **35**, 2719 (1993)

6.53   W.N.G. Hitchton, G.J. Parker, J.E. Lawler: IEEE Trans. **PS-21**, 228 (1993)

6.54   G.J. Parker, W.N.G. Hitchton, J.E. Lawler: Phys. Rev. E **50**, 3210 (1994)

6.55   D.G. Cooperberg, C.K. Birdsall: Plasma Sources Sci. Technol. **7**, 41 (1998)

6.56   Yu.M. Aliev, V.Yu. Bychenkov, A.V. Maximov, H. Schlüter: Plasma Sources Sci. Technol. **1**, 126 (1992)

6.57   Yu.M. Aliev: in *Microwave Discharges: Fundamentals and Applications*, ed. by C.M. Ferreira, M. Moisan (Plenum, New York 1993) NATO ASI Series, Series B 302, pp. 105–115

6.58   Yu.M. Aliev, U. Kortshagen, A.V. Maximov, H. Schlüter, A. Shivarova: Phys. Rev. E **51**, 6091 (1995)

6.59   H. Schlüter: in *Advanced Technologies Based on Wave and Beam Generated Plasma, NATO ASI, Sozopol, Bulgaria*, NATO ASI Ser. B, Subseries 3: High Technology, ed. by H. Schlüter, A. Shivarova (Kluwer, Dordrecht 1999) Vol. 67, p. 271

6.60   J. Krenz: Institut für Plasmaphysik, Univ. Hannover, BMFT Project Report No. 13N54069, 1987 (unpublished)

6.61   S.A. Self, H.N. Ewald: Phys. Fluids **9**, 2486 (1966)

6.62   K.-U. Riemann: J. Phys. D **24**, 493 (1991)

6.63   S.C. Brown: *Basic Data of Plasma Physics* (Wiley, New York 1959)

6.64   V. Martisovits: J. Phys. B **3**, 850 (1970)

6.65   C. Boisse-Laporte: in *Microwave Discharges: Fundamentals and Application*, ed. by C.M. Ferreira, M. Moisan (Plenum, New York 1993) NATO ASI Series, Series B 302, pp. 25–43

6.66   G.G. Lister, Y.-M. Li, V.A. Godyak: J. Appl. Phys. **79**, 8993 (1996)

6.67   Yu.M. Aliev, S. Grosse, H. Schlüter, A. Shivarova: Phys. Plasmas **3**, 3162 (1996)

6.68   A. Shivarova, Kh. Tarnev: J. Phys. D **31**, 2543 (1998)

6.69   Yu.M. Aliev, V.Yu. Bychenkov, M.S. Jovanovic, A.A. Frolov: J. Plasma Phys. **48**, 167 (1992)

6.70   S. Grosse, H. Schlüter, M. Schlüter: in *Electron Kinetics and Applications of Glow Discharges*, NATO ASI Ser. B: Physics, ed. by U. Kortshagen, L.D. Tsendin (Plenum, New York 1998) Vol. 367, p. 423

6.71   Y.A. Godyak, R.B. Piejak: Appl. Phys. Lett. **63**, 3137 (1993)

6.72   L.L. Alves, G. Gousset, C.M. Ferreira: Phys. Rev. E **55**, 890 (1997)

6.73   G. Mümken: J. Phys. D **32**, 804 (1999)

Chapter 7

7.1   D.R. Tuma: Rev. Sci. Instrum. **41**, 1519 (1970)

7.2   M. Moisan, C. Beaudry, P. Leprince: Phys. Lett. **50A**, 125 (1974)

7.3   J. Marec, E. Bloyet, M. Chaker, P. Leprince, P. Nghiem: in *Electrical Breakdown and Discharges in Gases*, ed. by E.E. Kunhardt, L.H. Luessen, NATO ASI Ser. B Vol. 89 (Plenum, New York 1982) p. 347

7.4   M. Moisan, C.M. Ferreira, Y. Hajlaoui, D. Henry, J. Hubert, R. Pantel, A. Ricard, Z. Zakrzewski: Rev. Phys. Appl. **17**, 707 (1982)

7.5   M. Moisan, Z. Zakrzewski: in *Radiative Processes in Discharge Plasmas*, ed. by J.M. Proud, L.H. Luessen (Plenum, New York 1986) p. 381

7.6   M. Chaker, M. Moisan, Z. Zakrzewski: Plasma Chem. Plasma Process. **6**, 79 (1986)

7.7   M. Moisan, Z. Zakrzewski: J. Phys. D **24**, 1025 (1991)

7.8   M. Moisan, J. Pelletier (eds.): *Microwave Excited Plasmas* (Elsevier, Amsterdam 1992)

7.9   C.M. Ferreira, M. Moisan (eds.): *Microwave Discharges: Fundamentals and Applications*, NATO ASI Ser. B, Vol. 302 (Plenum, New York 1993)

7.10   M. Moisan, J. Hubert, J. Margot, Z. Zakrzewski: in *Advanced Technologies Based on Wave and Beam Generated Plasma, NATO ASI, Sozopol, Bulgaria*, NATO ASI Ser. B, Subseries 3: High Technology, ed. by H. Schlüter, A. Shivarova (Kluwer, Dordrecht 1999) Vol. 67, p. 23

7.11   D. Grozev, K. Kirov, I. Koleva, K. Makasheva, A. Shivarova: in *Advanced Technologies Based on Wave and Beam Generated Plasma, NATO ASI, Sozopol, Bulgaria*, NATO ASI Ser. B, Subseries 3: High Technology, ed. by H. Schlüter, A. Shivarova (Kluwer, Dordrecht 1999) Vol. 67, p. 245

7.12   H. Schlüter: in *Advanced Technologies Based on Wave and Beam Generated Plasma, NATO ASI, Sozopol, Bulgaria*, NATO ASI Ser. B, Subseries 3: High Technology, ed. by H. Schlüter, A. Shivarova (Kluwer, Dordrecht 1999) Vol. 67, p. 271

7.13   M. Moisan, Z. Zakrzewski, R. Pantel, P. Leprince: IEEE Trans. **PS-9**, 55 (1975)

7.14   M. Moisan, C. Beaudry, L. Bertrand, E. Bloyet, J.M. Gagné, P. Leprince, J. Marec, G. Mitchel, A. Ricard, Z. Zakrzewski: 4th International Conference on Gas Discharges, Swansea 1976, IEEE Conf. Publication No. 143, 382 (1976)

7.15   M. Moisan, Z. Zakrzewski, R. Pantel: J. Phys. D **12**, 219 (1979)

7.16   Z. Zakrzewski, M. Moisan: *Seminar on Plasma Physics and Technology* (Academie des Sciences de Tchecoslovaqui 1979) rapport IPPCZ, p. 228

7.17   M. Chaker, P. Nghiem, E. Bloyet, P. Leprince, J. Marec: J. Physique Lett. **43**, 71 (1982)

7.18   M. Moisan, Z. Zakrzewski, R. Pantel, P. Leprince: IEEE Trans. **PS-12**, 203 (1984)

7.19   M. Moisan, M. Chaker, Z. Zakrzewski, J. Paraszczak: J. Phys. E **20**, 1356 (1987)

7.20   M. Moisan, Z. Zakrzewski: Rev. Sci. Instrum. **58**, 1895 (1987)

7.21   Z. Zakrzewski, M. Moisan: Plasma Sources Sci. Technol. **4**, 379 (1995)

7.22   M. Moisan, R. Grenier, Z. Zakrzewski, G. Sauvé: J. Microwave Power Electromagn. Energy **30**, 59 (1995)

7.23   M. Moisan, Z. Zakrzewski, R. Etemadi, J.C. Rostaing: J. Appl. Phys. **83**, 5691 (1998)

7.24   J. Rogers, J. Asmussen: IEEE Trans. **PS-10**, 11 (1982)

7.25   E. Räuchle: in *Proc. 3rd Int. Workshop on Microwave Discharges: Fundamentals and Applications, Abbaye de Fontevraud, 1997*, ed. by C. Boisse-Laporte, J. Marec, J. Physique IV, **8**, Pr7-99 (1998)

7.26   W. Lochte-Holtgreven: *Plasma Diagnostics* (North-Holland, Amsterdam 1968)

7.27   R.H. Huddlestone, S.L. Leonard: *Plasma Diagnostic Techniques* (Academic Press, New York 1965)

7.28   I.H. Hutchinson: *Principles of Plasma Diagnostics* (Cambridge University Press, Cambridge 1987)

7.29   D. Anciello, D.L. Flamm: *Plasma Diagnostics* (Academic Press, San Diego 1989)

7.30   H.R. Griem: *Plasma Spectroscopy* (McGraw Hill, New York 1967)

7.31   G.V. Marr: *Plasma Spectroscopy* (Elsevier, Amsterdam 1968)

7.32   S. Svanberg: *Atomic and Molecular Spectroscopy* (Springer, Berlin, Heidelberg 1991)

7.33   I.I. Sobelman, L.A. Vainshtein, E.A. Yukov: *Excitation of Atoms and Broadening of Spectral Lines* (Springer, Berlin, Heidelberg 1995)

7.34   Z. Zakrzewski, M. Moisan, V.M.M. Glaude, C. Beaudry, P. Leprince: Plasma Phys. **19**, 77 (1977)

7.35   M. Chaker, P. Nghiem, E. Bloyet, P. Leprince, J. Marec: J. Physique **43**, L71 (1982)

7.36   M. Chaker, M. Moisan: J. Appl. Phys. **57**, 91 (1985)

7.37   C. Boisse-Laporte, A. Granier, E. Dervisevic, P. Leprince, J. Marec: J. Phys. D **20**, 197 (1987)

7.38   C. Boisse-Laporte, A. Granier, E. Bloyet, P. Leprince, J. Marec: J. Appl. Phys. **61**, 1740 (1987)

7.39   J. Margot-Chaker, M. Moisan, M. Chaker, V.M.M. Glaude, G. Sauvé: in *Proc. ICPIG XIII*, ed. by T.W. Williams (Adam Hilger, Bristol 1987) Vol. 4, p. 602

7.40   J. Margot-Chaker, M. Moisan, Z. Zakrzewski, V.M. Glaude, G. Sauvé: Radio Sci. **23**, 1120 (1988)

7.41   K. Ivanova, I. Koleva, A. Shivarova, E. Tatarova: Phys. Scripta **47**, 227 (1993).

7.42   U. Kortshagen, A. Shivarova, E. Tatarova, D. Zamfirov: J. Phys. D **27**, 301 (1994)

7.43   K. Ivanova, I. Koleva, A. Shivarova: Plasma Sources Sci. Technol. **4**, 444 (1995)

7.44   J. Berndt, D. Grozev, H. Schlüter, A. Shivarova: in *Advanced Technologies Based on Wave and Beam Generated Plasma, NATO ASI*,

Sozopol, Bulgaria, NATO ASI Ser. B, Subseries 3: High Technology, ed. by H. Schlüter, A. Shivarova (Kluwer, Dordrecht 1999) Vol. 67, p. 519

7.45   J. Berndt, D. Grozev, H. Schlüter: J. Phys. D (1999), submitted

7.46   G. Golubyatnikov, A. Kostrov, A. Shivarova, E. Tatarova, D. Zamfirov: in *Microwave Discharges: Fundamentals and Applications*, *NATO Advanced Research Workshop, Vimeiro, 1992*, ed. by C.M. Ferreira, M. Moisan (Plenum, New York 1993), Abstracts, p. 57

7.47   S. Grosse: *Untersuchungen an einem mikrowellenangeregten Oberflächenwellenplasma in Argon*, Dissertation, Ruhr-Universität Bochum (1995)

7.48   A. Shivarova, E. Tatarova, D. Zamfirov: in *Europhysics Conference Abstracts, Xth ESCAMPIG, Orleans 1990*, ed. by B. Dubreuil (European Physical Society, Geneva 1990) Vol. 14E, p. 171

7.49   C.M. Ferreira, M. Moisan: Phys. Scripta **38**, 382 (1988)

7.50   U. Kortshagen, H. Schlüter, A. Shivarova: J. Phys. D **24**, 1571 (1991)

7.51   A.B. Sá, C.M. Fereira, S. Pasquiers, C. Boisse-Laporte, P. Leprince, J. Marec: J. Appl. Phys. **70**, 4147 (1991)

7.52   U. Kortshagen, H. Schlüter, A. Shivarova: in *Europhysics Conference Abstracts, Xth ESCAMPIG, Orleans, 1990*, ed. by B. Dubreuil (European Physical Society, Geneva 1990) Vol. 14E, p. 163

7.53   U. Kortshagen, H. Schlüter: J. Phys. D **24**, 1585 (1991)

7.54   U. Kortshagen: *Elektronenenergie-Verteilungsfunktionen in oberflächenwellenerzeugten Argonplasmen*, Dissertation, Ruhr-Universität Bochum (1990)

7.55   S. Grosse, H. Schlüter, E. Tatarova: Plasma Sources Sci. Technol. **3**, 545 (1994)

7.56   S. Grosse, H. Schlüter, E. Tatarova: Phys. Scripta **50**, 532 (1994).

7.57   A. Böhle, U. Kortshagen: Plasma Sources Sci. Technol. **3**, 80 (1994)

7.58   U. Kortshagen: Phys. Rev. E **49**, 4369 (1994)

7.59   H. Schlüter: in *Plasma '95*, ed. by H. Rotkaehl (Polish Academy of Sciences, Warsaw 1995) Vol. 2, p. 135

7.60   F.M. Dias, E. Tatarova, C.M. Ferreira: J. Appl. Phys. **83**, 4602 (1998)

7.61   S. Grosse: in *Advanced Technologies Based on Wave and Beam Generated Plasma, NATO ASI, Sozopol, Bulgaria*, NATO ASI Ser. B, Subseries 3: High Technology, ed. by H. Schlüter, A. Shivarova (Kluwer, Dordrecht 1999) Vol. 67, p. 517.

7.62   G.K. Vinogradov, Yu.A. Ivanov, Yu.A. Lebedev: *Methods of Contact Diagnostics of Nonequilibrium Plasma Chemistry* (Nauka, Moscow 1981) [in Russian]

7.63   S. Klagge: *Sondendiagnostik (Weiterbildungskurs Plasmaphysik)*, Universität Greifswald, Sektion Physik/Elektronik (1990)

7.64   V.A. Godyak: *Soviet Radio Frequency Discharge Research* (Delphic Association, Falls Church, VA 1986)

7.65   V.A. Godyak, R.B. Piejak, B.M. Alexandrovich: Plasma Sources Sci. Technol. **1**, 36 (1992)

7.66   K. Wiesemann: Phys. Lett. **25A**, 701 (1967)

7.67   K. Wiesemann: Ann. Phys. (Leipzig) **23**, 275 (1969)

7.68   H. Ameniya, K. Shimizu: Jpn. J. Appl. Phys. **13**, 1035 (1974)

7.69   R.L.F. Boyd, N.D. Twiddy: Proc. Roy. Soc. A **250**, 53 (1959)

7.70   Yu.M. Kagan, N.K. Nitrofanov: Sov. Phys. – Techn. Phys. **16**, 1636 (1972)

7.71   H.W. Rundle, D.R. Clark, M.J. Deckers: Can. J. Phys. **51**, 144 (1973)

7.72   T. Okuda, K. Yamamoto: J. Appl. Phys. **31**, 158 (1960)

7.73   Yu.A. Ivanov, Yu.A. Lebedev, L.S. Polak: Sov. Phys. – Techn. Phys. **21**, 830 (1976)

7.74   V.A. Godyak, R.B. Piejak: in *Proc. 3rd Int. Workshop on Microwave Discharges: Fundamentals and Applications, Abbaye de Fontevraud, 1997*, ed. by C. Boisse-Laporte, J. Marec, J. Physique IV, **8**, Pr7-241 (1998)

7.75   V.E. Golant: Sov. Phys. – Techn. Phys. **5**, 1197, 1961.

7.76   V.E. Golant: *Microwave Methods for Plasma Diagnostics* (Nauka, Moscow 1968) [in Russian]

7.77   M.A. Heald, C.B. Wharton: *Plasma Diagnostics with Microwaves* (Wiley, New York 1965)

7.78   G. Himmel, I. Koleva, H. Schlüter: in *Proc. 3rd Int. Workshop on Microwave Discharges: Fundamentals and Applications, Abbaye de Fontevraud 1997*, ed. by C. Boisse-Laporte, J. Marec, J. Physique IV, **8**, Pr7-327 (1998)

7.79   A. Ricard, J. Hubert, M. Moisan: in *Proc. XVIIth ICPIG*, Budapest, *1985*, ed. by Z.S. Bakos, Z. Sörlei (Roland Eötvös Physical Society, Budapest 1985) p. 712.

7.80   S. Piret, A. Granier, E. Bloyet, G. Gousset: in *Proc. XIIIth ICPIG*, ed. by T.W. Williams (Adam Hilger, Bristol 1987) Vol. 4, p. 611

7.81   M. Moisan, R. Pantel, A. Ricard: Can. J. Phys. **60**, 379 (1982)

7.82   R. Pantel, A. Ricard, M. Moisan: Beitr. Plasmaphys. **23**, 561 (1983)

7.83   A. Ricard, C. Barbeau, A. Besner, J. Hubert, J. Margot-Chaker, M. Moisan, G. Sauvé: Can. J. Phys. **66**, 740 (1988)

7.84   K. Ivanova, I. Koleva, A. Shivarova, E. Tatarova: *Proc. XIXth ICPIG*, ed. by J. Labat (Faculty of Physics, University of Belgrade, Belgrade 1989) Vol. 2, p. 434

7.85   S. Daviaud, C. Boisse-Laporte, P. Leprince, J. Marec: J. Phys. D **22**, 770 (1989)

7.86   S. Daviaud, G. Gousset, J. Marec, E. Bloyet: J. Phys. D **23**, 856 (1990)

7.87   J. Margot, M. Moisan, A. Ricard: Appl. Spectroscopy **45**, 260 (1991)

7.88   J. Berndt, U. Kortshagen, H. Schlüter: J. Phys. D **27**, 1470 (1994)

7.89   I. Koleva, Analyt. Lab. **1**, 6 (1994)

7.90   M. Moisan, R. Pantel, J. Hubert: Contr. Plasma Phys. **80**, 2 (1990)

7.91   L. Dountchev, I. Koleva, A. Shivarova: Plasma Sources Sci. Technol. **5**, 531 (1996)

7.92   A. Shivarova, Th. Tarnev: Phys. Scripta **T82**, 65 (1999)

7.93   A. Granier, C. Boisse-Laporte, P. Leprince, J. Marec, P. Nghiem: J. Phys. D **20**, 204 (1987)

7.94   G.S. Solntsev, P.S. Bulkin, M.M. Rakhman, L.I. Tsvetkova: Sov. J. Plasma Phys. **15**, 495 (1989)

7.95   A. Shivarova, E. Tatarova, D. Zamfirov: *Proc. XXIst ICPIG*, ed. by G. Ecker, U. Arendt, J. Böseler (Arbeitsgemeinschaft Plasmaphysik, Ruhr-Universität Bochum 1993) Vol. I, p. 111

7.96   H.-J. Wesseling: *Proc. XXIst ICPIG*, ed. by G. Ecker, U. Arendt, J. Böseler (Arbeitsgemeinschaft Plasmaphysik, Ruhr-Universität Bochum, Bochum 1993) Vol. II, p. 335

7.97   U. Kortshagen: J. Phys. D **26**, 1230 (1993)

7.98   I. Ghanashev, H. Sugai, S. Morita, N. Toyoda: Plasma Sources Sci. Technol. **8**, 363 (1999)

7.99   A. Durandet, Y. Arnal, J. Margot-Chaker, M. Moisan: J. Phys. D **22**, 1288 (1989)

7.100   Yu.M. Aliev, U. Kortshagen, A.V. Maximov, H. Schlüter, A. Shivarova: Phys. Rev. E **51**, 6091 (1995)

7.101   Yu.M. Aliev, I.D. Kaganovich, H. Schlüter: Phys. Plasmas **4**, 2413 (1997)

7.102   V.A. Godyak, R.N. Piejak, B.M. Alexandrovich: Phys. Rev. Letters **80**, 3264 (1998)

7.103   M. Moisan, C. Barbeau, R. Claude, C.M. Ferreira, J. Margot, J. Paraszczak, A.B. Sá, G. Sauvé, M.P. Wertheimer: J. Vac. Sci. Technol. B **9**, 9 (1991)

7.104   M. Böke: *Testwellendiagnostik und Emissionsspektroskopie an einem oberflächenwellenerzeugten Argonplasma*, Dissertation, Ruhr-Universität Bochum (1998)

7.105   M. Böke, G. Himmel, I. Koleva, M. Schlüter: J. Phys. D **32**, 2426 (1999)

7.106   C. Boisse-Laporte, J. Marec (eds.): *Proc. 3rd Int. Workshop on Microwave Discharges: Fundamentals and Applications, Abbaye de Fontevraud, 1997*, J. Physique IV, **5**.

7.107   G. Lister: in *Advanced Technologies Based on Wave and Beam Generated Plasma, NATO ASI, Sozopol, Bulgaria*, NATO ASI Ser. B, Subseries 3: High Technology, ed. by H. Schlüter, A. Shivarova (Kluwer, Dordrecht 1999) Vol. 67, p. 65.

7.108   M.E. Bannister, J.L. Cecchi: J. Vac. Sci. Technol. A **12**, 106 (1994)

7.109   C.F.M. Borges, M. Moisan, A. Gicquel: Diamond Relat. Mater. **4**, 149 (1995)

7.110   R. Claude, M. Moisan, M.R. Wertheimer: Appl. Phys. Lett. **50**, 1797 (1987)

7.111   A. Gahl, D. Korzec, J. Engemann, J. Voigt: in *Proc. XII Int. Conf. on Gas Discharges and their Applications, Greifswald 1997*, Vol. I, p. 312

7.112   L. Paquin, D. Masson, M.R. Wertheimer, M. Moisan: Can. J. Phys. **63**, 831 (1985)

7.113   S. Moreau, M. Tabrizian, J. Barbeau, M. Moisan, A. Leduc, J. Pelletier, T. Lagarde, M. Rohr, F. Desor, D. Vidal, L.H. Yhia: in *XXIVth ICPIG, Warsaw*, ed. by P. Pisarczyk, T. Pisarczyk, J. Wutowski (Space Research Centre, Polish Academy of Sciences, Warsaw 1999) Vol. 1, p. 51

7.114   G.M.W. Kroesen, E. Stoffels, W.W. Stoffels, G.H.P.M. Swinkels, A. Bouchoule, Ch. Hollenstein, P. Roca di Cabarrocas, J.-C. Bertolini, G.S. Selwyn, F.J. de Hoog: in *Advanced Technologies Based on Wave and Beam Generated Plasma, NATO ASI, Sozopol, Bulgaria*, NATO ASI Ser. B, Subseries 3: High Technology, ed. by H. Schlüter, A. Shivarova (Kluwer, Dordrecht 1999) Vol. 67, p. 175

7.115   A. Ricard: *Reactive Plasmas* (Societé Française du Vide, Paris 1996)

7.116   W. Demtröder: *Laser Spectroscopy*, 2nd edn. (Springer, Berlin, Heidelberg 1996)

7.117   F. Werner, D. Korzec, J. Engemann: J. Vac. Sci. Technol. A **14**, 3065 (1996)

7.118   E. Kaneko, T. Okamoto, S. Watanabe, Y. Okamoto: Jpn. J. Appl. Phys. Pt. 2 **37**, L170 (1998)

7.119   M. Nagatsu, I. Ghanashev, H. Sugai: Plasma Sources Sci. Technol. **7**, 230 (1997)

7.120   G. Sauvé, M. Moisan, Z. Zakrzewski: J. Microwave Power Electromagn. Energy **28**, 123 (1993)

7.121   I. Odobrina, J. Kúdela, M. Kando: Plasma Sources Sci. Technol. **7**, 238 (1998)

7.122   D.G. Cooperberg, C.K. Birdsall: Plasma Sources Sci. Technol. **7**, 41 (1998)

7.123   S. Morita, M. Nagatsu, I. Ghanashev, N. Toyoda, H. Sugai: Jpn. J. Appl. Phys. **37**, L468 (1998)

Subject Index

G

gas flow 251
– temperature 253
geometrical-optics approximation (WKB) 5, 32, 77, 78, 107, 114, 116, 126
group velocity 10

H

heat flux vector 25
heating
– noncollisional 4, 6, 15, 209, 218, 259
– Joule 14, 15, 25, 145, 211
– local 15, 27, 147
– nonlocal 14, 27, 147, 256

I

ionization
– direct 16, 22, 150
– step 5, 22, 150, 209
ionization nonlinearity 1, 3, 5, 139, 144

J

Joule heating 14, 15, 145, 164, 211
Joule losses 12, 69, 81, 83, 89, 101

K

kinetic numerical modelling 6, 205

L

Landau damping 4, 11, 50, 211
launchers 238
local kinetic model 15, 226
Lorentz approximation 13

M

maintenance field intensity 141, 146, 148, 154, 159, 163, 168
Maxwell's equations 7
mobility coefficient, electron 18, 21
mode interaction 48, 81, 87, 102, 105, 211

N

noncollisional heating 4, 6, 15, 209, 218, 222, 259
nonlinearity 2, 3
– ionization 1, 3, 5, 139, 144, 146
– thermal 5, 140, 146
nonlocal kinetic model 14, 206
normalizing
– conditions for the EEDF 13, 212
– field intensity (according to Schottky condition) 142, 157

P

partial waves 211
phase diagrams 32, 244
– homogeneous plasmas
–– cylinder 58
–– semi-space 37
–– semi-space with layer 48
–– slab 52
– inhomogeneous plasmas
–– cylinder 103, 217
–– slab 91
phase portrait of particle balance
– diffusion-controlled regime with recombination 153